SOIL MANAGEMENT
A World View of Conservation and Production

Suppose a brother or sister is without clothes and daily food. If one of you says to him, ''Go, I wish you well; keep warm and well fed,'' but does nothing about his physical needs, what good is it?

James 2:15 – 16

New International Version

SOIL MANAGEMENT

A World View of Conservation and Production

RAY L. COOK
Professor Emeritus of Soil Science
Michigan State University

BOYD G. ELLIS
Professor of Crop and Soil Sciences
Michigan State University

KRIEGER PUBLISHING COMPANY
MALABAR, FLORIDA
1992

Original Edition 1987
Reprint Edition 1992

Printed and Published by
KRIEGER PUBLISHING COMPANY
KRIEGER DRIVE
MALABAR, FLORIDA 32950

Library of Congress Cataloging-In-Publication Data
Cook, Ray Lewis.
     Soil management : a world view of conservation and production /
Ray L. Cook, Boyd G. Ellis.
         p.   cm.
     Originally published: New York : Wiley, 1987.
     Includes bibliographical references and index.
     ISBN 0-89464-682-6 (acid-free paper)
     1. Soil management. 2. Soil conservation.   I. Ellis, Boyd G.
     II. Title.
     [S591.C698 1992]
     631.4--dc20   91-38873
                                                            CIP

10   9   8   7   6   5   4   3   2

Leucaena leucocephala trimmed as a hedge at the Asian Vegetable Development Center in Taiwan. This leguminous tree, which may also be trimmed as a single trunk, is a very fast growing tree and an excellent source of firewood. Planted on the contour for erosion control and trimmed as a hedge, one row will supply nitrogen for several rows of maize on each side. It is adapted to the lowland tropics. It will not tolerate extreme acidity.

Preface

Soil Management for Conservation and Production was written after R. L. Cook had taught a course in soil management at Michigan State University for 30 years. During the decade that followed publication of that book, he also taught courses on the subject in three other countries. Furthermore, during that decade and the one that followed, he served as consultant on many UNDP/FAO and AID projects in Europe, Africa, Latin America, and Asia and for 7 years taught a course on exploring international agriculture. B. G. Ellis has taught courses in South Africa and served for periods of time in Europe, South America, and Asia.

These worldwide experiences pointed to the need for a book that would consider tropical as well as temperate-region crops and discuss the specific soil managment problems of subsistence farmers in tropical countries. Their needs for equipment, fertilizers, pesticides, better varieties, irrigation facilities, and agricultural know-how are appalling. We have attempted also to update soil classification terminology and take advantage of recent soil management research findings in our own country. We apologize for any late work that we may have missed.

East Lansing, Michigan

Ray L. Cook
Boyd G. Ellis

Acknowledgment

We thank our many colleagues in crop and soil sciences for the assistance they gave us during the writing of this revision of *Soil Management for Conservation and Production*. Special thanks go to Drs. Henry D. Foth and Eugene P. Whiteside for soil taxonomy advice.

During our consulting missions and other trips to developing countries, we learned much from foreign specialists and their counterpart nationals. Otherwise, our discussions on tropical agriculture would be much less valid.

In writing the chapters on world food problems, shifting cultivation, water for arid lands, multiple-cropping, and interplanting, we drew heavily on FAO and AID publications, reports from the International Research Centers, publications from the American Academy of Sciences, Worldwatch Institute bulletins, and many person books and bulletins. We have attempted to give credit for help received.

R. L. Cook is grateful to former Director Irving Wyeth and other members of the International Agricultural Institute in the College of Agriculture and Natural Resources at Michigan State University for the information they passed on to him during his 5 years as a member of the Institute.

R. L. C.
B. G. E.

Contents

Introduction

During the time this book was being written, B. G. Ellis spent 2 weeks in Taiwan (May 1982) consulting with their Bureau of Environmental Protection and R. L. Cook spent 5 weeks in Taiwan (March–April 1983) as a member of the Advisory Committee of the Food and Fertilizer Technology Center for the Asian and Pacific Region. The people of that island nation are to be commended for the marvelous progress they have made in agricultural development during the two-and-a-half decades since R. L. Cook was first there (1962) as a member of a U.S./AID institution building team.

By 1965, AID declared Taiwan to be developed, so it ended assistance. Now the country is experiencing some of the same problems that have faced us for many years — growing surpluses of food crops and labor prices so high that farmers do not receive enough for their crops to pay productions costs.

The people of the world look on the United States as an unfailing breadbasket, a country where grain production is unlimited. Sometimes it seems almost to be true. We have often wondered how much wheat and corn we could produce if our farmers were paid on a cost-plus basis, say cost plus $2.00/bu. The same could be said about certain other countries, but what would we do with millions of tons of grain that the hungry people of the world cannot buy?

No, overproduction in the developed countries is not the solution. The hungry people of the developing countries must produce their own food, fiber, and feed for farm animals. World prices, production costs, distribution costs, and dietary habits must be adjusted and balanced. Only poor people starve! As the family income increases from a low level, more food is purchased. Practically all of a relatively small increase goes for more food.

Governments must work together, the developed countries to supply emergency food as needed and to supply inputs and know-how to make it possible for developing countries to raise most or all of their own food. A final responsibility of the governments of the developing countries is to make it possible for the poor to buy the food that is available.

R. L. C.
B. G. E.

Soil Characteristics and Classification

Soils are destined to fall in a certain productivity range. Perhaps "predestined" is the proper term to use, because the characteristics of the soil, while it is immature, are dependent on the nature of the parent material. The forces that bring about soil formation from soil materials are commonly grouped under the single term **weathering**. Included are physical forces that result in rock disintegration, chemical reactions that dissolve minerals and change the composition of rocks, and biological forces that result in intensification of the physical and chemical forces.

Soils were not formed in past ages. They are **being formed** and are ever changing. As they approach maturity, even though that stage may be still far away, the original soil materials become more difficult to identify until mature soils are alike, regardless of the nature of the materials from which they were formed. There are very few locations in the world where mature soils exist. Soils that are presently forming from glacial materials are quite young, and in most nonglaciated regions, forces of wind, water, and gravity have moved soil materials during recent times to cover partly formed soils with new material. Then soil formation starts anew. Thus soils are kept young.

Soil Management Principle

Productivity of young soils reflects parent material—productivity of mature soils reflects soil management.

Within young soils, the nature of the soil materials has a marked effect on their productive capacity. Variations in texture and profile of the soil result from similar variations in the soil material. And topography will exert an additional influence on profile development. Thus, the soils are predestined to have certain production possibilities—to fall within a certain **range** in productivity. Correct management will place a particular area of a certain type of soil **at the top** of its range. Sometimes, of course, economy is such that it is better to aim for a production level somewhat below the possible top in the range.

Soils cause variation in the way production practices affect crop yields. As shown by Fig. 1-1, two untreated soils may fall fairly close together, from the standpoint of crop yields, but may be far apart when good production practices are followed. In other words, the "range" of production, as brought about by good management, is much wider with one soil than with another. Properly managed irrigation tends to minimize inherent soil differences (see Chapter 3). The soil characteristics discussed in this chapter are those that have a particular bearing on management, those that dictate the production practices necessary to move the yield figures to the top of their respective ranges.

PHYSICAL CHARACTERISTICS OF SOILS

Soil Management Principle

Topography will dictate the most critical area of management.

Topography

Perhaps the first and most important soil characteristic to be considered in soil management plans is topography. Level lands are often poorly drained. Lack of natural slope and inadequate outlets inhibit runoff of surface water, and infiltration and evaporation may be too slow for successful cropping. Many level soils lie in valleys and lake plains where fine-textured materials were deposited. Drainage from such areas is slow, even through closely spaced tile drains.

Sloping lands are subject to erosion by water. Erosion is today the world's most serious threat to the eradication of hunger. One has but to travel in mountainous areas to observe the results of this force. The Grand Canyon of the Colorado River is the result of erosion. Less spectacular but more disastrous, as far as agriculture is concerned, are the small canyons and gullies present everywhere in cropland areas. These are the result of gravitational forces producing overland flow of water on slopes to the lowlands. Water management is closely allied to soil management, and slope becomes a major factor in land use. The erosion-control expert considers three alternatives in water management. First, get the water into the soil where it falls; second, allow it to run off so slowly that it cannot carry soil particles; and/or third,

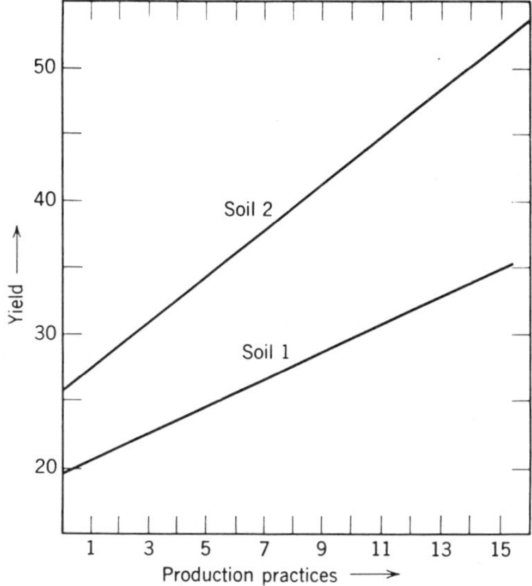

FIGURE 1-1 Soils are destined by their characteristics to fall in a fixed production range. Correct management raises them to the top of the range. Management may be defined as the application of certain production practices. Correct use of the practices makes more difference on soil 2 than on soil 1.

cover and tie the soil particles down so firmly that running water cannot dislodge them. Soil cover, either by living plants or plant residues is a principle means of effecting control by reducing velocity of surface flow and increasing infiltration of water.

Mechanical Properties

Fine-textured soils (sometimes called "heavy" soils) are those containing enough clay to make them sticky, plastic, and slow to drain. They must be worked when moisture is just right or a cloddy surface results. Tilth is easily ruined. The moldboard plow for loosening the soil is more often needed than it is on sandy soils.

Many fine-textured soils (i.e., those that contain 2 : 1 clay minerals) swell and shrink as moisture levels change. Infiltration is affected, and physical condition is changed. After saturation, drying soils may crack sufficiently to hasten evaporation and even to disrupt root systems.

Coarse-textured or sandy soils (sometimes called "light" soils) are those that contain so little clay and/or silt that sand particles are predominant. They drain readily, unless the sand is predominantly very fine or a tight horizon is close to the surface. Because such soils are not sticky or plastic, it is not so important that moisture and tillage operations be closely integrated.

Chemical Properties

Soil Management Principle

Soil chemical properties are easier to manage than are physical properties.

Clay and sand separates are by definition different only in particle size. Sand particles may be ground in a mortar until they are clay particles. Any handful of soil contains particles of all sizes, because weathering causes disintegration of some rocks and minerals faster than others. Thus the more easily weathered materials are reduced in particle size and add to the clay fraction. As a result, chemical variations develop. Clay minerals are not easily recognized by the naked eye, but the different clay minerals have very different properties that affect management. In contrast, sands are largely quartz and unweathered rock and mineral fragments that exert less direct influence on management.

The clay minerals are chemically active; sands are chemically inert. Clay may act as an acid if saturated with hydrogen ion or aluminum ions and, because of a negative charge, has the capacity to adsorb cations, a phenomenon that makes cation exchange possible. Partially decomposed organic matter (humus) also possesses this characteristic. Clay minerals vary in their capacity to adsorb cations (commonly termed **exchange capacity**).

Intelligent soil management makes use of the clay and organic matter collodial fraction as a storehouse or "bank" for available plant foods. A correct equilibrium between soluble, exchangeable, and fixed nutrients must be maintained.

GROUPING SOILS ON THE BASIS OF TEXTURE

Separates

Field experimental results, experiences, and observations show that soil management practices, particularly those that involve biological and chemical processes, must be varied with soil texture. Accordingly, soils have been given a textural classification. First, soil particles have been divided arbitrarily into the sizes called **separates**, as shown in Table 1-1.

Soil Class

A mechanical analysis determines the proportion between the separates in a given soil sample. Results are usually presented as percentages of sand, silt, and clay. These three percentages are used to designate the class in which the soil falls. The class name tells which separates impart their physical character-

TABLE 1-1 The Soil Separates

Name	Diameter (mm)
Very coarse sand	2.00 – 1.00
Coarse sand	1.00 – 0.50
Medium sand	0.50 – 0.25
Fine sand	0.25 – 0.10
Very fine sand	0.10 – 0.05
Silt	0.05 – 0.002
Clay	Below 0.002

istics to the soil. A clay soil is like the clay separate; sand and silt soils likewise resemble those separates and none other.

Many soils are not predominated by one of the three separates — clay, silt, or sand. The "feel" takes on characteristics of all three and is described as "loamy."

Loams do not contain equal proportions of all the separates. In fact, a loam may contain as little as 7 percent clay, as much as 50 percent silt or 52 percent sand, but not more than 27 percent clay. If a soil does contain more than 27 percent clay, it must have the word clay in its class name to show how dominant clay is in its characteristics.

Crop Adaptation

Soil class (texture), for many reasons, is probably the most important factor in determining crop adaptation. Without the use of fertilizers, nutrient supplies would be an important factor. Some crops can stand slow drainage and poor aeration. Grasses, for instance, may thrive where deep-rooted legumes fail. On the other hand, deep, loose sands may produce alfalfa better than grass.

Sugar beets need loose, friable soils, but they must be high in clay for top yields. Potatoes also do best on loose soils, but top yields may be produced on light, sandy loams where sugar beets are not adapted.

Moisture-holding Capacity

Soil moisture is held in the pores and as films around the particles. Coarse soils have a high bulk density and a correspondingly low total pore space, even though the pores are large. In fact, the large-size pores result in fast, excessive drainage. Large particles hold films with less tension than do small particles, so the films are relatively thin. The combined effects of these relationships is low moisture-holding capacity in coarse-textured soils. Drought limits production on these soils more frequently than on soils of finer texture.

Tillage

Tillage practices must definitely be associated with texture. Clay soils must not be worked while wet; but they develop a structure that resists wind erosion whereas very sandy soils blow readily. Cover crops and carefully timed tillage practices may prevent blowing.

Fertilizers

Experiments have shown that crops grown on sandy soils respond differently to fertilizers than do crops grown on clayey soils, even where chemical soil tests for available nutrients reveal no differences. This has resulted in the use of textural groupings as a part of the basis for fertilizer recommendations. The groupings are referred to again in Chapter 7.

GROUPING SOILS ON THE BASIS OF NATURAL DRAINAGE

Topographic variations bring about marked variations in natural drainage. Grass grows more readily in wet areas, and because high moisture levels cause poor aeration, decomposition of plant residues is slow. As a result, soil organic matter accumulation is directly proportional to soil moisture levels if temperature is held constant. Organic matter decomposition is slower with cooler temperatures, hence organic matter accumulation is inversely related to temperature if soil water is held constant.

Other soil formation forces, also affected by topography, are working concurrently with organic matter accumulation to bring about total soil differences greater than what may be attributed to simple organic matter accumulation. The total effect is that lowland soils high in organic matter should be managed much differently than upland soils in the same textural group.

Color

The key to identifying natural drainage and the characteristics that have resulted from the topographic locations of soils is color. Mineral soils have been placed in three natural drainage groups that are used in the designation of soil management groups.

1. Well-drained upland, light surface color, subsoil not mottled.
2. Imperfectly drained, intermediate topography and color, subsoil mottled as a result of variable rates of iron mineral oxidation.
3. Poorly drained lowland, dark surface color, subsoil gray because of lack of iron mineral oxidation.

These organic matter (color) groupings are being widely used as a basis for rotation and drainage planning and to some extent in irrigation and fertilizer recommendations. They are particularly useful in making yield predictions and land evaluations.

UNITS FOR SOIL CLASSIFICATION

Soil management discussions are greatly facilitated if the soils are named. Farmers discuss their cows by name, why not their soils? This question was asked many years ago. Individuals have spent a lifetime working out the names and making records as to where the named soils may be found, and the work is still going on. The extent of their progress is discussed further in Chapter 8.

The soil scientist may define soil as the portion of the crust of the earth that has been changed by soil-forming processes, the **solum** or the A and B horizons in the profile. We should remember, of course, that to a farm planner soil has three dimensions. Area and volume are involved. A soil may be considered as a section of the farm having a uniform profile. Such an area is called a **soil series**.

SOIL SERIES

The **series** name is usually taken from a town, county, river, or other geographic entity near where it was first identified or where it exists in extensive areas. Examples are the Miami soils from the Miami River area in Ohio and the Hillsdale soil named after Hillsdale County, Michigan.

The series is the classification unit that conveys the most meaning to farm planners. They immediately see a mental picture of topography, drainage conditions, profile, reaction (lime need), color, crop adaptation, and to some extent production capacity. In short, they see all but the texture of the surface soil, slope, degree of erosion, and such surface variations as stoniness, salinity, or acidity. They even know something about these latter features because usually these features do not vary widely within a series. For instance, Miami sand and Miami clay are not found within the same series. Widely different parent materials cannot develop similar profiles, so soils formed from them cannot fall in the same series. The soil profile is really the key to series identification. A sketch of a profile characteristic of cropped soils formed under forest vegatation in humid regions is shown in Fig. 1-2.

In soils formed under grass vegetation, the E horizon (formerly A_2) is less pronounced, and in drier regions there may be little evidence of a B horizon. Calcium and/or sodium accumulates in the solum of arid region soils.

The farm planner or the farmer is primarily interested in the series or in a still smaller division, the **soil phase**, but either may have passing interest at least in the soil order, of which the soil series is a subdivision. You are urged to consult local soil scientists at your state experiment station or at the local office of the U.S. Soil Conservation Service or to refer to H. D. Foth, *Fundamentals of Soil Science*, 7th ed., pp. 249–286 (Wiley, New York, 1984) or *Soil Taxonomy: A Basic System of Soil Classification for Mapping and Interpreting Soil*

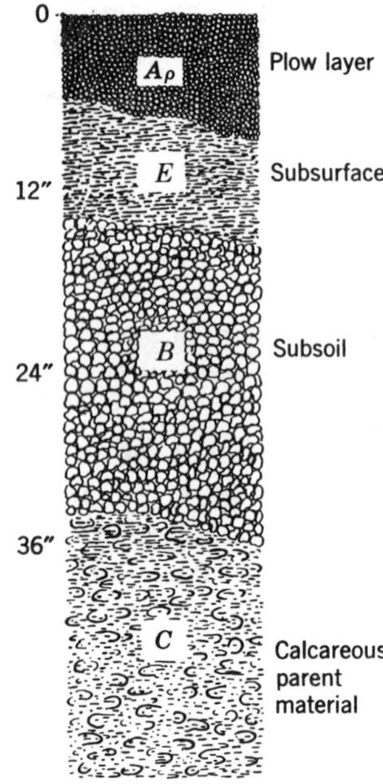

Most of the organic matter is in this zone because plant roots are largely confined to the plowed layer. Also, plowing determines depth to which residues are mixed.

Light-colored horizon where maximum eluviation and leaching have occurred. This layer is prominent in podzolic soils.

Horizon of maximum illuviation. Iron and organic matter or silicate clay minerals have accumulated. Darker color is caused by oxidation of iron compounds.

The weathered material above the parent material is commonly called the solum.

Less weathered, less roots, less clay, and more compact than the layers above. Also called the substratum.

FIGURE 1-2 Sketch of a soil profile characteristic of upland cropped soils formed under forest vegetation in humid regions. Photo by I. F. Schneider, Michigan State University.

Surveys USDA Soil Conservation Service, Agricultural Handbook 436, Washington, D.C. (December 1975).

SOIL PHASE

The farmer is directly concerned with a subdivision of the series called a **soil phase**. The phase name includes the series name and the texture (soil class) of the surface soil. Included also are degree of slope and erosion designations. Thus, there may be a term such as Miami loam B2 in which appear the terms

Miami = series (determined by the location where first mapped)

loam = class (determined by texture)

B = slope (2 – 6 percent)

2 = extent of erosion

The soil phase is essentially a mapping unit. On the detailed soil survey map it appears as a color or as a number on a black and white map. The term 450/B2, for instance, means the soil phase just described. It is the symbol the mapper uses in the field.

The series includes many soil phases, but their differences are not great. In most series only two or three, sometimes four, classes are included. Other variations that change the phase name include slope, extent of erosion, stoniness, etc. We may have, for instance, Miami loams with several slopes (A, B, or C) and with each slope variation, 1, 2, or 3 degrees of erosion, depending on how well the soil has been protected. Slope and extent of erosion are very important in farm planning. In fact, on the individual farm, these become as important as the series and texture of the surface. Thus it becomes important in soil management to have available the results of research on the soil phases present in the area. Research results have usually been presented on a soil-phase basis.

CHAPTER 2

Why World Hunger?

Soil is perhaps our most important natural resource, but it will not furnish sustenance to food crops without water in a suitable temperature and air environment. Scientific agriculture may actually be defined as mankind's manipulation of the interactions of soil, water, light, and solar energy. Science during this century has made fantastic progress, such as putting men on the moon and bringing them safely back to earth, but has not yet been able to provide, in all countries, sufficient food to eliminate starvation. *Why* have we failed to provide sufficient food for all the world's peoples?

First, water covers about 70 percent of the earth's surface. Only 30 percent is land, and if you look at a world map or globe you will see that much of the land is in the Northern Hemisphere too far north for efficient food-crop production. Furthermore, most of the land in the Southern Hemisphere lies rather close to the equator, where high temperatures bring about a need for rather specific management practices and where relatively little agricultural research has been done.

Only a small percentage of the world's land surface is arable. No more than 12 percent of China's land is arable, compared to 27 percent of the United States. The best agricultural area in the world, Hungary, is only 60 percent arable. The United States and China are about equal in size, but China has more than four times as many people. Is it any wonder, as population increases, that China is finding it necessary to import more and more food each year? China's low percentage of arable land results from the fact that a large part of the country is high, dry, and cold — the topography resulting from proximity to the Himalaya Mountains of South Asia. Even more seriously affected by these mountains are Tibet, Kashmir, Nepal, Sikkim, and Bhutan. In such areas, food production is very hazardous. When sloping lands are cultivated, erosion becomes so severe that topsoils are soon lost. In such areas, shifting cultivation with a long fallow period *can* result in a stable agriculture, but increasing population in many countries has made sufficiently long fallow periods impractical. This is discussed in detail in Chapter 18.

TOO MANY PEOPLE

Soil Management Principle

World population can only be substained by using correct soil management to increase food production.

As shown by Fig. 2-1, world population increases have been accelerating rapidly during this century. From the beginning of time until the year 1830, only 1 billion people had accumulated on the globe. After that it required only 100 years—until 1930—to bring the entire earth's population to 2 billion. Only 30 years later we had reached a population of 3 billion, and finally by 1975—after only another 15 years—we had reached 4 billion. The 5-billion mark was reached by 1985, and certainly by the year 2000 the number will fall between 6 and 7 billion, possibly *over* 7 billion.

In easier to visualize terms, this means we are currently adding more

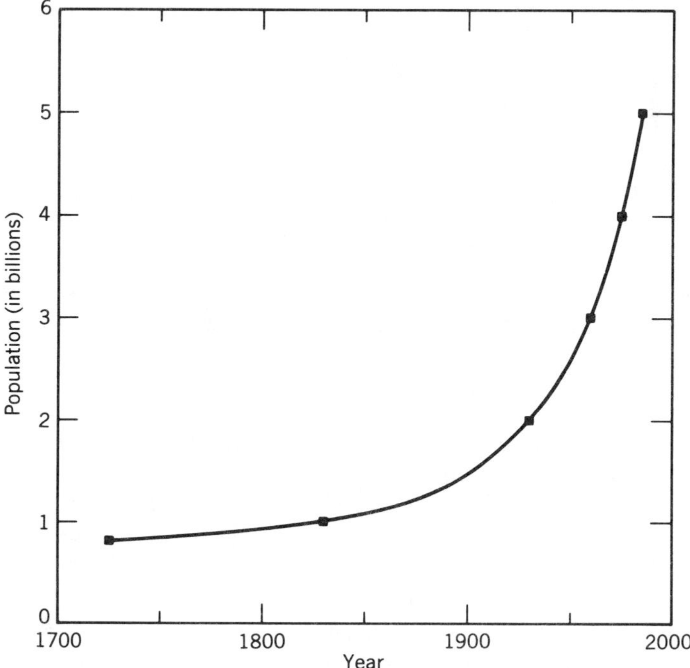

FIGURE 2-1 Changes in world population in the last 100 years.

people every year than the combined population of France, Belgium, and the Netherlands—or to put it another way, more people daily than are today living in the city of Flint, the third largest city in Michigan. The per hour increase is more than 8000 persons.

Why has population increased so rapidly during this century? It is *not* the result of an increase in birth rate but *rather* a decrease in death rate, the result of improved sanitation and disease control and/or eradication. Life expectancy in the United States in 1900 was about 50 years. Now it is over 71 years.

To make things even worse, population increases are still accelerating in some developing countries where food is scarce and must be imported. According to some estimates, by the year 2000, 80 percent of the world's population will be in countries now unable to raise all their own food. A further estimate is that 90 percent of the population *increase* that will occur during the last two decades of this century (1981–2000) will be in what are now considered developing countries. The population of India in 1981 was 684 million, an increase of 136 million during the decade of the 1970s, compared to an increase of 109 million during the 1960s—not a good showing for a great amount of work in the promotion of family planning. It is, however, a very creditable showing by the health and medical community, especially those working with child nutrition. In many developing countries, children are still considered as insurance against hunger during old age. Kenya, according to a World Development newsletter (1981) is increasing in population at the rate of 4.0 percent a year—a twofold increase in 16 years.

MOUNTAINS, SWAMPS, JUNGLES, ROCKS, DESERTS, PERMAFROST

Soil Management Principle

Much arable land lies in regions too cold for productive agriculture, and a lack of rainfall inhibits food production in vast areas.

Much of the land surface of the world can never produce crops. Some areas *could* be made arable, but at too great a cost under present standards of economy. By construction of terraces like those shown in Figs. 2-2, 2-3, and 2-4, some cropping is possible in mountainous areas, but the work is hard and hazardous. Intense storms may destroy crops and even the terraces. Large areas of Asia and Latin America are mountainous. The great Andes range extends the whole length of the continent of South America, occupying large portions of Peru, Chile, Bolivia, and Argentina.

Mexico is extensively mountainous (the Sierra Madres) and the northern

FIGURE 2-2 Pineapple in Okinawa. The land slope is almost 100 percent, too steep to walk down and carry a load, so the farmers lower the fruits by the overhead cables.

half of the country is largely desert, making it a country where agriculture is difficult. Extensive mountains also occur in the Central American countries.

Perhaps the best-known swamp and jungle area is the Amazon basin in Brazil. In many countries, as in Burma and Pakistan, some of the swamps cannot be drained off but are used to good advantage for paddy rice of the floating variety. Sometimes, however, excessive flooding ruins the crop.

Soils may be so thin, as the one shown in Fig. 2-5, that crop production cannot be satisfactory even if fertilizer and irrigation are available.

Much of the world's surface is desert. Seventy-four percent of the total area of Iran is arid or semiarid, having less than 250 mm of annual rainfall (see Fig. 2-6). In fact, the entire Middle East, perhaps we should say much of Asia, is either too dry or too cold for effective agriculture.

Vast regions of the north in America and Asia are too cold for production of food, even too cold for trees, the result of high altitude as well as latitude (see Fig. 2-7).

FIGURE 2-3 Tea may be grown on a steep hillside, but unless the slopes are carefully terraced, erosion will soon destroy the planting.

FIGURE 2-4 Terraced rice in the valley and terraced sugarcane on the hillsides, one row of cane on each terrace. All work must be done by hand. Literally, draft animals would fall out of the field.

FIGURE 2-5 Very thin soils over solid or coarsely disintegrated rock cannot be highly productive.

FIGURE 2-6 A desert area in Iran, quite typical of the entire Middle East. According to Foth and Schafer (1980), 36 percent of the world's land surface receives too little rainfall for crop production unless irrigation can be practiced.

FIGURE 2-7 An area in Alaska where it is too cold for efficient agriculture.

SUPPLIES AND EQUIPMENT LACKING

Subsistence agriculture depends on natural soil-formation processes for soil rejuvenation after a very short cropping period. In many of the developing countries, yields are low because fertilizer and pesticides are difficult to obtain and pests, including weeds, are rampant. Subsistence farmers are badly in need of such modern management inputs as fertilizers, pesticides, and improved crop varieties to help them produce the food and fiber so badly needed by their families.

The water buffalo (Fig. 2-8) is a dependable, powerful draft animal. It can work all day in the water (horses and oxen cannot), but it is slow. The garden tractor (Fig. 2-9) is much faster, but it is very expensive. It cannot be raised on the farm as can the buffalo.

In the more progressive rice-producing countries, such as Japan, Taiwan, the United States, and those in southern Europe, rice is now largely harvested with combines, but in most developing countries grains are still cut by hand and threshed by human foot power, as shown in Fig. 2-10.

Religion, customs, and traditions may be a hindrance to food production in some countries. Certain animals (cattle and monkeys in India) are sacred. They are allowed to eat food needed by the people, and/or they are allowed to destroy crops. Some people, even in our own country, refrain from the use of chemical fertilizers and pesticides, even though the penalty is much smaller yields of needed food. The National Academy of Sciences has recently estimated that total withdrawl of pesticides would result in U.S. consumers spending 30 to 40 percent of their income for food instead of the 17 percent

FIGURE 2-8 Most of the world's rice is grown in water. Low dikes hold the water in the "paddy," a term also applied to the crop until it is processed. The tireless, but slow water buffalo is the only draft animal that can tolerate working in the water.

FIGURE 2-9 In some countries, particularly Japan, the small two-wheel tractor is replacing the water buffalo. With the high cost of machine and fuel, even the more rapid operation may not be economical where the growers are short of funds.

FIGURE 2-10 Rice harvest in Taiwan in 1962. Where the paddy can be drained, much of the harvest is now done by small two-row combines.

they now spend. In many countries, pesticides are practically unknown by small farmers.

We do not mean to downgrade religion, but temples, images, and shrines cost a lot of money needed by the worshipers for family necessities (see Figs. 2-11 and 2-12).

WATER MAY BE SCARCE AND POLLUTED

According to *World Food Program News* of January–March 1981, half the world's people — over 2 billion — are without reasonable access to safe and adequate supplies of water. The World Health Organization (WHO) estimates that inadequate water and sanitation are responsible for 80 percent of the illness in the 100 or more food-short countries. Furthermore, the many hours required for replenishing the daily water supply cannot be used in growing food and the poor health resulting from drinking polluted water greatly lessens people's ability to work.

Most of the developing countries are in the tropics or subtropics where rainfall may be plentiful on a yearly basis but is seasonal. Cropping during the dry season may be impossible because pumps are not available and water storage and transportation systems are nonexistent. Thus, only one rain-fed crop a year can be grown. Diesel-powered pumps and dams such as the one shown in Fig. 2-13, will do a lot to facilitate year-round cropping with greatly expanded food production. Under adequate irrigation, a tropical climate may allow the production of four to six crops each year on the same land.

EROSION: A WORLD PROBLEM

Soil Management Principle

Erosion by water and wind, increasing throughout the world, is responsible for widespread crop failures. Hunger will be the result.

Considered worldwide, perhaps erosion by water and wind is still the greatest constraint to eliminating world hunger that faces us today. This, despite 60 years of intensive soil conservation research and teaching! Perhaps our own American farmers are most guilty of disregarding the lessons they should have learned just a relatively few years ago. Row crops are notably poor protectors of the soil, whereas small grain and hay crops keep more of the soil covered so that intensive rainstorms and high winds do not carry the soil away. Michigan farmers in 1959 planted only 30 percent of their land in row crops. Because row crops bring greater net returns, that percentage had grown by 1978 to 60. Erosion, of course, has increased.

Bulgaria is a mountainous, rolling country. The Bulgarians have very little land with less than 2 percent slope, and they have almost disregarded erosion control efforts since their present regime was established. They have paid dearly by serious erosion (see Fig. 2-14). In many countries, the desire to use big machines on more acres has discouraged such erosion-control practices as contouring and strip-cropping.

FIGURE 2-11 Some very religious people spend large amounts of money and time on churches, temples, and images. In a hungry world, perhaps more should go for fertilizers, good seeds, and pesticides.

FIGURE 2-12 Shrine in a hotel yard where food is always provided for the gods displaced by the hotel construction. Many homes have shrines in their backyards.

Population pressure has made it necessary for shifting-cultivation farmers to shorten the fallow period or increase the number of years of cropping. Increased erosion has resulted. The need for more crop acres has pushed cropping onto steeper soils, again setting the stage for increased losses of topsoil and eventually for lower yields.

The cutting of forests for firewood and lumber is leaving vast areas open for increased erosion. According to Douglas J. Bennet, Jr., former administrator of the Agency for International Development "The forests of the world are being cut down at a rate of 50 acres a minute." Much of the destruction is occurring in semiarid regions where tree growth is slow, so a reversal of the trend is difficult. Actually, one-third of the world's population depends on wood for cooking and heating. Just the boiling of water for drinking requires a continual search for fuel.

Fortunately, considerable research and teaching is now being done to increase the amount of firewood available in tropical and subtropical countries. Some of the work is published by the National Academy of Sciences (1980).

FIGURE 2-13 This reservoir provides for storage of irrigation water.

FIGURE 2-14 Serious gully erosion in Bulgaria. The country is very mountainous; there is little land with less than 2 percent slope.

HUNGER: THE RESULT OF POVERTY

Soil Management Principle

Government administrators can help. Only poor people starve.

Except for unusual cases, only poor people starve. In almost all countries, those people with money to buy will find food. Most of the people in developing countries live in rural areas. Their income is from agriculture. Population pressure has so increased the demand for land that prices put it beyond the reach of many. They must work by the day at pitifully small wages. All they earn is not sufficient for clothing, shelter, and proper nutrition.

The favored few who through village or family connections are fortunate enough to have possession of a hectare of land may be forced to do all the work by hand with machete and hoe because money is not available to purchase mechanical equipment. For such people, farming is very hard work at a subsistence level of living. Much of the responsibility for relieving poverty-induced hunger lies at the doors of government administrators. Agricultural research and extension has done and is doing much to increase food production, but problems pertaining to the local economy must largely be solved by public officials. Let us hope they accept their responsibilities and help those in need.

REFERENCES

Dewan, M. L., and J. Famouri (1964). *The Soils of Iran.* Food and Agriculture Organization, Rome.

Foth, H. D., and J. W. Schafer (1980). *Soil Geography and Land Use.* Wiley, New York, pp. 345–351.

Commission on International Relations (JH 205) (1980). National Academy of Sciences-National Research Council, *Firewood Crops: Shrub and Tree Species for Energy Production.* NTIS Accession PB81–150–716.

Popenoe, H., et al. (1981). *The Water Buffalo: New Prospects for an Underutilized Animal.* National Academy Press, Commission on International Relations (JH 217), National Academy of Sciences–National Research Council, Washington, D.C.

World Development Letter (1981). ISSN 0162–7007, p. 95. Newsletter Editor, AID/OPA, Room 4897, Washington, D.C. 20253.

Soil Water, Drainage, and Irrigation

Water is indispensable, nationally and on the individual farm. Water and soil are so intimately related that a management plan for one must include the other. In arid and semiarid regions and most of the time in humid areas, water supplies must always be managed. Methods vary according to soil and climate.

Soil Management Principle

Water and soil management plans must be closely linked — one without the other leads to disaster.

FORMS OF WATER IN SOILS

Farm planners should have a basic understanding of the relationships between soil and water. Otherwise, they will have difficulty in evaluating the importance of such water-control practices as drainage, tillage, mulching, and irrigation.

Water is held in soil as films around the particles and as free water or droplets trapped in the spaces between particles and/or aggregates. In addition, a small amount is held as vapor since the air within moist soil always has a relative humidity greater than 99 percent. Even dry soil particles are enclosed within a film of water, with the thickness varying inversely with the diameter of the particle and directly with the moisture content of surrounding air. Soil scientists call the soil "dry" when the surrounding air temperature is 110°C (oven dry). Soil water percentage is based on oven-dry weight. Thus, if an acre 15 cm (6 in.) of soil weighs 908,000 kg (2,000,000 lb) and it contains, at a given time, 20 percent water, the total weight of water in the acre 15 cm (6 in.) is 181,600 kg (400,000 lb).

Certain soil-water relationships are shown in Fig. 3-1. A **dry** soil, in the field, is one containing only the amount of water that it can absorb from the air — only hygroscopic water, which is unavailable to plants or soil organisms.

Moist soil varies in water content from the hygroscopic coefficient to field capacity, through the range of capillary water — that fraction which is available to plants and soil organisms. **Field capacity** has been defined as that

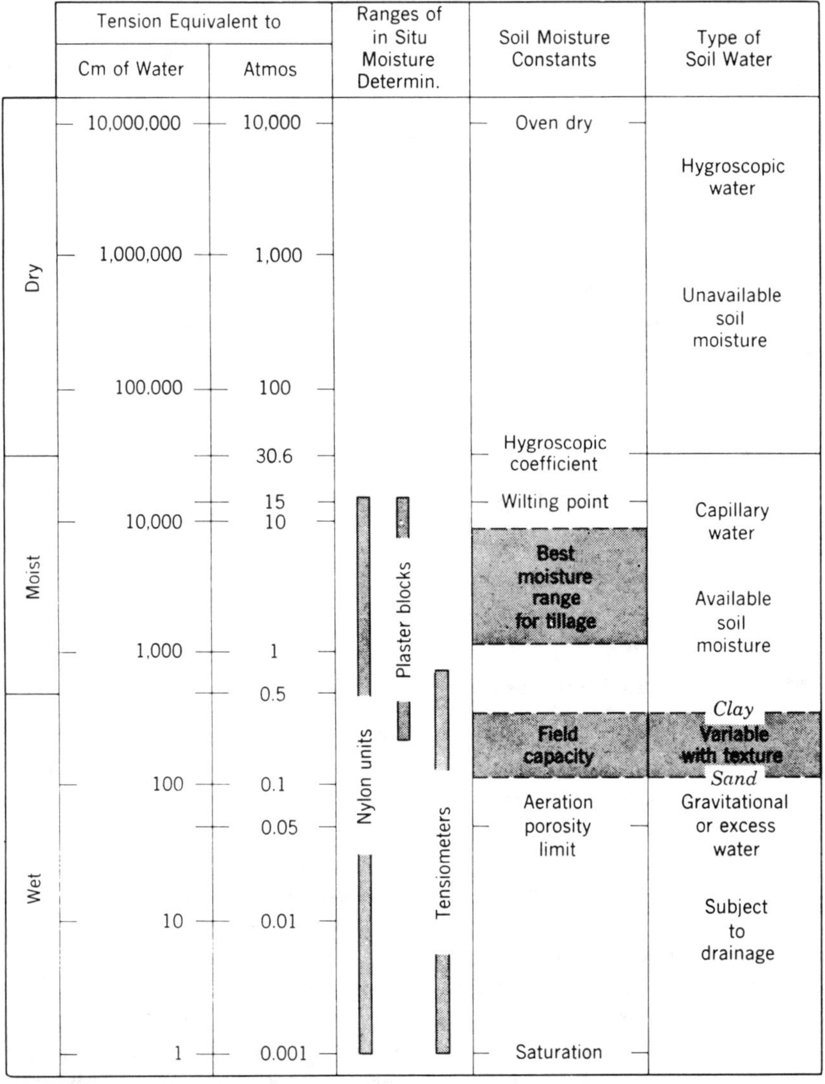

FIGURE 3-1 Soil-moisture relationships based on tension. (Somewhat modified, but in the main taken from H. Kohnke. "The practical use of the energy concept of soil moisture. *Soil Sci. Soc. Am. Proc.* 11:64–66, 1946.)

amount of water remaining in a soil profile 2 or 3 days after a saturating rain or irrigation, with no surface evaporation assumed.

Wet soil contains water in excess of field capacity — water subject to drainage or removal by force of gravity.

Tension

The force with which water is held in soil, commonly referred to as **suction** or **tension**, may be stated in bars, atmospheres, or centimeters of water. Figure 3-1 shows that tension decreases as moisture content increases. In other words, tremendous force (energy) is required to remove water from dry soil, whereas a very small amount of force will remove water from a wet soil.

A soil that contains only slightly more water than that which is hygroscopic in nature is unable to support plant growth, because the soil particles hold the water with such force that plants cannot remove it fast enough to avoid wilting. In fact, the soil-water matric potential must be reduced to about −15 bars before plants can survive. That potential has been defined as the **wilting coefficient**, a soil-water potential below which plants wilt permanently. It should be remembered that as soil water lessens, it becomes increasingly difficult for plant roots to absorb water. Then when transpiration rate is high, temporary wilting becomes a way of reducing the rate to avoid or postpone permanent wilting. Such temporary wilting occurs at considerably less than −15 bars matric potential and varies with atmospheric conditions.

Available water, the fraction between the wilting coefficient (−15 bars), and field capacity (−1/3 bars), is the important fraction as far as agriculture is concerned. Available water capacity is greatest in medium-textured soils and least in·those of coarse texture. Clay soils also decline in water capacity because of their very high wilting coefficient, a function of the extreme force with which clay particles hold water films.

Sandy soils have a low available water capacity because of the weak particle-film tension and the preponderance of large pores from which water drains readily.

SOIL WATER MOVEMENT

Optimum soil water is that fraction of the available water that ranges between suctions of −10 and −1 bars. Figure 3-1 shows that a water content below −1 bar may still be satisfactory for growth but too moist for tillage.

Soil Management Principle

Matric potential indicates the energy required to extract water — available water is the quantity that a plant can extract.

You are urged to keep in mind that matric potential measurements are not measurements of quantity of water, as on an acre-inch basis, but rather are energy measurements that indicate the suction necessary to remove water from the soil. Their use in soil-plant relationships is possible only through calibration to water percentages. Actual quantity measurements are possible only by sampling, drying, and weighing.

Water Movement

Water moves in the soil by percolation, capillarity, and vapor adjustment. **Percolation**, including infiltration, is caused by the pull of gravity. The objective of the soil conservationist is to get all precipitation to enter the soil where it falls. After water enters soil, its rate and extent of percolation depends on the texture, structure, organic matter content, and water content. Water movement by percolation, in arable soils, is usually complete within 2 or 3 days after a saturating rain. Field capacity measurements are usually made after about that lapse of time. Field capacity is often estimated in the laboratory by determining the soil water content at $-1/3$ bar by using a pressure apparatus. An impervious horizon would appreciably slow up percolation, whereas in coarse, uniform sands the rate would be much faster. Depth of the soil column to the free-water table determines how completely gravity may pull the water out of the soil. Tall columns exert more suction, which more completely dries the surface soil. To illustrate this principle, twist layers of saturated cheesecloth or a bath towel into a roll. Suspend the roll between the fingers, keeping it in a horizontal position until it stops dripping. It then illustrates a saturated **shallow** column of soil. Now turn the roll to a vertical position. Water will immediately start dripping from the lower end. The roll now illustrates a **deep** soil column. The increased suction causes the upper end to lose much of the water that it held while in the horizontal position.

The same principle may be illustrated with two **porous-bottom** flowerpots filled with saturated soil. After they cease draining, place one pot on top of the other. Drainage will start again from the lower pot because, in effect, the soil column has been doubled in height.

Capillarity refers to the movement of the capillary water by film adjustment. When gravitational water is not present in the profile, capillary movement may be in any direction. It is not affected by gravity, but rather is the result of adjustments caused by suction, the force that holds water in surface films.

Moisture films tend to be uniform in thickness over the entire surface of a single particle, assuming no interference from surrounding particles. All particles of equal diameter would have equally thick films. As water is removed from one side of a soil particle, perhaps by a root or by evaporation, flow starts in that direction and continues until water withdrawal ceases and uniform suctions are reestablished (see Fig. 3-2).

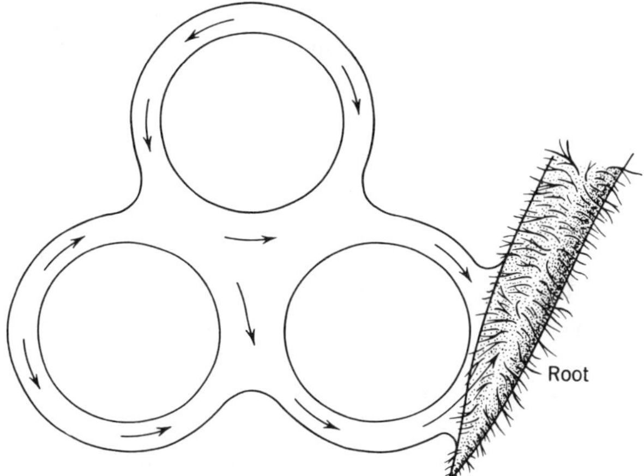

FIGURE 3-2 When water is taken in by a root, the thinning of films at that point results in a movement of water from around the particle and from surrounding particles where films are continuous. Such water movement is called **capillary adjustment.**

Vapor adjustment accounts for small but sometimes quite effective soil moisture movement. The air within a soil with more moisture than the hygroscopic coefficient is always at 99+ percent relative humidity. The resulting vapor pressure varies with temperature changes that bring about air movement and often moisture precipitation or internal evaporation.

Water as liquid or vapor is constantly moving in soil. It flows through the pores between soil particles and aggregates as long as gravitational water is present, then through capillary films to equalize tension, as film water is withdrawn by roots or through evaporation at the air surfaces.

Since water is held most strongly by fine soil particles (clay), it does not readily move from a fine soil layer to a coarse layer. Likewise, a thin layer of clay under a medium-textured topsoil will slow up penetration of water from above until the soil above the clay layer is almost saturated, in other words, to the extent of formation of a "perched" water table. A plow pan may actually bring about this effect.

<hr>

Soil Management Principle

Artificial soil profiles should be constructed of soil materials that are uniform in texture.

<hr>

Landscapers have been known to use coarse sand as fill material, then cover it with fine topsoil. If the surface is sloping, this means simply that the fine soil will act as an impervious roof to prevent the sand subsoil from becoming moist, an especially bad medium for trees, perhaps even worse than for the subsoil to be clay. The solution, of course, is to mix the soil materials until the result is a homogeneous soil profile.

WHY DRAINAGE IS NECESSARY

Soil conservation, synonymous with soil building in its fullest sense, is possible only in soils that are favorable for plant and microorganism growth. Saturated soils, or those near saturation, are poorly aerated and so cannot support activities requiring oxygen. Figure 3-1 shows that soils with suctions of less than about 50 cm of water are below what some have called the aeration porosity limit. The large pores, through which air may move, are so completely filled with water that oxygen movement practically ceases. This condition must be corrected by drainage. Several improvements from drainage may be listed.

1. Time is saved because farmers would otherwise be waiting for soils to dry sufficiently for tillage.

2. Lower soil water lets soils warm up more rapidly.

3. Aeration is improved by drainage. When a medium-textured soil has its pore space filled half with air and half with water, growth conditions are ideal.

4. Root respiration is dependent on oxygen supply, and nutrient uptake is dependent on respiration.

5. Deeper, faster root development occurs in well-drained soils. This minimizes later injury from drought. Shallow rooting induced by early season wet soils results in starvation of plants for nutrients that are sometimes quite plentiful in lower horizons. This is especially true of phosphorus in some soils.

6. Drained soils develop better structure. Drying, due to surface tension of water, is a powerful aggregating force.

   Standing water destroys aggregates. This action is simply a slaking effect. The individual particles are pushed so far apart by the expanding water films that structure of the entire granule becomes weak.

   Structure of poorly drained soils is further jeopardized by farmers who do not, and sometimes cannot, wait for sufficient drying before they start tillage operations or before they turn cattle into pasture fields, thus causing much damage.

7. Winter killing of wheat, alfalfa, clover, and other perennial or fall-planted crops is less in well-drained soil. Free water at the soil surface is largely responsible for winter heaving of plants.

8. Drainage increases the effective size of the farm enterprise by bringing more land under cultivation.

9. Drained soils harbor fewer diseases.

10. Weed control is easier on drained soils.

11. Drainage makes it possible to leach saline soils.

KINDS OF SOIL NEEDING DRAINAGE

Drainage systems are expensive, so they should be carefully planned. Many soils do not need artificial drainage. Others cannot be adequately drained. Tile may be successful in some soils, not in others. Over large areas, both tile and open ditches are quite satisfactorily employed.

Fine-Textured Soils

Fine-textured, slowly permeable soils occupy lake plains, valleys, and potholes. Surface waters usually soak away too slowly to depend on percolation as a means of drainage. In most instances, early season water tables are high in such soils. Thus they may be quite permeable and still need artificial drainage.

Color can be used as an indication of drainage condition. A dark surface indicates poor natural drainage — that poor drainage conditions prevailed during most of the period of soil formation. Uniformly red to brown subsoils indicate a high state of oxidation and therefore good drainage. Mottled red, yellow, and gray subsoils indicate intermediate drainage, and gray subsoils are indicative of poor drainage where little oxidation has occurred.

Some soils are so slowly permeable that drainage with tile is not practical. Such an area is shown in Fig. 3-3. Only a very narrow strip was benefited by each tile line, as shown by the height of the alfalfa. Donahue, Miller, and Shickluna (1983) suggested a laboratory method of determining the possibility of tile-draining a slowly permeable soil. The determination is based on the assumption that water held at $-1/3$ bar closely approximates field capacity. Then

$$\% \text{ drainage capacity} = \% \text{ soil volume that is pore space} -$$
$$\% \text{ water-filled pore space at field}$$
$$\text{capacity, on a volume basis}$$

It is assumed that pore space filled with water at field capacity cannot transmit water into tile drains. Soils with low drainage capacity cannot be tile-drained.

Drainage Guide

The Michigan Experiment Station and Cooperative Extension Service and the U.S. Soil Conservation Service have published *Recommended Standards for Drainage of Michigan Soils*. Listed are the soil types that are generally in need of drainage with recommendations as to methods to be employed. You are urged to consult local authorities for similar information on soils of your area.

FIGURE 3-3 A new alfalfa seeding on Brookston clay loam soil in the Michigan Saginaw Valley area. The soil was so slowly permeable that drainage by tile was effective only a few feet each way from the line. Note the height of the alfalfa directly over the tile.

Sandy Soils

When sandy soils overlie impervious subsoils or hard pans, they are very difficult, frequently impossible, to drain. Sometimes a subsoiler is effective. This is true if the pan is thin and is underlain by permeable soil. On the other hand, subsoiling will do little permanent good if the loosened soil is still underlain by impermeable material.

Tile drainage in this kind of soil is not likely to be satisfactory if the tile must be laid in slowly permeable clay loam or clay. Such soils are even more difficult to drain than is soil that is fine-textured right at the surface, because surface-water removal is more difficult from the sand surface.

Inherent productive capacity should be considered before money is spent on sandy-soil drainage. The wise manager may decide to use the land for permanent grass, a crop that does well on wet soils.

Irrigated Lands

Irrigation may intensify the need for drainage in humid areas. Tight, impermeable soils usually grow shallow-rooted crops that suffer quickly during drought periods. Thus, a need for irrigation is always evident. Rain is always expected in humid areas, but one never knows when it will come. Irrigations may be followed immediately by heavy rains. Tile drains may then carry away the excess water before injury results.

In semiarid and desert regions, soluble salts are usually a problem. It is necessary to apply enough water to cause drainage or salts will accumulate.

The need for underdrainage is at once evident. Leaching, which is discussed in Chapter 16, is a common practice in the reclamation of saline soils.

DRAINAGE PRACTICES

The goal of drainage is to remove water that is in excess of field capacity. Drainage practices must be tailored to the type of soil involved, particularly with respect to topography and texture profile. Local situations pertaining to the nature of the total watershed, land ownership where outlets are involved, crops to be grown, value of the land, and technical assistance available must be considered.

Drainage Systems

A **natural drainage** system is planned for an area having rolling to hilly topography where the problem is largely the removal of excess water from valleys and potholes. In many respects, such a system is more difficult to engineer than is one designed for flat country. Outlets may be difficult to reach and are often natural creeks too shallow for adequate fall in the tile lines. Furthermore, in such an area many kinds of soil are involved.

We start at the outlet to observe how much fall is available. Mains must empty into the creek at a level **above** the normal creek water level. If the tile outlet is under the water level, we "shoot" back to the lowest point in the area to be drained and decide how near the surface we must place the upper end of the tile. It is simple arithmetic then to calculate how much fall is available between the upper and lower ends of the tile.

In this kind of a system, tile lines are seldom straight and there are no rules regarding distance apart. A single line may be carrying the runoff from an entire watershed. Experienced drainage engineers should be consulted. They have graphs and tables for determining tile size and minimum slope, and they are able to give advice on special structures that may be needed. In organized Soil Conservation Districts, U.S. Soil Conservation Service engineers are available for such advice.

Systematic drainage systems are laid out in relatively flat lands where all the soil needs drainage. Most lands have at least a little slope that may be used to advantage. Some lands, however, are practically flat. Mains and laterals are varied with the location of the outlet, natural slope, and the dimensions of the farm. Laterals cannot be too long, and they require more fall because they are smaller. Eight-inch (20-cm) mains may be successfully laid without fall where it is necessary. Figures 3-4, 3-5, and 3-6 are suggested systematic drainage patterns.

Cut-off or hillside patterns are for the purpose of drying up a hillside seep so the water will not spread out in the lowland. Drains thus installed sometimes remove the need for more extensive patterns, thus effecting considerable savings.

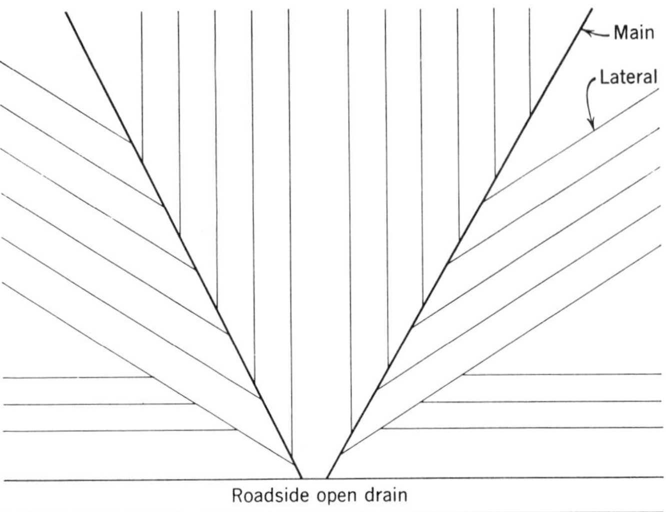

FIGURE 3-4 An actual tile pattern being used on a farm in Saginaw County, Michigan. It has operated successfully for more than 30 years. The rather unusual pattern was the result of an attempt to take advantage of the topography. The two mains were placed in natural waterways. This made it possible to obtain more fall in the laterals. The pattern is really a combination of two herringbones.

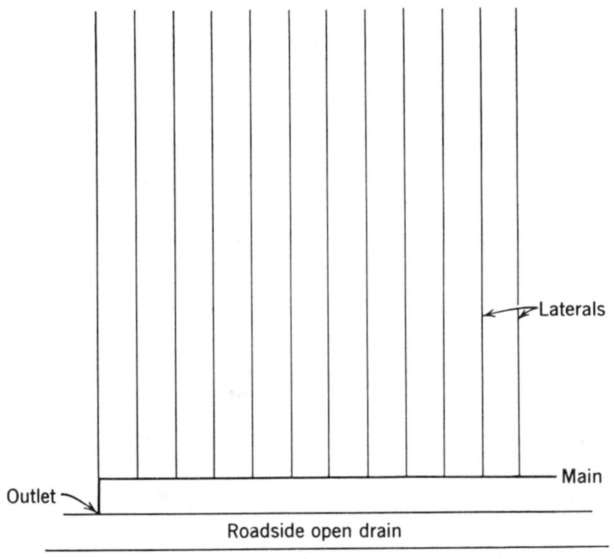

FIGURE 3-5 A grid drainage pattern with one main near the road ditch outlet. One tile outlet serves the entire farm. The size of the main would be dependent on the area of the farm. Laterals should empty into the top of the main.

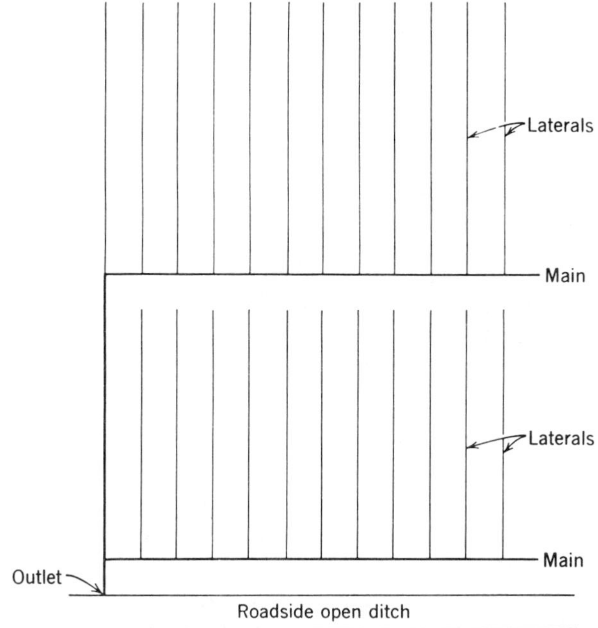

FIGURE 3-6 A grid drainage pattern designed for an area where lateral length is limited by lack of fall. Rather than use larger tile as laterals, to compensate for lack of fall, an extra main is installed to carry the water from the back half of the farm. Otherwise, the plan is like the one shown in Fig. 3-5.

Surface drains are the last resort for soils that cannot be drained with tile. Inadequate outlets may prohibit the use of under drainage, or soils may be too slowly permeable. Perhaps money for tile is not available. Whatever the reason, a decision to rely on surface drainage does not eliminate the need for planning.

Surface drains should be combined with a system of ridging and bedding, with broad channels constructed so that they may be readily crossed with farm implements. The ridges and channels (beds) should be parallel with the natural slope, so water may be conducted into natural waterways or dredge ditches. Slight contouring may be used to prevent erosion if slopes are too steep for parallel arrangement.

Deep open ditches at the boundaries or through the area are necessary for outlets. Water must be carefully transferred from a shallow surface channel into a deep open ditch. Erosion will start where water falls over a bank unless masonry or rock construction is used. Sumps may be used to lower the water before it enters the ditch, because even a few inches of abrupt fall will start a gulley if the falling water strikes soil.

Surface and under drainage are often combined. Even in fairly permeable soil, under drainage may be too slow to remove all the water, especially during flood periods where large watersheds are involved. A few large open ditches, which may also be used for tile outlets, and carefully planned, grassed waterways or diversion channels are necessary to prevent delays in the spring and serious crop injury at other times.

Mechanics of Tile Drainage

Space does not permit a full discussion of a subject as involved as tile drainage, but a few points will be discussed. Emphasis has already been placed on the importance of soil permeability. Even quite permeable soils may become difficult to drain with tile if soil structure is allowed to deteriorate.

Outlets

The mechanics of tile drainage should start with the outlet, usually a stream or open ditch. It should have sufficient capacity that the water level will drop soon after the peak flow from a heavy storm.

Pump outlets are often used, but they are expensive. An intensive type of agriculture, such as vegetable production, can support pumping to remove excessive quantaties of water. Pumping is necessary where outlets ditches are too shallow.

Tile Size

Several factors must be considered in estimating the size of tile needed for a certain job.

1. Area of watershed.
2. Vegetative cover on the watershed. Complete, permanent cover usually means relatively little runoff. Bare soil means much runoff and that it will be fast. Tillage practices should be considered if the watershed is under cultivation.
3. Slope of the watershed. If the hillsides are steep, the runoff will be fast and tile must be larger than if the slopes are gentle.
4. Nature of the soil on the watershed. Sandy, permeable soils mean more infiltration and less runoff, so tile size may be minimal.
5. Amount of fall available in the tile line. Tile size may be reduced where fall is adequate.
6. Tile should be 10 cm (4 in.) or larger in diameter, regardless of amount of water to be carried. Those smaller than 10 cm fill up too fast to be worthwhile. Lines longer than 402 m (80 rods) should be laid with 12.5-cm tile. The minimum diameter in organic soils should be 15 cm (6 in.), and the minimum tile length should be 60 cm (2 ft).

Tile Spacing and Depth

Recommended tile line spacings in mineral soils range from 10 to 35 m (2 to 7 rods). The narrow spacing is suggested for fine-textured, slowly permeable soils where high-value crops are to be grown. Twenty meters (4 rods) is usually considered a minimum for common field crops.

Tile are usually placed deeper in organic soils to allow for settling, and permeability is faster than in most fine-textured mineral soils. As a result, spacings as great as 40 m (8 rods) may be permissible. Water levels in outlet ditches should be controlled and adjustable in organic soils so that water tables may be kept high during nonproductive seasons, and so lessen subsidence.

The depth of tile should vary with soil type and topography. Recommendations for mineral soils vary from 75 to 120 cm (30 to 48 in.). Shallower depths are recommended for less-permeable soils, and then closer spacing of the lines is required. Depths as great as 120 cm may sometimes be accompanied by spacings as great as 35 m.

Lack of sufficient fall makes it sometimes necessary to place the upper ends of laterals at shallower depths. It is then wise to back-fill to the extent of mounding to give additional protection to the tile so that they are less likely to be broken by frost, heavy machinery, or subsoilers.

Blinding

It is always a good practice in clay soil, sand soil, or organic soil to cover the tile with an organic material to exclude soil particles. This practice is referred to as **blinding**. Some drainage experts feel that topsoil contains enough organic matter to be used directly on the tile. This may be true for medium-textured soils very high in organic matter, but additional organic material is generally recommended. Hay or straw, enough to fill the trench several inches when packed, is very satisfactory, because in the bottom of a trench, such material will last a long time. Coarse gravel may be used where straw and hay are not available.

Fall

It is very important that drainage tile be laid with sufficient fall to keep them from filling with sediment.

Diameter of Tile	Recommended Minimum Grade per 25 m (83 ft)
10.0 cm (4 in.)	2.5 cm (1 in.)
12.5 cm (5 in.)	1.75 cm. (0.7 in.)
15.0 cm (6 in.)	1.25 cm (0.5 in.)

Tile larger than 15 cm (6 in.) may be laid with still less fall where it is necessary. However, it is better to hold to the 1.25-cm grade.

Too much fall is dangerous. Where it exceeds 0.1 percent (2.5 cm/25 m), there is danger of tile washing out. Sometimes mains must be laid with still greater slope; if so, they should be cemented together. Bell tile are safer, although much more expensive.

Plastic Tubing

Corrugated plastic tubing has gained favor in the United States because it is better suited to machine installation. It comes in variable lengths (100 m or more) and in sizes ranging upward from 10 cm (4 in.). Placement machines are able to unroll the tubing as they dig the trench and move along (see Figs. 3-7 and 3-8).

The tubing is an improvement over tile in that it is cheaper (because less labor is needed) and alignment may be more easily maintained. Also, there is less chance of a washout if excess gradient is being used. A little special care is necessary, however, when plastic tubing is being installed. The trencher should leave a groove in the bottom of the trench to take the tube as it is laid. This ensures better alignment and provides support against bulging. When water comes into the trench soon after digging, some care must be taken to keep the tube from floating or back-filling may push it out of line. The machine shown in Fig. 3-8 was provided with spring teeth to immediately pull topsoil into the trench (Fig. 3-9). Further blinding with gravel is recommended.

FIGURE 3-7 Plastic tubing for drainage being laid by machine in Swaziland.

FIGURE 3-8 Plastic tubing for drainage being laid by machine in the United States.

Miscellaneous Suggestions

Tile should pass the quality specifications of the American Society of Testing Materials (ASTM). Clay tile may be used in any soil. Concrete tile may not be satisfactory in strongly acid soils, particularly organic soils.

Tile should be laid with 1/8-in. spaces between joints in mineral soils and with 1/4-in. spaces in organic soils. In loose soils and well-decomposed organic soils, the joints should be covered with tar building or roofing paper. Cradling on boards is necessary to preserve alignment in very sandy soils and in some organic soils.

Relief wells, junction boxes, and sediment traps are at times necessary for best results. Questions regarding their use and other questions regarding drainage practices should be referred to local drainage engineers or to local cooperative extension personnel.

SCIENTIFIC IRRIGATION

Fifty-five percent of the land area of the earth is arid or semiarid, with less than 50 cm (20 in.) of annual precipitation, and another 20 percent is subhumid, with 50 to 100 cm (20 to 40 in.) of precipitation. These figures show that water deficiency is a problem over at least 75 percent of the land area of the world. Indeed, short but sometimes very injurious periods of drought occur in the areas having more than 100 cm (40 in.) of precipitation. Nowhere can precipitation patterns be accurately predicted.

Modern-time irrigation agriculture was started by the Mormons in 1847 near Salt Lake City, Utah. It has grown until today more than 29 million acres in the 17 Western states are irrigated. Further expansion is hampered by a shortage of satisfactory water and by increased use of water for domestic and industrial purposes.

FIGURE 3-9 Spring teeth to cover plastic tubing after placement in a trench.

Water

Irrigation requires tremendous volumes of water. An experiment in New Mexico showed that 200 cm (80 in.) of water was required for top yields of alfalfa. It is necessary to supply the fraction lost by evaporation as well as the amount needed by the crop. In addition, there are frequently serious losses from seepage along flumes and ditches. This all means that the supply of water at peak use periods must be great. Some growers have found to their sorrow that "water right" means only the right to purchase and is no guarantee that water will be for sale.

Water quality is important. In arid regions, soluble salts may be too high. A complete discussion of this subject is presented by Thorne and Peterson (1954). They classified waters in six groups on the basis of conductivity. Class 1 includes waters ranging up to 250 micromhos.[1] Such water is not likely to cause trouble except on very low permeability soils where leaching is nil. It is well to obtain professional advice before using waters having a higher conductivity than 250 micromhos. Also, the sodium content of the water should be low. Obtaining advice on this point is also recommended. In recent years, successful irrigation has been accomplished with relatively high-salt water on coarse-textured soils using large quantities of water. Also, a contrasting technique has been "pinpoint" irrigation, where low volumes of water are placed in the immediate root zone of a plant. This conserves water, but salts may still accumulate.

Humid area water is very low in soluable salts. However, it is usually high in calcium and magnesium, and sometimes iron. Such water is commonly called "hard." It does not puddle soil as does saline water high in sodium, but it will act as a liming agent and may actually improve the structure. An old irrigation saying is, "Hard water makes soft land and soft water makes hard land." We do know, of course, that sodium-saturated water is soft, whereas calcium water is hard.

The liming effect of hard water used for irrigation in humid areas is seldom a serious problem because the total amount added is generally small relative to the amount of lime needed to appreciably raise the pH of the soil.

Most irrigation waters in the Western states contain boron. Trouble will eventually occur where the boron content of the water exceeds 2 ppm. Among the plants tolerant to boron are sugar beet, cauliflower, garden beet, alfalfa, lettuce, and cabbage.

Sodium chloride is in some of the deep well waters in the Great Lake states. Where the concentration is high, the water should not be used for irrigation.

---

[1]Micromhos, more completely means "micromhos/cm at 25°C." A micromho is a reciprocal microohm. One micromho is 1/1000 of a millemho or 1/1,000,000 of a mho. In SI units, 1 mmho/cm = 1 dS/m.

Water Distribution

Arid Regions

Furrow irrigation and flooding are commonly employed in the Western states. Such methods require careful soil leveling and sloping and an extensive system of flumes and canals. When economically feasible, permanent canals should be lined with concrete or plastic to avoid seepage. The canals are usually built on ridges so that water may be turned or siphoned into the rows or strips. Improperly shaped surfaces result in erosion and unequal distribution of the water. Figure 3-10 illustrates furrow irrigation; Fig. 3-11 illustrates an irrigation ditch.

Sprinkler irrigation has increased during recent years in both arid and humid regions. The increase has been rapid since aluminum pipe came on the market. Sprinkling is particularly advantageous for very permeable soils because of the lesser amount of water required. The cost of a sprinkler system is usually greater than the cost of land-forming for flood irrigation, and the labor involved in moving pipes and sprinklers is greater than for flooding. These are perhaps the primary reasons for the predominance of flood irrigation in the Western states.

Throughout the Midwest (eastern Michigan to western Nebraska) center-pivot sprinklers are used and are increasing in popularity. One unit that is 394 m long can revolve around the pivot in 1 day or less and apply 1/2

FIGURE 3-10 Furrow irrigation. Two-inch gated siphon tubes set up on a new irrigation ditch, every furrow being irrigated. This entire farm is irrigated by the use of these tubes. Reuben Saver farm, Jerome, Idaho. (U.S. Soil Conservation Service photograph. Courtesy of Eldo Bectke and Harold Harris.)

FIGURE 3-11 Permanent, concrete-lined irrigation ditch built on a ridge so water may be turned or siphoned into this rice field in Taiwan.

in. of water on about 54 ha (132 acres) of land. Such a unit is carried on 13 pairs of wheels, each pair powered by an electric motor. Each motor is electronically started and stopped so the 13 pairs of wheels stay in a straight line. Incidentally, the outside motor runs all the time. Figure 3-12 shows an area in Michigan on the Muskegon County Wastewater farm where center-pivot sprinklers were operating. Other sprinkler systems include solid set, linear traveler, and big gun. The selection of the best one for a particular site may depend on economics, water supply, and site characteristics.

Trickle (drip) irrigation is gaining in favor where water is especially costly or in short supply, as in arid regions. Recently developed equipment makes it possible to move water directly to such individual plants as tomatoes, berry bushes, and fruit trees.

Moisture barriers, in some places made of asphalt, are very effective in keeping irrigation water from percolating below the root zone of deep sand soils. This subject is discussed in detail in Chapter 21.

Humid area soils are not adapted to furrow or flood irrigation. The more intensive weathering has formed topsoil that would be lost in a leveling procedure. Also, the saturation from flooding would cause puddling of clay soils, and the sandy soils are too permeable. Water sources are commonly ponds or rivers at considerably lower elevations than are the fields. Thus pumps and pipe are necessary to force the water to the fields. It then seems logical to continue the distribution by pipe and sprinklers.

FIGURE 3-12 Center-pivot irrigation system in operation on the Muskegon County Michigan Wastewater farm.

When to Irrigate?

In an arid region where rain never falls during the growing season, the question of when to irrigate is rather simple. One has but to determine the consumptive use of water for the crop to be grown and the water-holding capacity of the soil to the depth of rooting. These data are available at most state experiment stations. Then, if it is assumed that irrigation should be started when available water has fallen to 50 percent, a determination of water in centimeters (inches) will show when to apply the amount of water necessary to bring the water to 100 percent available moisture.

Soil Management Principle

Need for and amount of irrigation should be based on soil moisture measurements.

Bouyoucos Block Method of Moisture Determination

The determination of available moisture by electrical conductivity has greatly simplified the answer to the question "When to irrigate?" Plaster of

Paris blocks (see Fig. 3-13) in which two electrodes are imbedded are placed in soil previous to or at the time of planting. The blocks should be in the region where most of the roots will be, or at any depth where one expects to control moisture levels. Soon the moisture contents of the soil and the blocks become equal.

Wire leads from the electrodes are extended to any convenient place in the field. This may be a stake in the row, the fence at the field border, or even a central shelter. When a moisture reading is desired, the leads are attached to the Bouyoucos moisture meter for a direct reading in terms of percentage available moisture remaining in the soil. Only a few seconds are required for a reading. Figure 3-1 shows the range over which the blocks are effective.

When the meter shows that available moisture is down to 25 percent, watering should be started and continued until the meter shows 100 percent available water. It is wise to have blocks at two levels. We can then tell how fast the irrigation water penetrates and can thus estimate rather accurately when the water should be turned off.

The Tensiometer Method of Moisture Determination

The **tensiometer** is a porous cup at the end of a tube filled with water (long enough to reach the zone where moisture is to be measured) and with a vacuum gauge at the top of the tube. As water moves out through the porous cup and into the soil, a vacuum is formed at the top of the tube. The range of

FIGURE 3-13 Plaster of Paris block and improved Bouycoucos moisture meter. The meter is calibrated to be read as percentage available moisture remaining in the soil. The block and meter make scientific irrigation easy.

tension that can thus be measured is much less than with the resistance blocks. Figure 3-1 shows that the limit of measurement would be between -0.5 and -1.0 bars. Such soil would still be above the best moisture range for tillage.

Other Methods of Moisture Determination

Neutron moisture probes were introduced a number of years ago as a rapid means of measuring soil moisture in the field. Access tubes (generally 5-cm aluminum pipes) are established in the field to a convenient depth. When the moisture measurement is to be made, a neutron source is lowered to a measured depth in the access tube and a measurement of the back-scatter of the neutrons is made. This measurement is related to the moisture in the soil, so an accurate measurement of moisture content can be made once the instrument is calibrated. This method has several advantages. It is very rapid once the access tubes are installed, and a single measurement can be made in about 1 minute. Measurement can also be made easily with depth, so one gains information about the moisture status of the entire soil profile. The major disadvantages for a farmer are that the instrument is expensive and that it requires a license to own and operate because it contains a radioactive source. However, consulting companies may offer this service in the future.

Irrigation may be scheduled based on evapotranspiration and rainfall data. Microcomputers make it possible for a farmer to have a program that will predict the time to irrigate based on weather data. This is still in the developmental stage, but it appears to offer promise in the future.

How Much Water is Needed?

The rule-of-thumb method of adding a certain amount of water each week is not good. Too many people add "2 in./week." Water is expensive; in some regions it is scarce. Too much water may wash plant nutrients below the root zone. Leaching may be beneficial in arid soils, but not in the Eastern humid soils or even in most subhumid soils.

Growers should determine the location of most of the roots of their crops. Soil phase will cause this to vary, so an exact depth cannot be given. Those who use the Bouyoucos electrical conductivity method of moisture measurement should place one moisture block at the low point of the root mass and another at about the middle of the root mass. Enough water should be applied to make the high block read 100 percent and the lower block somewhat less, perhaps 90 percent, to be sure that water is not wasted. Moisture measurements are particularly important in sandy soils because total water-holding capacity is relatively low and permeability is high. With good water control, such soils can be made as productive as soils of finer texture. Those who farm *sand* soils should consider the use of moisture barriers (see Chapter 21).

Research and Irrigation

Research has made scientific irrigation possible. The work is continuing, and new and more complete information will be forthcoming. Anyone who plans an irrigation system will be well repaid for seeking out the latest information from those best qualified to give it. The U.S. Agricultural Research Service at Beltsville, Maryland, and at other stations across the nation, and many state experiment stations, are engaged in this work. You are advised to seek answers to local problems from your extension service.

REFERENCES

American Society of Testing Materials (1955). Specifications for Drain Tile. ASTM C4–55. The Society, Philadelphia.

Donahue, R. L., R. W. Miller, and J.C. Shickluna (1983). *Soils, an Introduction to Soils and Plant Growth*, 5th ed. Prentice–Hall, Englewood Cliffs, N.J.

Drablos, C. J. W., E. H. Kidder, M. L. Palmer, et al. (1973). *Corrugated Plastic Drain Tubing*. University of Illinois at Urbana-Champaign, College of Agriculture Cooperative Extension Service Circular 1078.

Gardner, W. R. (1979). How water moves in the soil. *Crops and Soils*. November; 13–18.

Kohnke, H. (1946). The practical use of the energy concept of soil moisture. *Soil Sci. Soc. Am. Proc.* 11:64–66.

Richards, L. A., and S. J. Richards (1957). Soil moisture. *In Soil — the 1957 Yearbook of Agriculture*. U.S. Department of Agriculture, Washington, D.C., p. 49–60.

Thorne, D. W., and H. B. Peterson (1954). *Irrigated Soils, Their Fertility and Management*, 2nd ed. Blakiston, New York.

Vitosh, M. L. (1977). *Irrigation Scheduling for Field Crops and Vegetables*. Cooperative Extension Service, Michigan State University, Bulletin E-1110.

CHAPTER 4

Organic Matter, Microorganisms, and Soil Structure

Organic matter makes up a small but extremely important fraction of all mineral soils. It is common knowledge that dark-colored soils are productive soils. Such knowledge is reflected in the frequent advertisements of black topsoil, commonly called "black dirt." Homeowners purchase such soil with assurance that it will grow better lawns than will the light-colored soil over which it is spread. They know furthermore that gardening is easier and growth faster in such soil, and since it does not bake and crust, water soaks in readily and seeds are more likely to germinate and produce seedlings.

Corn[1] yields, other factors being equal, are always higher on soils with higher organic matter content. Bronson[2] grouped 791 fields in two canning plant areas in Minnesota according to content of soil organic matter. Then he converted each yield to percentage of the yield average for the plant area. The percentages of organic matter plotted against yields of sweet corn resulted in a nearly straight-line relationship. Any single factor that can produce so positive a correlation with yield is surely very important and worthy of considerable study.

ORGANIC MATTER ACCUMULATIONS

Soil organic matter accumulates during soil formation. A flowerpot filled with crushed granite, standing on the outer windowsill, will not furnish sustenance for a geranium. It is not a soil. Let time elapse, however, and the mass of granite will become a soil. The accumulation of organic matter plays a

---

[1]Corn (*Zea mays*), so called in the United States, is known in the rest of the world as *maize*. In the British Isles, the term *corn* is applied to any commonly raised grain. To keep you mindful of worldwide terminology, both terms for *Zea mays* will occasionally be used.

[2]Roy Bronson, unpublished data obtained while employed by the Green Giant Co. in Minnesota.

significant and necessary role. Such an accumulation starts, no doubt, from the introduction of dust and bacteria from the air. Seeds are dropped by birds. Carbon dioxide from the air combined with water to form carbonic acid, which reacts with the minerals in the granite to release mineral nutrients. Bacteria fix atmospheric nitrogen for use by higher plants. Eventually, a soil is formed and the geranium grows.

Climate

Rate of organic matter accumulation depends on several factors. Over wide areas, climate is important. Jenny showed that within the climatic factors, temperature is the most important. The balance between growth of higher plants and decomposition of organic matter by soil organisms determines the extent of organic matter accumulation. Jenny showed that higher temperatures stimulate soil organism activity to a greater extent than they do plant growth. As a result, soil organic matter varies in the subhumid prairie region from 9 percent in northern Canada to less than 2 percent in the southern United States, through an annual temperature range of 0° to 21°C. Throughout the world, the highest organic matter percentages in mineral soils are found in cold climates where the annual growth of vegetation is small. In such regions, soils are frozen during much of the year and relatively cool the rest of the time. Organisms work slowly in decomposing organic matter under these conditions.

Soil moisture also affects organic matter accumulation. Jenny showed that humid prairie soils contain about 2 percent more organic matter than do those in the subhumid area at the same temperature. Desert soils, of course, contain still less. Virgin soils of Arizona were reported to vary in organic content from 0.1 to 1.0 percent. Even those low percentages, however, represent a very important fraction of Arizona soils.

In fact, Fuller stated that, "The continued productivity of the soils in Arizona will depend largely upon the replenishment and maintenance of the soil organic constituents." He likened the small amount of organic matter in the semiarid soils of the U.S. Southwest to the amount of water in a horse tank. It does not matter to the horse how much water is in the tank, just as long as replenishment is continuous.

Local Factors

Vegetation: Grasses result in marked organic matter accumulation whereas forests do not. The differences resulting from vegetation become less as annual temperature increases. Jenny showed that at annual temperatures of 21°C, forest soils and grassland soils approach each other in organic matter content; whereas in regions of 7°C annual temperatures, they vary greatly in this respect.

The annual organic matter contribution of trees is largely in the form of leaves that accumulate on the surface. Under the trees they remain wet much of the year, so decomposition is rather rapid. Tree roots are deep; they add little to the organic content of the surface soil.

Grasses produce extensive fibrous root systems in the surface soil. Over a long period of time (centuries), the roots are primarily responsible for organic build-up under this type of vegetation. In the American prairies, burning destroyed much of the top growth, but accumulation of soil organic content occurred, nevertheless. Roots must have been the source.

Drainage is probably the most important local factor in organic matter accumulation. Well-drained soils are relatively dry much of the year. Drought sometimes limits vegetative growth, but there is always sufficient moisture for organisms to function. Also low moisture results in good aeration. Decomposition is in reality a burning (oxidation) process. Increasing aeration has the same effect as opening the draft in the furnace — the fuel burns more quickly. We should remember, however, that an open draft is necessary for more heat. To obtain the good from organic matter, it must be allowed to decompose. Good management will provide replacement, and thus soil depletion will be avoided.

Soil Management Principle

Organic matter maintenance is essential for high production on either fine- or coarse-textured soils.

The upland, well-drained soils are light in color because they are low in organic matter. Decomposing organic matter darkens the soil. Soils on intermediate slopes are medium in color, whereas those in the lowlands are dark.

Where shallow water actually covers the soil surface during much of the year, surface organic accumulations result and organic soils (Histosols) are formed. They represent the extreme in organic matter accumulation.

Nutrients affect organic matter accumulation. Plant growth, other factors being equal, is in proportion to nutrient levels, especially the level of nitrogen. Modern soil management is taking the matter of nutrients into consideration. Where nutrients can be added to bring about greater yields of grains, tubers, or roots, more residues are returned to the land, and organic levels may be increased. The result is effective soil conservation.

Texture affects organic accumulation, because it affects drainage and aeration. Fine-textured soils contain roughly twice as much total organic matter as do sandy soils in the same region. A fine-textured soil, however, with 3 percent organic matter may be as badly in need of an addition of fresh material as would be a sandy soil with only half that amount. To be effective, organic matter must be in a state of active decomposition.

Topography, aside from drainage effects, plays a role in organic matter accumulation through its effect on temperature. Soils on north slopes contain more organic matter than do those on south slopes, and the lower temperatures at high altitudes slow decomposition to effect faster accumulation.

Moisture affects organic matter accumulation. On north slopes in the British Isles where moisture is seeping down the slope, peat soils (Histosols) actually develop in much the same manner as they do in potholes in Michigan, Wisconsin, or Minnesota.

FUNCTIONS OF ORGANIC MATTER

Organic matter admixes so completely with the mineral fraction of a soil that it is considered an integral part of the soil body. It is so necessary, for good production, that the organic fraction be maintained that we should take a look at the functions of soil organic matter.

Aids in Water Management

Residues or plants protect the soil surface from beating rain, resist wind action, and thus greatly aid in erosion control. Decomposing organic matter causes soil aggregation, which aids infiltration and increases pore space in clay soils. Thus water-holding capacity is increased, even beyond the absorptive capacity of the organic matter. Decaying roots leave channels through which water may move more readily. Thus drainage of fine-textured soils is improved.

Partially decomposed organic matter fills the noncapillary pores of sandy soils, thus making them capillary in size and better able to hold water against the pull of gravity. Organic matter additions are the only means of making some sandy soils economically productive.

Food for Soil Organisms

Soil Management Principle

Organic matter feeds microorganisms that release nutrients from organic matter and minerals that improve soil structure and increase water-holding capacity.

A living soil is one with active microflora. The by-products of organism activity, mucilagenous in nature, stabilize soil aggregates and thus greatly improve moisture and air relationships. The carbon dioxide given off by the organisms and the production of certain organic acids increase the solution capacity of the soil liquids to dissolve minerals and release the potassium, calcium, and magnesium needed by higher plants. Within the possibilities of economical procurement of organic matter, a farmer should "feed" soil organisms for maximum activity. This means frequent additions of easily decomposed organic matter. In some soils, the stimulated activity of macroorganisms (worms, ants, etc.) is also important.

Increases Exchange Capacity and Buffer Capacity

The phenomenon of cation exchange has been said to rank next to photosynthesis in importance to agriculture. Well-decomposed organic matter (humus) has a very high cation-exchange capacity that adds to the buffer capacity of the soil. This is especially important in sandy soil, which without organic matter is poorly buffered. Mineral nutrients and ammonium would be loosely held in such soil, and thus easily lost by leaching. A good supply of soil organic matter makes it safe to apply rather large applications of fertilizer at planting time and thus avoid a need for a second application. If lime is needed on a well-buffered soil, rate of application must be increased to bring the soil to the desired pH.

Minimizes Leaching Losses

Organic substances have the ability of holding against leaching substances other than cations. Nitrate, for instance, may be filtered from a solution by passing the solution through activated wood charcoal. Likewise, other nutrients may be absorbed and thereby saved for future use by plants.

A Source of Nutrients and Growth-Promoting Substances

Organic matter serves as a very important source of plant nutrients. Nitrogen exists almost entirely in the organic fraction. If there is very little stored in the soil, a farmer may need to make several applications during a growing season. The growing of leguminous organic matter actually adds nitrogen to the soil.

Mineral nutrient availability may be improved by good organic matter management. Organic phosphorus, for instance, makes up more than half the total phosphorus in some prairie soils. Mineralization is increased if the pH is raised from 5.5 to 6.5 by liming because organism activity is increased. Of course, the solubility of iron and aluminum phosphates are at the same time increased.

Most of the potassium taken up by a wheat or corn crop is left in the straw or stover. Residue management is very important. Such materials should be uniformly spread before they have time to leach, and varieties should be grown that are high in proportion of residue to grain.

Trace elements may be very satisfactorily supplied by decomposing organic matter. This is especially true during the production of crops that need certain trace elements in limited quantities. For instance, it may be necessary to supply boron in the fertilizer for alfalfa on a boron-deficient soil, because its need for the element is quite high. The following corn crop, on the other hand, needs only a small amount of boron and it is easily injured by a direct application of borax. It is much more satisfactory to allow the corn to take its boron from the decomposing alfalfa residues.

Other trace elements, known and unknown, may be likewise furnished from decomposing organic matter. The extra available moisture in a highly organic soil may be quite important in keeping nutrients in a form readily taken up by plants. There is some evidence that plants absorb some nutrients in organic form—perhaps one reason why organic soils are so highly productive.

It has long been known that plants derive certain unexplained benefits from decomposing organic materials. Hormones or growth-promoting and regulating substances valuable to plants may be produced by organisms that decompose soil organic matter.

Decomposed organic matter (humus) possesses chelating properties. These properties bring about covalent bonding between the organic matter and ions of copper, zinc, manganese, and iron. In alkaline soils or in acid soils after liming, such metallic nutrients remain in solution and in a state of availability to plants.

A Stabilizer of Soil Structure

Indirectly, organic matter is largely responsible for aggregation in humid temperate region soils. No soils of such areas have better structure than do those formed under grass vegetation, Mollisols (prairie soils). Grasses produce an extensive, fibrous root system, and root tissue is continually sloughing off. Decomposition is always going on in an old sod. The grass uses so much water that frequent drying and wetting occurs. The result is a very stable structure. A biological explanation of aggregation is offered in the section entitled Microorganisms and Soil Structure later in this chapter.

SOURCES OF ORGANIC MATTER

Generally, organic matter must be grown rather than purchased. In that respect, it is different from other soil amendments.

Plant Roots

The value of roots as a source of organic matter is ably demonstrated by the high organic content of grassland soils. Small grains and corn (maize) produce similar root systems that may greatly be stimulated by application of phosphorus. It has been shown that very high levels of soil phosphorus result in extreme proliferation of roots. Where a main root penetrated a band of fertilizer, there developed a dense mass of fibrous roots. It seems that phosphorus serves as a rooting medium, perhaps causing root cells to send out side branches rather than to remain as simply vegetative cells.

We have observed masses of fibrous roots in fertilizer bands, almost without exception. There is no doubt that phosphorus causes this intensified production. It may be good soil conservation to apply more phosphorus than top growth seems to justify, simply to bring about greater root production and thereby increase soil organic matter. This would have special value where corn or other row crops are being grown continuously and where success hinges on organic matter maintenance. On the other hand, excessive phosphorus leads to increased phosphorus levels in drainage waters, which is undesirable from an environmental quality viewpoint. Soil testing allows one to maintain high levels of phosphorus in soils to obtain excellent root growth and at the same time avoid applying phosphorus when soil test levels exceed high values. Plant breeders should be selecting for greater root development, thereby adding to soil organic matter. Such a characteristic would also enhance water and nutrient uptake by the plant.

Crop Residues

On most farms, crop residues are an important source of soil organic matter. Generally, grain straw and corn stover yields are slightly more than grain yields if the entire above-ground portions are included. This ratio varies somewhat with climatic area, variety, and soil nutrient balance, but as an average a 55–45 ratio is rather close for humid regions. This ratio may be wider in semiarid regions.

The U.S. Soil Conservation Service *Ready Reference Guide* allows full protective credit (against erosion) for residues if the quantity equals 2 tons or more an acre. On the basis of these ratios, yields of 55 bushels of wheat and 100 bushels of oats would just about produce the required amount of straw for full protection. A 100-bushel corn crop produces about 3 1/2 tons of residue. The cobs would supply an additional 1400 pounds of organic matter. See Fig. 4-1 for an illustration of the residues from a 100 bushel corn crop. Such an amount of organic material would be sufficient for protection to the soil.

Too little attention is being paid to the idea of selecting species, varieties, and fertilizers that produce higher residue yields. It is known that organic-matter depletion finally causes reduced yields or makes it necessary to change

FIGURE 4-1 The residue from last year's 100 bu corn crop was sufficient to provide good water permeability in this soil. There can be considerable slope without danger of erosion, perhaps up to 6 percent.

cropping systems. Why not, then, delay or prevent such a necessity by growing more organic matter in the form of residues? It might even eliminate the need for growing a cover crop or a green manure crop some place in the rotation or cropping sequence.

Care should be taken that residues do not inhibit the growth of the next crop, either as a result of moisture relationships or immobilization of nitrogen. Even spreading, timely plowing, and adequate and timely application of nitrogen fertilizers help to avoid such injury. Where residues are left on the surface in conservation tillage systems, special planters may be necessary to ensure good seed germination.

Green Manures

Effective cropping systems should include green manure or cover crops wherever they may be effectively and economically worked in. Legumes are preferred (see Figs. 4-2, 4-3, and 4-4) because they add nitrogen to the soil. In many systems, however, nonlegumes fit better into the rotation because they are usually easier to establish and generally produce more total carbon in a short period than do legumes. The nitrogen problem may easily be taken care of by use of fertilizer nitrogen, although it is now a rather expensive fertilizer.

Cover crops are planted to furnish soil protection, to pick up soluble nutrients that might otherwise be leached, and to deprive fruit trees of nitrogen so the wood will mature and thus avoid winterkill. The crops also add organic matter, even though they may be dead when plowed under.

Animal Manures

The value of animal manures in organic matter maintenance is universally recognized. They were once a part of every farm operation. But tractors do not produce manure, and modern technology and economics have favored large animal production centers for beef, dairy, and poultry production. This has drastically changed the availability and distribution of manure. Nonetheless, long-time field experiments have shown that annual applications of manure help to maintain organic matter levels. At the Sanborn field experiments in Missouri, continuous corn on untreated plots for 50 years left the soil containing 1500 lb of nitrogen in the surface 7 in. as compared with 2480 lb in soil similarly cropped where 6 tons of manure had been applied annually.

Other studies conducted over the same period at Sanborn field showed that manure is no more effective in organic matter maintenance than are certain other practices. Fertilized plots, for instance, in the same field, where a 6-year rotation had been used, contained even more soil nitrogen, 2530 lb/acre. No doubt, much of the organic matter maintenance quality of manure lies in the effect of the nutrients. This was indicated by results from the Broadbalk field at Rothamstead, where chemical fertilizers over a period of 100 years were just as effective as manure for continuous wheat production. The soil is fine sandy loam.

Animal manure is particularly valuable as a source of organic matter and

FIGURE 4-2 This sweetclover was seeded in corn at the time of the last cultivation. It served as an ideal winter cover and was furnishing green manure for another corn crop. Sims clay loam soil, Saginaw County, Michigan.

as a protective cover on irrigated land where a shortage of water makes it unwise to grow green manures. At Scotts Bluff, Nebraska (see Chapter 16) 6 tons of manure annually resulted in maintenance of nitrogen levels during 30 years of irrigated cropping. On adjoining plots, 4-ton applications were about 65 percent effective.

Many other data could be presented to show that animal manures are valuable as soil amendments and that organic matter may thus be maintained. However, the data show further that, generally, at least 6 tons annually are required to do the job. Also, wherever comparisons have been made, commercial fertilizers in conjunction with rotations have been equally effective. This points then to the advisability, on all but a few farms, of using animal manures for special purposes. Most farmers are not able to cover all their land more often than once in 4 or 5 years. Many, of course, have none at all.

Special Uses for Manures

Animal manures are, on most farms, most profitably used as amendments for "poor spots" and as protective covers or mulches. The effect of manure on runoff and soil loss have been reported from two corn plots at Zanesville, Ohio. The soil is Muskingum silt loam (Typic Dystrochrepts) with 12 percent slope. One plot was topdressed with manure; one was not. Runoff from the topdressed plot, from June until harvest, was less than one-half that from the unmanured plot, and soil loss figures were 1.4 tons and 41 tons, respectively.

This protective value of a topdressing of manure has been recognized by the U.S. Soil Conservation Service. In their *Ready Reference Guide* they rate 6 tons of manure to be equal to 2 tons of crop residues in this respect. Thus it is possible to use the manure to replace residues in the farm plan, as for instance on a field where corn is used as silage, or after a crop that leaves inadequate residues.

Manures are also effectively used as mulch material. This was shown in an Ohio experiment where strawy manure spread as a mulch after the first cultivation increased corn yields more than did an equal amount of manure plowed under before planting.

Manure mulches are particularly valuable for wind erosion control. On coarse-textured soils, and on most soils in the Great Plains, careful consideration should be given to this use.

Farmers are continually plagued with "poor spots" in their fields. There may be sandy, droughty spots or areas very high in clay or badly eroded

FIGURE 4-3 Sweetclover roots and nodules show why the crop is good for green manure. These plants were taken from those being plowed under in Fig. 4-2.

FIGURE 4-4 This good stand of sweetclover is the right size to be turned under for corn. The man is holding a yardstick to show the height of the clover.

hillsides or hilltops, sometimes relatively small compared to the entire field. A heavy application of manure will usually do wonders for such an area. An experiment at Ohio showed, for instance, that manure was more valuable when used on soil that had the surface removed than when applied to plots that had the normal thickness of topsoil.

Animals manures are especially useful for such intensive agriculture as vegetable growing, where land may be too valuable for green manure production or where fall vegetables are taken off too late to grow a cover crop. In some sections of the country, dairying and poultry production are concentrated on small areas of land. To such enterprises, manure is a by-product, a source of income. Growers of special crops are taking advantage of such opportunities. In 1957, 250,000 tons of dried animal manures were sold annually in the United States. Near Los Angeles alone, the dried manure sales amounted to $3.5 million each year; now the value would be several times this figure.

MICROORGANISMS AND SOIL STRUCTURE

Microorganisms may, in one sense, be considered as a source of soil organic matter. Their bodies, in a fertile soil, make up as much as 2 tons per acre, perhaps only 2 or 3 percent of the total organic matter, but a very active portion.

These organisms are so numerous that a handful of good soil may contain a larger number than there are people in the world. The bulk of the bacteria and fungi are heterotrophic. They obtain their energy, as do animals, by oxidizing (decomposing) organic matter. Their numbers greatly increase after fresh organic matter is added. The rate of decomposition is dependent on the existing biological conditions. Chief among these are temperature, moisture, oxygen, and nutrients. Conditions favorable for the growth of higher plants are generally suitable for the life processes of soil microorganisms. They are often limited by low temperatures, lack of oxygen, and a shortage of nutrients, particularly nitrogen, phosphorus, and calcium.

The organic matter decomposers are highly beneficial, largely because of the by-products of their activities. These include not only ammonia (NH_3) and the mineral nutrients including sulfur but also carbon dioxide (CO_2) and certain gummy substances that hold soil particles together to form soil aggregates. In other words, the soil organisms are really the soil structure forming agents in our soils.

Grasses, grown continuously, are the best-known plants to build soil structure (see Table 4-1). The reason is obvious. They produce tremendous masses of roots near the surface where aeration is good and decomposition is continuous. Thus, there is always a supply of decomposition by-products to promote aggregate formation and stability. There is yet another factor. Water molecules are readily adsorbed by clay or organic colloids. As dehydration occurs, water films become thinner, which draws attached colloids closer together, thus aggregation is more intensive. Surface tension also plays a role in forcing particles together.

Clay particles have natural coherence when they are placed in close contact with one another. Any force that tends to pull dispersed clay particles closer together helps in aggregate formation. Among the early by-products of microbial decomposition are materials called **microbial gums** or **mucilages**. These exhibit a high degree of "stickiness" because of their great attraction for water molecules and because of an uneven distribution of electrical charges in the large molecules of which they are composed. These, in effect, serve as electrical binders to draw clay particles together. Their

TABLE 4-1 The Effect of Grass on Stability of Soil Aggregates

Soil[a] and history	Percent of soil in stable aggregates over 0.25 mm
Sims, alfalfa and grass for 10 years	51.5
Sims, depleting crops for 10 years	36.4
Brookston, grass for at least 50 years	55.1
Brookston, depleting crops for 50 years	23.8

[a]The Sims soil is sandy clay loam (mollic Haplaquepts); the Brookston is clay loam (Typic Argiaquolls). Grasses are very effective in bringing about the formation of large, water stable aggregates. Such a state of aggregation ensures ready intake of water and good aeration.

cementing action is particularly effective in aggregate formation during the early stages of decomposition of fresh plant residues. The goal of the soil manager should be to furnish a continuous supply of "food" for the willing workers that prepare these electrical binders.

Once formed, the stability of the clumps of mineral particles that we call aggregates is promoted by conditions or materials that tend to reduce the forces of attraction between adjacent clumps. Some of the later products of decomposition of plant residues are relatively inert. When these are deposited over the surfaces of aggregates previously formed, they serve as insulators to weaken the electrical forces of attraction between them. To the extent that this insulation is effective, the aggregates become stable; they do not tend to run together again when the soil becomes saturated with water during periods of wet weather. Although these insulating materials are relatively resistant to decomposition, they do decompose slowly. As the soil becomes depleted of these materials, aggregate stability is lost and soil structure deteriorates. A continuing supply of raw plant material is as necessary to maintenance of these organic insulators as it is in the maintenance of the electrical binders.

Published reports have shown a need for a continuous state of organic matter decomposition in order that soils may remain in an aggregated condition. Maximum aggregation occurred within 1 week after ground wheat straw or sucrose was mixed with soil; no further aggregation occurred during 12 weeks of incubation. In other words, the aggregating effect of the organisms was very fast.

Some microorganisms effect soil aggregation, but others may destroy the effective substances exuded by the aggregating organisms. Among a group of 34 soil isolates (33 fungi and 1 actinomycete), 19 were quite ineffective in promoting aggregation when wheat straw was the source of carbon, whereas all were effective when sucrose was used. With the sucrose, however, the isolates varied greatly in *degree* of effectiveness.

The 19 isolates that were ineffective in promoting aggregation would be among those organisms that would destroy the effective substances exuded by the aggregators. In other words, they would be responsible for making soil aggregation a temporary phenomenon.

Alfalfa is known to stimulate more aggregators than does straw. Other sources of organic matter probably affect various genera of fungi differently than does alfalfa, which leads to the conclusion that more continuous aggregation might be effected by several sources of carbon. Mixing would, of course, be impractical, but during a cropping sequence or rotation we can use several materials.

One percent mixtures of either sucrose or straw have been reported to be as effective as larger concentrations. Other studies have shown that relatively small amounts of green alfalfa are very effective as microorganism stimulators. It would thus seem logical to suggest frequent, even though small, additions of organic matter to bring about continuous aggregation and thus guarantee the benefits of good soil structure.

Organic Matter Calendar

In a passage entitled "Looking Ahead" written in 1954, Melsted stated, "The realization that the more a soil produces the better is the care it is receiving will influence future soil management practices and tillage operations." Now we see the truth of these words. We cannot consistently obtain high yields without being good to the soil. Fertilizer applications are being geared to expected yields. To expect high yields, we must "feed well."

Again quoting Melsted, from his last sentence, "For the moment it must be assumed, for lack of definite data to the contrary, that maintaining soil organic matter and supplying residues for active decomposition are desirable and economical practices and that the restoration and maintenance of soil tilth are necessary for good soil management."

Note the reference to "active decomposition." This means frequent additions, not once in a rotation but every year, perhaps oftener than once a year. What is the schedule?

Soil Management Principle

Soils should never be allowed to "rest"; they should be always growing something.

Growing Plants

Our prarie soils prove that growing plants add organic matter. Roots are continually sloughing off, and mature plants are disintegrating. High fertility promotes fast, extensive growth. Soils cannot be allowed to rest, they must never be left bare! This means plowing under a cover or green manure crop, or the aftermath of the previous crop, and planting immediately. The idea works well with minimum tillage (see Chapter 9). By means of soil tests, we must always be sure the growing plants are well supplied with nutrients, especially nitrogen, the limiting nutrient in so many efforts at producing organic matter. Growing plants probably are the best source of what Melsted called "residues for active decomposition."

Residues

Straw from small grains, grasses cut for seed, beans, or peas and stover from maize, sugercane, or sorghum should be carefully returned to the soil in the manner recommended for the soil and climate. In the humid and subhumid regions, or on irrigated lands, cover crops can often be under- or interseeded. In some cases, the residues may be worked into the soil before the cover crop is seeded. Thus, two additions of organic matter are made in one year.

Animal Manures

These are fine sources of organic matter that decompose quickly. They should be applied, whenever possible, on land that does not already have a good plant or residue cover, thus furnishing another "feeding" for the soil organisms.

Organic-matter management should be based on a schedule that includes frequent (two or three times a year) additions of easily decomposed material. Thus the soil microorganisms may be constantly working to improve the soil and provide more take-home pay for the farmer.

REFERENCES

Anderson, M. S. (1957). Farm manure. In *Soil, The Yearbook of Agriculture,* U.S. Department of Agriculture, Washington, D.C., pp. 229–237.

Fuller, W. H. (1951). *Soil Organic Matter.* Arizona University Agricultural Experiment Station Bulletin 240.

Jenny, H. (1930). *A Study on the Influence of Climate upon the Nitrogen and Organic Matter Content of the Soil.* Missouri University Agriculture Experimental Research Bulletin 152.

McCalla, T. M., F. A. Haskins, and E. F. Frolik (1957). Influence of various factors on aggregation of Peorian loess (albic glossic natraqualfs) by microorganisms. *Soil Sci.* 84:155–161.

Melsted, S. W. (1954). New concepts of management of corn belt soils. *Advances in Agronomy* 6:121–142.

Ohlrogge, A. G. (1958). *How Roots Tap a Fertilizer Band. Plant Food Rev.* 4 N. (2 and 3).

Smith, G. E. (1942). *Sanborn Field, Fifty Years of Field Experiments with Crop Rotations, Manure, and Fertilizers.* Missouri University Agricultural Experiment Station Bulletin 458.

Nitrogen Problems and Rotations

Nitrogen is the key to successful organic-matter management and thereby to successful soil management. More than 99 percent of the nitrogen in a soil at any one time exists as a part of complex organic materials in various stages of decomposition. Where no fresh organic matter is mixed with soil for a considerable period, organism activity falls to a very low ebb and practically all the nitrogen is a part of the humus. Such nitrogen is not rapidly available to plants, and the productive capacity of the soil is temporarily low. In other words, a soil relatively high in total nitrogen may be quite unproductive, possibly because of unwise organic-matter management, or perhaps unwise microorganism management. Perhaps the "willing workers" below the surface (the microorganisms) have not been well treated.

SOIL NITROGEN VERSUS ORGANIC MATTER

The average percentage of nitrogen in a large number of soils multiplied by the factor of 20 is roughly equal to their average percentage of organic matter. Individual soils, however, vary considerably from this relationship owing to variations in the state of decomposition of the organic materials. Variations from the average become less as the easily decomposed organic fraction is reduced by organism activity.

Soil Management Principle

Humus undergoes change in a soil — but at a very slow rate.

Humus is a term applied to soil organic matter that has reached a rather advanced stage of decomposition. It is still undergoing change, but at a slow rate. Extracted from soil, it is dark, viscous material composed of 40 to 45 percent lignin, 30 to 35 percent proteins, and the remainder of waxes, fats, and other resistant materials. The lignin, waxes, and fats are directly from plant materials, whereas the proteins are synthesized by microorganisms. Several mineral plant nutrients are held by the humus as exchangeable cations. Chief among these are calcium, magnesium, potassium, manganese, and iron. Nitrogen, phosphorus, and sulfur are chemical constituents of the protein.

From these characteristics of humus, we would think it should be an easily available source of nutrients for higher plants, particularly nitrogen, since its nitrogen content is about 5 percent. Such is not the case. The high percentage of lignin makes it quite resistant to the attack of decomposing organisms, commonly called **ammonifiers.** It is fortunate, of course, that this is true, otherwise humus would soon disappear, would exist in name only, and the valuable soil characteristics imparted by humus would be unknown. Silicate clay soils would be untillable. We would be forced to confine crop growing to sands and sandy loams, and they would of necessity be handled much differently than at present.

Carbon : Nitrogen Ratio

Soil organic matter is decomposed by the great mass of organisms that obtain energy through decomposition, through the oxidation of carbon. The CO_2 thus formed plays the important role of dissolvent to break down the soil minerals and so release such nutrients as potassium, magnesium, and calcium. It is in the order of natural processes that about 65 percent of the carbon in organic matter should be so dissipated.

The nitrogen contained in organic matter is liberated as ammonia to be used as a source of energy by two groups of autotrophic bacteria called **nitrifiers.** They change the ammonia to nitrite, then to nitrate (NO_3). The nitrate (or the ammonia) may be used by plants to synthesize protein, or by the decomposing organisms to build their own bodies.

This dissipation of 65 percent of the organic carbon, during the process of humus formation, results in a narrowing of the carbon:nitrogen (C : N) ratio to about 10 : 1. This ratio is reasonably constant the world over. The total amount of carbon in the organic matter also regulates organism activity. If the organic matter being decomposed is lacking in nitrogen, as is the case with straw, corn stalks, or sawdust, the soil organisms turn to any soluble nitrogen that may be in the soil and/or they attack the humus or any partially decomposed products that may be present. The result of their utilizing all soluble nitrogen is commonly termed **nitrogen immobilization.** Plants growing at that time suffer from nitrogen starvation.

If the organic matter being decomposed is high in carbon and low in nitrogen (a wide C : N ratio), organism activity slows to that made possible by nitrogen from the soil or ceases entirely. At such a time, nitrogen fertilizer is badly needed, especially if a crop is growing. Microorganism activity also slows, of course, when there is too little carbon in proportion to nitrogen (too narrow a C : N ratio). It is actually the carbon that the organisms desire as a source of energy.

The C : N ratios of several organic materials commonly added to soils are shown in Table 5-1. The most desirable C : N ratio is between 20 : 1 and 30 : 1. A narrower ratio material furnishes too much nitrogen and may result in an attack on the soil humus and a total loss of organic carbon from the soil. A ratio of 30 : 1 is safe as far as immobilization of soil nitrogen is concerned. Such material usually contains about 1.5 percent total nitrogen.

The plowing under of materials containing less than 1.5 percent of nitrogen sets up a nitrogen problem. Either nitrogen must move from the soil to the decomposing organisms or nitrogen must be applied as fertilizer. If the soil is very fertile, the former may present no problem. If the soil is not very fertile, however, growing crops will be deprived of nitrogen until decomposition is complete and the microbial bodies die and decompose, or in other words, until the nitrogen is again mineralized. An alternative is to wait to plant the crop after the organic matter is decomposed. This is **not** a recommended practice, however, as the soil is in the meantime left open to packing, leaching, and erosion. Modern tillage practices usually call for planting immediately after plowing (see Chapter 9).

Nitrogen Fertilizers

Soil nitrogen should be supplemented with fertilizer nitrogen when the C : N ratio in the added organic matter is wide, when the nitrogen content is below

TABLE 5-1 The Carbon : Nitrogen Ratio
of Organic Materials
Commonly Added to Soils

Material	C : N ratio
Sweetclover (young)	12
Barnyard manure (rotted)	20
Alfalfa (blossom stage)	20
Clover residues	23
Green rye	36
Cane trash	50
Corn stover	60
Straw	80
Timothy	80
Sawdust	400

Source: Except for the alfalfa figure, these data are from H. D. Foth, *Fundamentals of Soil Science*, (7th ed.) Wiley, New York, 1984.

1.5 percent. Simple arithmetic gives a practical answer. Wheat straw contains 0.5 percent nitrogen or 10 lb/ton. To come up to the 1.5 percent required, if immobilization is to be avoided, it will be necessary to add 20 lb/ton of straw.

Corn (maize) residues vary greatly in nitrogen content. If the corn was well fertilized with nitrogen (100 lb nitrogen or more per acre), the residues may contain as much as the required 1.5 percent. No additional nitrogen would be needed for humus formation. On the other hand, if nitrogen fertilizer was not applied for the growing corn and the soil was relatively infertile, the residues may be as low as the 0.5 percent in wheat straw. Then, it would be wise to apply an additional 20 lb of nitrogen per ton of residues when they are turned under or make doubly sure the next crop does not suffer. Sometimes the latter alternative is preferable.

LEGUME VERSUS COMMERCIAL NITROGEN

Legume Nitrogen

Legumes behave differently than do nonlegumes as far as response to nitrogen fertilizer is concerned. We have long known this to be due to a symbiotic relationship that exists between these plants and certain soil bacteria called **rhizobia.** The various strains of the organism are quite specific with respect to the particular legumes with which they associate. Inoculation is always advisable to be sure that an effective strain is present.

Soil Management Principle

Nitrogen fixation by root-nodule bacteria begins within a few weeks after the organism enters the root.

Nitrogen fixation by root-nodule bacteria begins within a few weeks after the organism enters the root. The bacteria secrete enzymes that dissolve a portion of a root hair wall. Then the organisms enter the cell, and the root hair becomes the nodule in which the bacteria perform their work. The growing legume is able to use nitrogen as it is fixed, and other plants (grasses and weeds) growing in the immediate vicinity use the nitrogen from sloughed off nodules and roots. This is why grasses such as timothy, bromegrass, and bluegrass do so well in legume mixtures.

Symbiotic nitrogen-fixing organisms (**rhizobia**) are much more effective than are the free-fixing organisms (**azotobacter** and **clostridia**), the latter groups accounting for 10 to 50 lb of nitrogen per acre per year. Most of the nitrogen in our grassland soils can be credited to these free-fixing organisms. A little, of course, came down with precipitation, mostly the result of fixation by electric discharge (lightning) or from nitrate originating from industrial or automobile emissions.

In comparison, in Michigan on May 6, about the time of plowing for corn, alfalfa *(Medicago sativa)* tops and roots to the depth of plowing contained 132.9 lb of nitrogen per acre. Assuming one-third came from the soil, the crop was adding about 90 lb from the air. The corresponding figure for sweetclover *(Melilotus alba)* was 62.0 lb/acre. The amounts in both crops increased as the season progressed.

The *Book of the Rothamsted Experiments* (Hall, 1905) shows that over a 20-year period, the presence of 25-percent legumes in the crops on a soil resulted in an increase of 1364 lb of nitrogen per acre more than was added to a soil where only grasses were grown. This amounts to an increase of 68 lb/year from only 25-percent legumes in the vegetation. Incidentally, the production of grasses alone had resulted in an increase of 50 lb of nitrogen per acre per year, mostly owing to the free-fixing organisms. The data were based on tests of samples taken at the beginning and at the end of the period without plant removal.

Lyon and Bizzel (1934) grew legumes and nonlegumes for a period of 10 years. The soil was a mixture of 60-percent Dunkirk silty clay loam (Glossoboric Hapludalfs) and 40-percent fairly clean sand. It was carefully mixed, a wheelbarrow load at a time, and placed in 48 frames. Lime, phosphorus, and potassium were liberally supplied, but no nitrogen as fertilizer or manure was added during the 10 years. The average nitrogen content of the soil was 0.084 percent. By analyzing all the crops and the soil at the conclusion of the experiment, they found that continuous alfalfa resulted in an apparent fixation of 268 lb of nitrogen per acre per year; alfalfa in rotation with grain, 241 lb; sweetclover alternating with grain, 163 lbs; and red clover *(Trifolium pratense)* alternating with grain, 146 lb. In a Michigan experiment, the three crops, alfalfa, sweetclover, and red clover, ranked in this order in their beneficial effect on yields of following crops.

Numerous other investigators have shown that legumes may profitably be used as a way of fixing atmospheric nitrogen for use by crops. Many early agronomists, chief of whom perhaps was Cyril G. Hopkins of the University of Illinois, thought it possible to supply all nitrogen with legumes. Hopkins based his soil-building formula on "lime, raw rock phosphate, and clover." Such a philosophy, seemingly sound at the time, would have obtained fewer followers in an area where soils were less well supplied with organic matter.

Today, with corn yields averaging 90 to 100 bu/acre and with some farmers getting 200 or more bushels, one cannot depend only on the nitrogen added to the soil by the last year's legume crop.

Commercial Nitrogen

Thomas Robert Malthus (1766–1834), an English economist, stated in 1798 that population would increase faster than food supplies and, accordingly, many people would starve unless wars and diseases reduced their numbers. Low crop yields of that day were partly responsible, no doubt, for that attitude. From 1850 to 1950, wheat yields in England rose from 20 to 35 bu/acre. Similar increases occurred on the continent of Europe and in the United States.

During the latter part of the nineteenth century, it became evident that nitrogen as a fertilizer was badly needed, that if humanity perished because of lack of food, nitrogen shortage would be the cause. Then around the turn of the century, synthetic nitrogen-fixation processes were developed. The arc process, still used in Norway, was the first. It was soon followed by development in Germany of the cyanamid process, used since 1909 in the Canadian plant at Niagara Falls.

The Haber-Bosch ammonia process is now used in manufacturing a large part of the synthetic nitrogen produced in the world. In the United States, the 1955 yearly consumption of nitrogen in mixed fertilizers and for direct application was 1,960,536 tons. By the year 1980, that figure had increased to 11,399,508 tons. The corresponding figures for Michigan were 36,521 and 236,482. That is almost a 6 times increase for the United States as a whole and a 6.5 times increase for Michigan. In Ohio, Indiana, and Missouri, nitrogen consumption trends were almost identical to those for Michigan, but in Wisconsin, Iowa, Illinois, Minnesota, South Dakota, and North Dakota, the respective amounts consumed in 1980 were 11, 15, 15, 18, 22, and 73 times what they were in 1955.

Several factors were responsible for this rapid increase in nitrogen consumption in the United States and an even more rapid increase in the general area represented by the just mentioned states. First, the Haber-Bosch ammonia process made cheaper ammonia possible. The price of nitrogen fertilizer was continuing to decrease until the mid-1970s, when natural gas suddenly became much more costly, a direct result of increased petroleum prices.

Secondly, the 25-year-period 1955 to 1980 saw a marked change in farming systems in those portions of the United States where corn was a main crop. As an illustration, 30 percent of Michigan's cropland was in row crops in 1950. By the year 1978, that percentage had increased to slightly more than 60 percent. A similar change occurred in the other states of the region. The changes in the Northern Plains were even greater. Actually, corn accounted for much of the increase in row-crop acreage throughout the North Central states.

During the period of rapid transition to row crops (largely corn in Michigan), state average yields almost trebled. Hybrid corn, largely replacing the old open-pollinated varieties, is much more responsive to nitrogen fertilizer. The much higher yields induced our farmers to greatly expand acreage. Corn has become a profitable cash crop at the expense of livestock production on many Michigan farms. As of 1980, about 25 percent of all corn grown in the United States was exported. Many farmers like a farming system that gives them free time during the winter months. We call it the "Winter in Florida" farming system. It sounds good, but it is resulting in increased soil loss because, since hay and pasture are no longer needed (hay is not a good cash crop), legumes and grasses may not be grown. Instead, the needed nitrogen is purchased as fertilizer, often as much as 200 lb/acre/year. This is a very expensive input.

Nitrogen consumption has increased worldwide. In 1980 it was close to 60 million tons, almost as much as the tonnage of phosphate (P_2O_5) and potash (K_2O) together. Thus we see that worldwide, nitrogen is by far our most important plant nutrient, and now the most expensive.

We have already shown that carbon is the critical element for humus maintainance or building. With pencil and paper, we can readily compare the cost of nonleguminous carbon plus commercial nitrogen against the cost of producing leguminous carbon and nitrogen in the form of alfalfa, clover, or other legumes. There is some evidence, as has already been mentioned, that the organisms that decompose legumes do a better job in stabilizing structure than do those that decompose nonlegumes. Whether or not this characteristic should tip the balance in favor of producing legumes as a source of nitrogen is still questionable. Soil type and climate will make a difference. You are urged to keep posted on the current research being conducted by local experiment stations on this problem. In tropical, semiarid regions, leguminous trees are a good prospect. *Leucaena leucocephala,* called "Ipil-Ipil" in the Philippines, has been known, according to AID advisor Mike Benge, to fix 590 kg of nitrogen per hectare per year. Its nitrogen can furnish all the nitrogen needed by a nonlegume growing nearby.

ROTATIONS VERSUS CONTINUOUS CROPPING

Crop rotation may be defined as a cropping system that provides for the production of two or more harvestable crops in succession on the same land. The term is usually used in contrast to "continuous cropping," a practice of growing the same crop during consecutive years on the same land. In this regard, a crop followed by fallow may be considered a rotation in dryland farming.

Many data obtained from all parts of the country may be assembled to prove that rotation farming results in higher yields than does continuous cropping. We believe that rotations are *not* outdated. The advantages outweigh the disadvantages.

Soil Management Principle

Rotations are a valuable tool in soil management today.

Advantages of Rotations

1. Soil-building crops are distributed over all fields.
2. Rotation provides for more uniform distribution of stable manure and fertilizers to all fields.
3. Well-planned rotations provide for more continuous vegetative cover.

4. Soils are given time to recuperate from the effects of a crop that may have extra-heavy demands for certain nutrients.

5. Crops vary in the feeding range of their roots. Rotation provides for more complete use of the soil profile.

6. Crops vary in the way they affect following crops. Many examples could be quoted showing the effects of rotations. For example, in Rhode Island, potatoes did poorly after millet, rutabagas, or potatoes and best after rye, oats, and onions. The very lowest yields were after millet and the best after rye with a significant difference of 95.9 bu/acre. In the same experiment, onion yields were also the highest after rye and the lowest after rutabagas — 604.0 as compared to 115.9 bu/acre.

7. Rotations favor weed control.

8. Rotations favor disease and insect control.

9. Insofar as it means the production of a larger number of crops, a rotation system usually results in a broader distribution of labor and income. Diversification provides some assurance of a more steady income.

Advantages of Continuous Cropping

1. A soil may be particularly well adapted to the production of a certain crop or at least better adapted to the production of one crop than to any other. For instance, a certain area may be too hilly for rotation cropping. Grasses may be the only crop that may be grown without excessive soil loss, or it may be ideal for corn. If climate is favorable, that is the most profitable crop to grow, particularly if corn is a desirable crop in the total farm-management plan. Soils on intermediate slopes may be safely cropped to small grains but not to cultivated crops. On such soil a "modified" rotation, one that includes only small grains and meadow, may be suitable.

2. Markets and climate in a certain area may be such as to favor a one-crop agriculture. On the best corn land in portions of Iowa, Wisconsin, Illinois, and Indiana, there are few crops more profitable than corn. The same reasoning applies for wheat on the Palouse silt loam (Ultic Haploxerolls) of Washington and for cherries on Michigan's Leelanau Peninsula.

3. Machinery and building costs are less in a one-crop system.

4. A farmer may prefer the production of a single crop. Thus he can become a specialist and perhaps work on his farm only a few days during the year. The owner of an orange grove in California finds it relatively easy to hire specialists to prune and spray, and sells his crop on the trees. By such management, he may be employed elsewhere or may spend his time in recreation, as he chooses.

5. Few people keep themselves well-enough informed to do an expert job in the production of a large number of crops and in the raising of livestock to consume the forage and feed grains.

SUGGESTED ROTATIONS

Are Rotations Outmoded?

On farms where the soil is relatively uniform, and where more than one crop is to be grown, rotations are highly recommended. Classic among rotation experiments are the Morrow plots. They were started in 1876 in an area now included on the campus of the University of Illinois. Three of the original 10 plots remain. Results from the experiment were discussed in detail by Millar (1955).

Three principles are now widely recognized: first, that rotation of crops is better than continuous cropping even though the crops are all nonleguminous or depleting in nature (corn compared with corn-oats); second, that to be highly effective a rotation system must include a legume; and finally, rotation alone is not sufficient to maintain productivity. Extra plant nutrients are needed.

As mentioned in Chapter 4, the value of the legume in the rotation is not simply that nitrogen is added to the soil but that soil structure is improved by the organisms that decompose the leguminous organic matter. It has been shown, for instance, that extra nitrogen for corn on Paulding clay (Typic Haplaquepts) in northern Ohio was less effective in increasing corn yields than was a legume in rotation with the corn. Soil structure made the difference. They could plow the legume plots when water was standing on the continous corn plots.

A Modern Rotation Experiment

One of our most up-to-date rotation experiments was located on Sims clay loam (Mollic Haplaquepts) in the Saginaw Valley of Michigan (see Fig. 5-1). The plots were established in 1940 to furnish answers to some questions asked by people interested in promoting beet-sugar production in the eastern sugar-beet area (Wisconsin, Michigan, Illinois, Ohio, and Ontario). Accordingly, seven 5-year rotations were established.

1. Alfalfa-brome, alfalfa-brome, corn, beets, barley (brome refers to *Bromus inermis*).
2. Alfalfa-brome, alfalfa-brome, beets, corn, barley.
3. Alfalfa-brome, alfalfa-brome, beans, beets, barley.
4. Alfalfa-brome, corn, beets, barley, oats (changed in 1956 to alfalfa-brome, corn, beets, beans, barley).
5. Clover-timothy, corn, beets, barley, oats (changed in 1951 to alfalfa-brome, alfalfa-brome (plowed down), beets, corn, barley). Timothy is *Phleum pratense.*
6. Beans, wheat, corn, beets, barley.

FIGURE 5-1 Ferden rotation plots, Saginaw County, Michigan. The soil is Sims clay loam. The work began in 1940.

7. Beans, wheat (G.M.),[1] corn (Sw. Cl.),[2] beets, barley (G.M.).

You will notice that this is a crop *sequence* as well as a crop rotation study; in fact, it is actually a *system of farming* study, as there are certain variations other than crops. Plots in rotations 1, 2, and 3 received 10 tons of manure, those in rotations 4 and 5 received 7 tons, and those in the cash-crop rotations 6 and 7 did not receive manure, instead all crop residues were returned. Residues were removed from plots in all other rotations.

Fertilizers were applied at two levels, the same in all rotations, a so-called "low" rate on one-half of the plot and a "high" rate on the other half. In addition, the subplots were divided again for two nitrogen treatments, planting time only, and planting time plus a side or topdressing.

Two crops appeared in all the rotations and beans in all but one.

Corn yields were highest where the crop followed alfalfa. Figure 5-2 shows that corn grown in the rotation without a legume produced a smaller number of ears than did that grown in rotation with alfalfa. Furthermore, the ears were small and poorly filled with kernels when grown on soil not improved by growing legumes, which was clearly the result of nitrogen starvation. Red clover-timothy hay and mixed-legume green manure were about equal in the way they affected corn yields. Corn after sugar beets yielded 5 to

[1]G.M. = green manure mixture of sweetclover and alsike, June, and Mammoth clovers.
[2]Sw. Cl. = sweetclover green manure *(Melilotus alba)*.

7 bu less than after alfalfa. During the years 1951 to 1957, all corn yields were considerably higher than during the previous 10 years except in rotation 6 where there was no legume. This was an interesting development. It is believed that the differences during the two periods were largely due to cultural changes. Improved varieties were planted, plant population increased, minimum-tillage practices adopted. Without the advantage of a legume, however, the corn was not able to take advantage of the cultural improvements.

Sugar-beet and barley yields were likewise the lowest where there was no legume in the rotation. Sugar-beet yields were the highest where the crop followed beans. During the later years, this yield advantage ranged from 1.0 to 3.3 tons/acre. It is particularly interesting because the beet plots in the bean rotation always looked especially good throughout the season.

There is a little doubt that much of the value of the legumes came from the nitrogen added to the soil. With all three crops, the extra nitrogen tended to even the yields during the 1951 to 1957 period. In fact, where sugar beets followed beans and alfalfa (rotations 3 and 2, respectively), extra nitrogen actually reduced the yields. Apparently, the legumes had fixed all the nitrogen the sugar beets needed.

The applied nitrogen was not sufficient, however, to make rotation 6 yields (no legume) equal to the others. This indicates that some of the benefits from the legumes were from effects other than nitrogen, probably soil struc-

FIGURE 5-2 Corn grown on the Ferden rotation plots shown in Fig. 5-1. The left pile was from a plot where there had been no legume grown since the experiment started. Note the short, poorly filled ears (nubbins). The right pile was produced where alfalfa was grown 1 year out of 5, immediately before the corn. There were more and much longer ears from an equal-sized area.

ture. If space permitted, many additional data could be assembled to prove that crop rotation is advantageous, especially when soil-building legumes are included either as meadow or green manure crops, for various farming situations.

Rotations Illustrated

We remind you that these are general suggestions. Soil series and degree and length of slope should be considered. A cropping system that will keep soil loss below permissible amounts for the series must be selected (see Chapter 6). The broad, general suggestions presented in Table 5-2 assume soil losses will be within these limits. Cover crops and green manure crops are included, and it is assumed that adequate lime and plant nutrients are to be supplied. Further, it is assumed that farm manure will be applied once in rotations that include meadow crops, and that all crop residues not needed for bedding will be returned to the soil.

A MODERN CONCEPT OF ROTATIONS

Continuous cropping is not the sin it was once thought to be. Modern methods of disease, insect, and weed control have greatly reduced the hazards of growing the same crop year after year. There are in fact many advantages to such a system, as has already been pointed out. Chief among them perhaps is soil-crop adaptation. This may be accomplished though while still following a rotation.

The early concept of rotations was to grow the same crops on fields of equal size. Thus farmers would harvest about the same amount of each crop each year, facilitating, of course, their livestock program and giving them a rather constant cash-crop income each year.

Contouring and strip-cropping have ruled out uniform-sized fields. Specialization has reduced the number of crops on many farms. As a result of these changes, it now seems advisable to arrange a rotation for each field without too much regard for the balance between acreages of different crops during any particular year. It might, for instance, be that a farm plan might include 10 acres of corn one year and 30 the next. The low corn year might be a high wheat year. Such a system requires more storage space for hay and silage but not necessarily for grains.

Continuous cropping and rotations may well be practiced to advantage on the same farm. For instance, alfalfa on hilly land should be grown as long as the stand persists. Intermediate slopes may be used for a meadow–small-grain rotation (perhaps meadow-meadow-grain-grain) and relatively flat lands for continuous corn (maize). Where feasible, all manure may be applied to the corn land. Liberal use of fertilizer on the upland fields could bring sufficient residues to maintain organic matter and total farm production at very high levels. Land class should be considered before such a plan is attempted.

TABLE 5-2 Recommendations for Rotations That Should Maintain Productive Capacity

Corn belt	
Alfalfa Corn (cover crop) Corn (cover crop) Grain (seed alfalfa)	For livestock farm on fine-textured nonerodable soil. On sandy and/or erodable soils, grow 2 years alfalfa and 1 year corn.
Clover seed Corn (cover crop) Soybeans or corn (cover crop) Grain Grain	For cash-crop farms and on fine-textured soil.
Northern parts of Michigan and Wisconsin	
Alfalfa Alfalfa Alfalfa (rye after first cutting) Potatoes	For livestock and potato farm on sandy soils. Omit rye and grow corn and/or sunflowers for silage if potatoes are not desired
Southern parts of Michigan and Wisconsin and northern Ohio, Indiana, and Illinois	

Alfalfa
Alfalfa
Corn or beans
  Cover crop in grain if spring
    planted
Grain (seed alfalfa)
	For general farms on fertile, sandy loam and loam soils. Permissible to drop a year of alfalfa or add a year of row crop on permeable clay loam soils.
	Add a year of alfalfa and omit the row crop on very light sands or impermeable clay soils.

Alfalfa Alfalfa Corn (cover crop) Grain Wheat (sweetclover) Corn (cover crop) Grain Wheat (seed alfalfa)	For general farms on highly fertile, fine-textured soil where alfalfa does better the second year than the first but where the area of alfalfa desired is only 25 percent of the acreage. If organic matter is especially needed, seed Mammoth clover with grains.
Great Plains	
Northeastern Plains	
Alfalfa Corn Wheat Wheat Corn Cowpeas or soybeans Wheat (cowpeas cover)	Leave alfalfa more than 1 year if more hay is needed. Substitute a second-year corn for first-year wheat on better soil. Grow a cover crop during the corn year where moisture is available. Cowpeas or soybeans should be cut for hay.

TABLE 5-2 Recommendations for Rotations That Should Maintain Productive Capacity

Corn belt	
	Southern Plains
Wheat, sorghum, or fallow	Adjust sequence according to moisture
Wheat, sorghum, or fallow	content of the soil and presence of residues
Wheat, sorghum, or fallow	at planting time. Wheat does not do well after sorghum.
Southwestern States (Irrigated)	
Alfalfa	For sandy soil in warmer sections. Use
Alfalfa	Papago peas as cover crop during winter.
Alfalfa	Winter vegetables may be substituted for
Cotton (cover crop)	cotton or grain. Then grow Sesbania or
Cotton (cover crop)	Guar as summer cover crops. Substitute
Grain (cover crop)	sorghum for cotton in cooler sections.
Grain (cover crop or seed alfalfa)	A year less alfalfa and an extra year of cotton is permissible on fine-textured soils.
Southern States	
Legume meadow (hay or pasture)	Use one or more row crops according to
Cotton, tobacco, corn, or peanuts	capacity of the soil. Crotalaria should be
Grain	grown in corn as green manure, either between rows or in alternate rows. Grow hairy or smooth vetch as cover crop after peanuts. Annual lespedeza may be seeded in small grain.
East-Central Uplands	
Alfalfa or clover	Use winter legumes (Crimson clover or
Cotton (cover crop)	vetch) as cover crops after cotton. Cotton
Grain	may follow corn where a longer rotation is
Tobacco	desired. Use ryegrass or fescue cover crop
Corn or cotton (cover crop)	seeded in grain. Avoid legumes before
Grain (cover crop or seed legume)	tobacco.
Northeastern States	
Alfalfa	Omit corn on steep slopes. Grow corn
Alfalfa	continuously with cover crops on level
Alfalfa	fields in the valleys. Grow clover and
Corn (cover crop)	timothy on adequately limed soils.
Grain (seed alfalfa)	Substitute potatoes for corn and add wheat after oats where desired.

REFERENCES

Allaway, W. H. (1957). Cropping systems and soil. In *Soil—the 1957 Yearbook of Agriculture.* U.S. Department of Agriculture, Washington, D.C., pp. 386–395.

Davis, J. F., and L. M. Turk (1944). The effect of fertilizers and the age of plants on the quality of alfalfa and sweet clover for green manure. *Soil Sci. Soc. Am. Proc.* 8:298–303.

Hall, A. D. (1905). *The Book of the Rothamsted Experiments.* Dutton, New York.

Lyon, L. T., and J. A. Bizzell (1934). A comparison of several legumes with respect to nitrogen accretion. *J. Am. Soc. Agron.* 26:651–656.

Millar, C. E. (1955). *Soil Fertility.* Wiley, New York.

Odland, T. E., and J. B. Smith (1948). Further studies on the effect of certain crops on succeeding crops. *J. Am. Soc. Agron* 40:99–107.

National Academy of Sciences (1977). *War on Hunger.* 11(2), Washington, D.C.

National Academy of Sciences (1979). *Tropical Legumes: Resources for the Future.* Washington, D.C.

CHAPTER 6

Fitting Crops to the Soils

The word "crops" in this chapter title is intentionally plural because the management of any piece of land should always, or almost always, involve more than one crop. Perhaps the only exceptions are when the land is reforested or set aside for permanent grass pasture. It is not enough to grow a single annual crop that occupies the land for only a few months.

Soil must always be "busy," always growing something. Idle soil is depreciating, wasting away. Growing crops build soil, and when all or most of the dry matter stays on and in the soil, greater yields result in faster building. To keep the soil forever busy, it is necessary to produce a crop that grows over a 12-month period or to grow more than one crop. In many systems, two crops keep the soil covered the year round. We may grow one for harvest, the other for cover and organic matter.

Note also that this chapter title refers to "soils," again plural, but with the article "the" indicating **certain** soils. Usually farmers can select their crops, but they are confined to a certain group of soils. They can, however, select crops adapted to their soils and improve the soils within the **limits of their capabilities.** This chapter deals with crop and soil limitations to effect maximum production on a sustained basis.

PRODUCTIVITY INDEX

Selection of a cropping system that will result in effective soil conservation and at the same time provide an acceptable living for the farm family necessitates a knowledge of the conserving or depleting effects of the crops to be grown. History shows that in temperate regions grasses build deep, highly productive soils. Few persons, however, care to produce only grass. Furthermore, they cannot wait hundreds of years for grasses to build a productive soil.

The question arises then, how do such crops as corn (maize), small grains, and legumes affect soil? Perhaps the best answers to this question were provided by long-time experiments conducted at the Ohio Agricultural Experiment Station (reported by Salter, Lewis, and Slipher, 1938). Their results have been supplemented by additional, more recent information. They found that crop yields over a 29-year period varied directly with changes in soil nitrogen content. As the latter dropped to 41.2 percent of the beginning nitrogen levels, corn yields dropped to 45.6 percent of initial yields.

Unfertilized, continuous corn plots at the end of a 32-year period contained 840 lb of nitrogen per acre, whereas continuous wheat plots had dropped only to 1315 lb, and plots that had produced a 5-year rotation of corn, oats, wheat, clover, and timothy still contained 1546 lb. The nitrogen content of the soil at the beginning of the trials was 2176 lb/acre. *Losses* over the 32-year period were as follows.

Continuous corn	1336 lb nitrogen/acre
Continuous wheat	861 lb nitrogen/acre
Continuous oats	751 lb nitrogen/acre
Corn, oats, wheat, clover, timothy	630 lb nitrogen/acre
Corn, oats, clover (over 29 years)	396 lb nitrogen/acre

These data show that continuous corn production in Ohio without use of fertilizer or manure over a 32-year period resulted in a loss of almost two-thirds of the soil nitrogen, and that losses during the production of small grains over the same period amounted to about one-third the soil nitrogen. The introduction of a clover crop once in 5 years reduced overall nitrogen losses, and production of the clover once in 3 years reduced it still further.

It has since been reported that 40 years of cropping to continuous cotton resulted in the loss of two-thirds of the organic matter and 50 percent of the nitrogen in an Oklahoma soil, a figure surprisingly close to that obtained with corn in Ohio.

These and other data from Ohio, Michigan, and other states have led to the soil-productivity indexes (P.I.) shown in Table 6-1. The indexes are stated in terms of percentage changes effected by one season's growth. As shown previously, they are based on the effect the crop has on soil nitrogen content and the data that show that productivity is proportional to nitrogen content. The truth of this latter relationship is strengthened by some data from Minnesota that found actual corn yields to be very close to those predicted by calculations from nitrogen levels determined by the Kjeldahl method.

The indexes shown in Table 6-1 apply on adequately limed soil, to which the crops are climatically adapted. For instance, in the South, crimson rather than red clover is grown. Good yields are necessary or soil-building effects cannot be taken as credit. Likewise, credit for residues varies with crop yields and should be based on about 2 tons for the + 0.25 credit. Credit for fertil-

TABLE 6-1 Soil-productivity Indexes for Scoring Soil Management and Cropping Program[a]

Crop or land use	Soil productivity index (percentage)
Row crops—corn, potatoes, sugar beets, field beans, soybeans for grain or hay	−2.0
Orchards, clean cultivated (plus factor for interplanted crops or cover crops)	−2.5
Wheat, barley, oats, rye, spelt, buckwheat (straw not returned)	−1.0
Millet, sudan grass, etc. (hay or pasture)	−1.0
Field peas (canning, grain, or hay)	−0.5
Crop land—bare or summer fallow	−2.0
Crop residue (2 tons) left on field (field bean, soybean, or small grain, straw, cornstalks, beet tops)	+0.25
Perennial grasses for hay	0.0
Perennial grasses for pasture	+0.5
Alfalfa, for change effected by end of first hay year	+2.5
Alfalfa, for change effected during second year	+0.5
Alfalfa, for change effected during third year and after	0.0
Red, mammoth, and alsike clovers—as hay or pasture	+2.0
Red, mammoth, and alsike clovers plowed under in spring following seeding	+1.75
Clover-timothy—as hay or pasture (second year)	+1.25
Sweetclover, seeded in small grain, plowed green in April or May	+2.5
Sweetclover, seeded in corn, plowed green in April or May	+1.5
Soybeans or vetch, plowed down bloom stage or later	+1.5
Rye, wheat, oats, or buckwheat—green manure	+0.5
Application of 40 units of commercial fertilizer	+0.15
Application of 1 ton of protected manure	+0.15
Application of 1 ton of manure not protected	+0.125

Correction for erosion.

Degree of erosion and percentages of slope often involved.

Erosion	Slope	Factor[b]
Little or none	Less than 1%	0.0
Slight	1–2%	0.25
Moderate	3–5%	0.5
Severe	6–10%	1.0
Very severe	10%	2.0

[a]This plan and most of the values are suggested by the Ohio Experiment Station. In some cases, the values have been changed to a figure that, it is believed, more nearly represents correct values for the broad Midwest region.

[b]Increase all negative indexes by this factor, i.e., corn on a 4% slope has a minus 3 index unless minimum tillage is practiced. Then, the erosion factor should be reduced to 60% of these figures, in this case, 60% of 0.5 = 0.3, or the index becomes −2.6.

izers should be taken only to the extent that yields are increased (harvested and residue portions). *Higher rates simply to increase residues are justified.* We freely admit that many assumptions are made. In other words, the experiments on which these values are based were not broad enough nor specific enough to furnish certain proof that all indexes are exact. However, they have been used so long and so widely that reliance now seems justified. Furthermore, methods of calculating safe cropping systems worked out by U.S. Soil Scientists and State Experiment Station and Extension personnel verify the reliability of the productivity index philosophy.

The use of productivity indexes is simple. Corn grown continuously, without fertilizer or other ammendments on level land, will deplete the soil at the rate of 2 percent/year. A 100 bushel yield in 1979 would be followed by 98 bushels in 1980, by 96.04 bushels in 1981, 94.13 bushels in 1982, and so on. Small grains are one-half as depleting as a row crop, and clover is soil-building to the extent of 2 percent/year. Thus a corn-oats-clover rotation scores $-2 - 1 + 2 = -1$ for 3 years or would be depleting to the extent of 0.33 percent/year. Nitrogen losses under such a rotation just about verify this figure. Such a rotation can be made soil-building by returning residues and/or manure and by applying commercial fertilizer.

Productivity Index Illustrated

Perhaps the best way to explain the use of productivity indexes is to calculate a proposed cropping system — in effect, to determine whether or not a certain rotation is safe, whether it will increase or decrease productivity.

Assume a five-field farm with unequal-sized fields and a total of 61 tillable acres.

Year	Field 1 9 acres	Field 2 14 acres	Field 3 16 acres	Field 4 10 acres	Field 5 12 acres
1	Alfalfa	Corn	Corn	Oats	Wheat
2	Corn	Corn	Oats	Wheat	Alfalfa
3	Corn	Oats	Wheat	Alfalfa	Corn
4	Oats	Wheat	Alfalfa	Corn	Corn
5	Wheat	Alfalfa	Corn	Corn	Oats

Erosion, all fields — slight factor 0.25

Fertilizer — 400 lbs 6-24-12 and 200 lbs 45-0-0 corn
      400 lbs 6-24-12 oats
      400 lbs 6-24-12 wheat

Note: The fertilizers contain slightly more than 40 units of plant food per 100 lb, so credit of +0.15 may be taken for each 100 lb applied.

Green manure — Seed rye in each corn crop (+0.5)

Manure — 8 tons for each corn crop (+0.15)

Residues — Corn stover, plow under (+0.25). Straw will be taken off for bedding.

Calculate the productivity index for each year and for the 5 years.

			$-$	$+$
First year Alfalfa	9 acres $\times +2.5 =$			22.5
Corn	14 acres $\times -2.0 =$		28	
Corn	16 acres $\times -2.0 =$		32	
Oats	10 acres $\times -1.0 =$		10	
Wheat	12 acres $\times -1.0 =$		12	
			-82	$+22.5$

Erosion $\quad -82 \times 0.25 = -20.5$	
Fertilizer $\quad 6 \times 0.15 = 0.90 \times 30$ acres corn $=$	27.0
$4 \times 0.15 = 0.60 \times 10$ acres oats $=$	6.0
$4 \times 0.15 = 0.6 \times 12$ acres wheat $=$	7.2
Rye green manure 0.5×30 acres corn $=$	15.0
Manure $8 \times 0.15 = 1.2 \times 30$ acres corn $=$	36.0
Residues 0.25×30 acres corn $=$	7.5
Total positive	121.2

$121.2 - (82 + 20.5) = +18.7$

$18.7/61$ acres $\qquad = +0.306$ P.I.

Likewise the P.I. for each of the other 4 years may be calculated. The indexes for the 5 years and the average follow.

$$\text{First year} = +0.306$$
$$\text{Second year} = +0.347$$
$$\text{Third year} = +0.211$$
$$\text{Fourth year} = +0.537$$
$$\text{Fifth year} = +0.499$$
$$\text{Average} = +0.380$$

In actual calculation of a management plan, the 5-year average is sufficient, provided the same rotation and soil treatments are adhered to in all fields. It is calculated as follows.

			$-$	$+$
5 years Alfalfa	61 acres $\times +2.5 =$			152.5
Corn	122 acres $\times -2.0 =$		244	
Oats	61 acres $\times -1.0 =$		61	
Wheat	61 acres $\times -1.0 =$		61	
			-366	$+152.5$

Erosion $\quad -366 \times 0.25 = -91.5$	
Fertilizer $\quad 6 \times .15 = 0.9 \times 122$ acres corn $=$	109.8
$4 \times .15 = 0.6 \times 61$ acres oats $=$	36.6
$4 \times .15 = 0.6 \times 61$ acres wheat $=$	36.6
Rye green manure 0.5×122 acres corn $=$	61.0
Manure $8 \times 0.15 = 1.2 \times 122$ acres corn $=$	146.4
Residues 0.25×122 acres corn $=$	30.5
Total positive	$+573.4$

573.4 − (366 + 91.5) = 115.9
115.9/61 × 5   = +0.38

Note: 61 × 5 equals the number acres involved in 5 years. The 5-year calculation checks with the average of the single-year figures.

Variable Rotations or Soil Treatments

Soil variations on some farms make it wise to follow different rotations and
use different soil treatments in each field. We must then guard against excessively depleting some fields to build others. Under such a soil management
plan, a productivity index is calculated for each field through the rotation
followed on that field. For one field of continuous corn, only 1 year is involved and would look like this:

Corn	−2.0	
Fertilizer, 600 pounds 40% fertilizer		+0.9
Residues		+0.25
Sweetclover green manure		+1.5
	−2.0	+2.65

Whereas, the remainder of the farm might be handled as in the previously discussed rotation, or, it might be too hilly for row crops and be used
only for small grains and alfalfa. Thus the need for calculating each portion
separately is plainly evident.

A positive productivity index means that productive capacity will gradually increase. There is, of course, a limit where it will level off regardless of the
extent of building practices employed. That limit is dependent on certain
factors beyond the control of the operator, such as moisture, temperature,
physical crowding, and supply of carbon dioxide.

SLOPE-LENGTH LIMITATIONS

As slope increases, erosion rapidly comes into play as a force to be considered.
In the discussion of productivity index, the slight erosion did not increase the
negative (depleting) side of the ledger enough to rule out use of the rotation.

Suppose we take a look at an identical proposition for an area with a
moderate erosion hazard. The modifying factor is 0.5, and the productivity
index becomes + 0.08, a figure too small to be considered safe. Several
alternatives are available. First, erosion may be reduced by minimum tillage,
contouring, strip-cropping, or terracing. Another possibility is the use of
sweetclover instead of rye as green manure, and/or it may be wise to limit
corn to 61 acres in 5 years. Further benefits from the use of fertilizer might be
possible by an application for the green manure crop.

Soil Management Principle

Erosion may be reduced by conservation tillage, minimum tillage, contouring, strip-cropping, or terracing.

The U.S. Soil Conservation Service and various state experiment stations have experimented widely with the effect of degree and length of slope on water and soil losses from different types of soil.

The particular practices that may be necessary to effect water conservation and control erosion on a certain soil area are largely determined by the percent and length of slope, tillage practices including residue management, and the cropping system, already referred to as rotation. In many areas and on many farms, the most profitable crops are those that are least conserving of water and are least protective to the soil. Accordingly, wise management necessitates the establishment of a **safe** rotation for each area on the farm. Such a rotation, referred to in these pages as a "minimum rotation" is defined as the least protective rotation that will provide for sustained productivity for an indefinite period. The need for soil protection increases rapidly as slopes become more severe. Experiments have shown that soil loss increases 2.4 times as slope steepness is doubled and 1.4 times as length is doubled. The U.S. Soil Conservation Service soil-loss tables are based on these relationships.

The U.S. Soil Conservation and Agricultural Research Services personnel and representatives from the State Cooperative Extension and Experiment Station staffs of the North Central Region formulated an easy system of calculating a minimum rotation on a specified soil. For instance, a farmer may wish to know whether or not a row crop-row crop-oats-meadow rotation is advisable in a certain field, or whether it will result in soil depletion and lowered productivity. In other words, would such a rotation be using the soil beyond its capability? If we assume the previously mentioned rotation is not advisable on the particular soil [perhaps one with B slope (2 – 6 percent) and class 2 erosion], what is the minimum rotation that *is* permissible?

Relative Soil Loss

The first step in working out this system of calculating a minimum rotation is to determine relative soil losses from various rotations. Experimental data from runoff plots in various sections of the country (Coshocton, Ohio; La-Crosse, Wisconsin; and others) were used for this purpose. The percentages (see Table 6-2) are based on continuous corn, conventionally grown without

TABLE 6-2 Relative Soil Loss and Protectiveness Exerted by Individual Crops Under Various Cropping Systems[a]

Crop	Relative soil loss	Relative protectiveness
Continuous row crop	100	0
Row crop after 1 year meadow	40	60
Row crop after 2 or more years meadow	35	65
Row crop after small grain	90	10
Row crop after row crop after small grain	100	0
Row crop after row crop (or grain) after 1 year meadow	80	20
Row crop after row crop (or grain) after 2 or more years meadow	70	30
Third year row crop after 1 year meadow	95	5
Fourth or more year row crop after 1 year meadow	100	0
Spring grain after 2 years row crop (or grain) after catch crop	35	65
Spring grain after row crop (or grain) after 1 year meadow	30	70
Spring grain after row crop (or grain) after 2 or more years meadow	25	75
Spring grain after second year row crop (or grain) after 1 year meadow	35	65
Spring grain after second year row crop (or grain) after 2 or more years meadow	30	70
Spring grain after 3 or more years row crop (or grain)	40	60
Spring grain after 1 year meadow	15	85
Spring grain after 2 or more years meadow	10	90
Meadow crops	1.0	99
First year Korean lespedeza (no companion crop) after row crop (or grain)	40	60
Lespedeza hay	4.0	96

Soil loss from winter grain is the same as that from spring grain in Kentucky, southern Illinois, Indiana, and Missouri; elsewhere it is 0.7 that of spring grain.

[a]Fertility treatments sufficient to produce good yields based on experimental results and farm-planning recommendations are assumed in these values. This table was taken from the U.S. Soil Conservation Service ready reference guide.

special management practices as 100 percent. At the other extreme from the zero protection offered by continuous corn is the almost complete protection offered by continuous meadow, where the relative soil loss drops to 0.5 percent of that which occurs under continuous corn.

Following is a list of management practices and factors to show how each practice reduces soil loss.

	Practice	Factor
Residues	(2 tons/acre or more) left on the surface through year.	0.50
Residues	(Moderately grazed) left on surface through year.	0.75
Residues	(2 tons/acre or more) left on surface until planting time.	0.80
Winter cover crops	Plowed at planting time.	0.80
Mulch	(Manure 6 tons or more) applied immediately after planting.	0.60
Plow-planting	A minimum of cultivation and packing (minimum tillage.)	0.60
Use of field cultivators on meadow	Instead of plowing	0.50

To determine the effect of the practice, multiply the relative soil loss shown in Table 6-2 by the factor. For instance, plow-planting reduces soil loss from a continuous row-crop area to 60 percent of what it would be without the practice. If residues are left on the soil until planting time, and the last-years crop was sufficient to leave 2 tons of stover, the relative soil loss becomes $0.60 \times 0.80 \times 100 = 48$ percent of what it would be without the two practices.

A further reduction in soil loss in continuous corn culture may be effected by growing a winter cover crop to be plowed at the time of planting the next crop. The relative soil loss of 48 percent, obtained as has been shown, is then further reduced (0.80×48) to 38.4 percent, almost down to one-third the maximum without management practices. It is immediately evident that the combination of practices reduces the hazard of slope and makes it possible to use a less protective cropping sequence. In other words, use of the practices just listed makes it possible to hold soil losses to levels consistent with good soil management.

The use of "productivity index," "relative soil loss," and "relative protectiveness" furnishes us with a simple way of determining what effect a certain cropping system may have on the long-time capacity of a soil. We see further how a given cropping system may be improved by certain conservation practices and/or how a system may be changed so it is permissible on a certain soil. Erosion by water or wind is the culprit that must be controlled. Soil scientists have long studied ways of estimating erosion so it might be possible to determine how intensive a cropping system may be followed without depleting the soil. Success has been attained.

Predicting Water-Erosion Losses

As long ago as 1932, Professor A. R. Whitson had some of his graduate students (R. L. Cook was one of them) studying possible ways of estimating the extent to which soils might be depleted by water erosion. To our knowledge, his students did not make any lasting contributions. But others were more successful. A. W. Zingg in 1940 published an equation relating soil loss to certain soil characteristics. In 1941, D. D. Smith added crop and conservation-practice factors and the concept of a specified soil-loss limit to Zingg's equation. In 1947, G. M. Browing and coworkers added soil erodability and management factors and prepared a set of tables for use in Iowa.

Several other soil scientists, including D. M. Whitt, G. W. Musgrave, C. A. Van Doren, L. J. Bartelli, and, finally, W. H. Wischmeier, made important contributions to the development of what is now termed "The Universal Soil-Loss Prediction Equation." The equation is explained in excellent detail by Wischmeier and Smith in USDA Agricultural Handbooks 282 and 537.

THE UNIVERSAL SOIL-LOSS PREDICTION EQUATION

In the remainder of this chapter we present a brief summary of the high points of USDA Agricultural Handbooks 282 and 537, just enough to convince you that soil-loss prediction is possible. Please note that consideration is given to all the forces that may affect soil management practices.

The equation:

$$A = RKLSCP$$

where,

A = computed soil loss in tons/acre[1]

R = rainfall factor

K = soil erodibility factor

LS = topographic factor (length and steepness of slope)

C = crop and management factor

P = erosion control factor

The Rainfall Factor (R)

For a specific storm, the erosion index unit is the product of the kinetic energy of the storm measured in hundreds of foot-tons per acre times its maximum 30-min intensity in inches per hour. The energy intensity products (EI) can be computed from recording rain-gauge data with the aid of a published rainfall energy table. The rainfall factor is the average yearly sum of the erosion index units during 1 year. The yearly average is determined over a period of years. In the continental United States and Hawaii, the factor

---

[1]Because the discussion on soil-loss prediction is taken largely from Agricultural Handbooks 282 and 537, we take the liberty of using the old English unit terminology as did the authors of those publications. Conversions to the metric system are possible.

R varies from less than 20 in the mountainous areas of the Far West to more than 500 along the Gulf of Mexico.

In Michigan, the factor varies from 70 near the Straits of Mackinac to 155 in Berrian County. Most agronomists work in a relatively small area so they need deal with only one or two rainfall factors. See Fig. 6-1 for the factors in Michigan's 83 counties. Note that seven adjoining counties at the northern tip of the lower peninsula have 70 as a rainfall factor. Also, all the counties adjoining Saginaw Bay have the factor 75. We see then that an agricultural extension worker, in most cases assigned to a single county, would work with only one rainfall factor.

FIGURE 6-1 Rainfall factor, R, for Michigan counties. Although this factor varies across the state, it may be considered uniform over a countywide area. (Taken with permission from Research Report 310, Michigan Agricultural Experiment Station, 1976.)

The Soil Erodibility Factor (K)

The soil erodibility factor is experimentally determined on a unit plot 72.6 ft long with a uniform lengthwise slope of 9 percent. The plot must be in continuous fallow for at least 2 years before measurements are taken, and it is tilled up and down the slope. During the period of soil-loss measurements the soil is plowed and fitted as for conventional production of corn. It is tilled when necessary to prevent vegetal growth and serious crusting. When all these conditions are met, each of the factors L, S, C, and P has a value of 1.0 and K equals A/EI (EI = energy intensity product.)

K values have been determined for a number of soils. Those included in Handbook 537 are shown in Table 6-3. Local research and USDA Soil Conservation Service personnel have determined K values for other soils. Rain-makers are used for rapid determination of K values. In Fig. 6-2, Earl Erickson of Michigan State University is working with one.

TABLE 6-3 Computed K Values for Soils on Erosion Research Stations

Soil	Source of data	Computed K
Dunkirk silt loam	Geneva, N.Y.	0.69[a]
Keene silt loam	Zanesville, Ohio	0.48
Shelby loam	Bethany, Mo.	0.41
Lodi loam	Blacksburg, Va.	0.39
Fayette silt loam	LaCrosse, Wis.	0.38[a]
Cecil sandy clay loam	Watkinsville, Ga.	0.36
Marshall silt loam	Clarinda, Iowa	0.33
Ida silt loam	Castana, Iowa	0.33
Mansic clay loam	Hays, Kans.	0.32
Hagerstown silty clay loam	State College, Pa.	0.31[a]
Austin clay	Temple, Tex.	0.29
Mexico silt loam	McCredie, Mo.	0.28
Honeoye silt loam	Marcellusss, N.Y.	0.28[a]
Cecil sandy loam	Clemson, S.C.	0.28[a]
Ontario loam	Geneva, N.Y.	0.27[a]
Cecil clay loam	Watkinsville, Ga.	0.26
Boswell fine sandy loam	Tyler, Tex.	0.25
Cecil sandy loam	Watkinsville, Ga.	0.23
Zaneis fine sandy loam	Guthrie, Okla.	0.22
Tifton loamy sand	Tifton, Ga.	0.10
Freehold loamy sand	Marlboro, N.J.	0.08
Bath flaggy silt loam with surface stones >2 in. removed	Arnot, N.Y.	0.05[a]
Albia gravelly loam	Beemerville, N.J.	0.03

Source: W. H. Wischmeier and D. D. Smith, *Preceding Rainfall Erosion Losses: A Guide to Conservation Planning.* Reproduced with permission from USDA Agricultural Handbook 537, 1978.

[a]Evaluated from continuous fallow. All others were computed from rowcrop data.

FIGURE 6-2 Dr. Earl Erickson, soil physicist at Michigan State University, using a rain-maker in soil-erosion research.

The Topographic Factor (LS)

Considered together for convenience in the field, LS refers to length and steepness of slope. It "is the expected ratio of soil loss per unit area from a field slope to that from 72.6 foot length of uniform 9 percent slope under otherwise identical conditions." (Wischmeier and Smith, 1958). This ratio, for a specific area, may be taken from the chart (Fig. 6-3) and is the LS factor to be used in the soil-loss prediction equation. For instance, if the slope is 8 percent and is 200 ft long, LS equals 1.4.

Let us assume at this point that the 8 percent slope-200 ft area is in central Michigan where the rainfall factor is 97, LS equals 1.4, the soil has been fallow for 2 years, and no erosion-control practice is in effect. Then factors C and P would each equal 1. The soil might well be a sandy loam with a K value of 0.28. Then,

$$A = R \times K \times LS \times C \times P$$

$$A = 97 \times 0.28 \times 1.4 \times 1 \times 1 = 38 \text{ tons/acre/year}$$

Contouring, a practice to be discussed in a later chapter, would reduce soil loss on such a slope (8 percent-200 ft) to 60 percent of maximum. The P becomes 0.60 and

$$A = 97 \times 0.28 \times 1.4 \times 1 \times 0.60 = 22.8 \text{ tons/acre/year}$$

The Crop and Management Factor (C)

Any crop on a soil will furnish some protection and will reduce erosion. In fact, growing plants with a massive root system and canopy effect *can* reduce soil loss to almost zero. C then would be very small. It all depends on the crop mass produced (yield), the dates of soil fitting and planting with reference to when the rains occur, tillage practices, cover crops grown, and crop residue management. An illustration of C-factor calculation was given by Wischmeier and Smith and, with their permission, is reproduced here as Table 6-4.

The crop rotation in their example was meadow, corn, corn, wheat. Yields were high and residues were well managed. The 4-year C value is

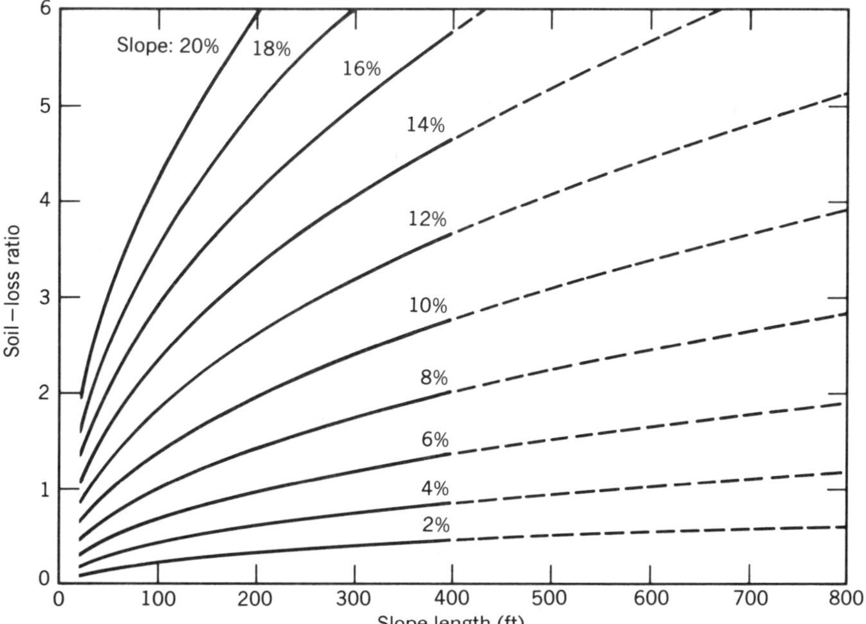

FIGURE 6-3 Slope-effect chart (topographic factor, LS) Reproduced with permission from W. H. Wischmeier and D. D. Smith (1965). *Predicting Rainfall Erosion Losses from Cropland East of the Rocky Mountains.* USDA Agricultural Handbook 282. Agricultural Research Service – U.S. Department of Agriculture.

TABLE 6-4 Sample Working Table for Derivation of a Rotation C Value

(1) Event	(2) Date	(3) Table 6, area 16	(4) Crop- stage period	(5) EI in period	(6) Soil-loss ratio[a]	(7) Sod factor	(8) Crop-stage C value	(9) Crop year
Pl W	10/15	92	SB	0.03	0.27(132)	0.95	0.0077	
10 percent c	11/1	95	1	0.03	0.21	0.95	0.0060	
50 percent c	12/1	98	2	0.12	0.16	1.0	0.0192	
75 percent c	4/15	10	3	0.46	0.03		0.0138	
Hv W	7/15	56	4	0.28	0.07(5C)		0.0196	0.066
Meadow	9/15	84		1.26	0.004(5B)	1.0	0.0050	0.005
TP	4/15	10	F	0.05	0.36(2)	0.25	0.0045	
Disk	5/5	15	SB	0.10	0.60	0.40	0.0240	
Pl C	5/10	—						
10 percent c	6/1	25	1	0.13	0.52	0.40	0.0270	
50 percent c	6/20	38	2	0.14	0.41	0.45	0.0258	
75 percent c	7/10	52	3	0.40	0.20	0.50	0.0400	
Hv C	10/15	92	4L	0.05	0.30	0.60	0.0090	0.130

Chisel	11/15	97	4c	0.17	0.16(46)	0.60	0.0163	
Disk	5/1	14	SB	0.11	0.25(48 and 61)	0.80	0.0220	
Pl C	5/10	—						
10 percent c	6/1	25	1	0.13	0.23	0.80	0.0239	
50 percent c	6/20	38	2	0.14	0.21	0.85	0.0250	
75 percent c	7/10	52	3	0.40	0.14(48)	0.90	0.0504	0.138
Hv C and pl W	10/15	92						
Rotation totals				4.0			0.3392	
Average annual C value for rotation							0.085	

Source: Reproduced with permission from W. H. Wischmeier and D. D. Smith. *Predicting Rainfall Erosion Losses: A Guide to Conservation Planning.* USDA Agriculture Handbook 537, 1978.

[a]Numbers in parentheses are line numbers in table 5.

[b]Abbreviations: c, canopy cover; C, corn; hv, harvest; pl, plant; TP, moldboard plow; W, wheat, SB, seed bed; F, fallow.

[c]Table 5 and 6 refer to Wischmeier and Smith (1978).

0.085. By adding such a C value to the foregoing illustration for Central Michigan we have

$$A = 97 \times 0.28 \times 1.4 \times 0.085 \times 0.60 = 1.94 \text{ tons/acre/year}$$

an amount well below the permissible amount for that soil.

Soil Conservation Service personnel working with local agronomists have determined C values for the various cropping and management systems within their areas. They are glad to make the figures available to extension agronomists working with farmers.

Soil Management Principle

Erosion, either by wind or water must be understood and controlled.

PREDICTING WIND-EROSION LOSSES

Generally speaking, wind erosion is a problem in the less-humid areas, any place where soils are loose, dry, and finely granulated, where the soil surface is relatively smooth, and where vegetation is sparce or nonexistent.

It is possible to predict soil losses by wind action by solving the equation:

$$E = f(I', K', C', L', V)$$

where,

E = the soil loss in tons per acre per year
I' = the soil erodibility index, a characteristic dependent on texture, density, and structural stability. Organic and sandy soils erode readily.
C' = the climatic factor, related to wind velocity and soil moisture
K' = soil roughness (smooth or ridged)
L' = field length along the prevailing wind direction
V = the equivalent quantity of vegetative cover
f = "a function of" since the factors are not simply multiplied, as with the water erosion prediction equation.

Skidmore and Woodruff (1968) have published charts, tables, and maps necessary for the solution of the wind-erosion prediction equation. A computer solution was made available by Skidmore, Fisher, and Woodruff (1970).

Woodruff and Siddoway explained the use of the wind erosion equation in 1965, and Troeh, Hobbs, and Donahue presented an excellent review of the subject in 1980.

Practices for the control of water and wind erosion are presented in Chapters 7 and 16.

REFERENCES

Browning, G. M., C. L. Parish, and J. A. Glass (1947). A method for determining the use and limitation of rotations and conservation practices in control of soil erosion in Iowa. *J. Am. Soc. Agron. J.* 39:65–73.

Grava, J. (1958). *Nitrogen Availability Measurements in Soils.* Proceedings of the Minnesota Nitrogen Conference, University of Minnesota, St. Paul, February 20–22.

Reed, L. (1959). Quoted in Cropping a field for 40 years without fertilizer. *Crops and Soils* 11:19, February.

Salter, R. M., R. D. Lewis, and J. A. Slipher (1938). Our Heritage—the soil. Ohio Agr. Serv. Bull. 175.

Salter, R. M., R. D. Lewis, N. P. Woodruff, and F. H. Siddoway (1965). A wind erosion equation. *Proc. Soil Sci. Soc. Am.* 29(5).

Skidmore, E. L. P. S. Fisher, and N. P. Woodruff (1970). Wind erosion equation: Computer solution and application. *Proc. Soil Sci. Am.* 34:931–935.

Skidmore, E. L., and N. P. Woodruff (1968). *Wind Erosion Forces in the United States and Their Use in Predicting Soil Loss.* USDA Agricultural Handbook 346, April.

Slipher, J. A. (1938). *Our Heritage—the Soil.* Ohio Agricultural Extension Service Bulletin. 175.

Smith, D. D. (1941). Interpretation of soil conservation data for field use. *Agr. Engineering* 22:173–175.

Troeh, F. R., J. A. Hobbs, and R. L. Donahue (1980). *Soil and Water Conservation for Productivity and Environmental Protection.* Prentice–Hall, Englewood Cliffs, N.J.

Wischmeier, W. H., and D. D. Smith (1958). Rainfall energy and its relationship to soil loss. *Am. Geophys. Union Trans.* 39:285–291.

Wischmeier, W. H., and D. D. Smith (1965). *Predicting Rainfall-Erosion Losses from Cropland East of the Rocky Mountains.* USDA Agricultural Handbook 282, Agricultural Research Service in cooperation with Purdue Agricultural Experiment Station.

Wischmeier, W. H., and D. D. Smith (1978). *Predicting Rainfall Erosion Losses: A Guide to Conservation Planning.* USDA Agricultural Handbook 537.

Woodruff, N. P., and F. H. Siddoway *Conservation Tillage.* Proceedings of a National Conference, March 28–30, 1973, Des Moines, Iowa. Kansas Agricultural Experiment Station, Department of Agronomy Contribution 1332.

Zingg, A. W. (1940). Degree and length of land slope as it affects soil loss in a runoff. *Agr. Engineering* 21:59–64.

CHAPTER 7

Water Conservation and Erosion Control

One who has stood on the rim of the Grand Canyon of the Colorado River in Airzona is both awed and appalled at the sight. It is hard to believe that erosion was the cause. Equally impressive is the realization that a great butte was carved by the erosion process or that great logs once covered by enough earth to exert the pressure needed for petrifaction were laid bare by erosion.

Equally devastating, although less spectacular, is the continuous soil loss from our agricultural lands (Figs. 7-1, 7-2, and 7-3). Soil develops slowly, the result of a combination of forces (weathering) acting on the crust of the earth. The surface layer weathers most rapidly and increases most rapidly in organic matter. Thus, per unit mass it becomes the portion most productive of plant life. This is also the portion from which soil particles are removed by erosion, both wind and water.

Soil Management Principle

Vegetative cover prevents soil erosion.

Nature customarily protected land from excessive soil loss through erosion by establishing a vegetative cover, usually grasses or forests. Lack of moisture has prevented adequate cover establishment in desert regions. The shifting of desert sands as a result of wind action would stop if some form of vegetation could be established. The silting of the Colorado River, for instance, could be avoided if the watershed could be covered with grass. Much of the area is too dry for grass to survive, so erosion continues, and streams are polluted and reservoirs filled. Thus, it is clear that soil erosion is a national

FIGURE 7-1 Serious sheet erosion in sandy loam soil in central Michigan. This washing resulted from one summer rain while the soil was fallow. Table beets were occupying the more level land at the top of the slope. The boulders show the depth to which soil was removed. This kind of erosion destroys the valley soil by covering it with subsoil from the slope.

FIGURE 7-2 Serious gully erosion.

FIGURE 7-3 Unchecked gully erosion destroys land.

and world, as well as a local, problem. In fact, excluding economic, political, and cultural problems, erosion is presently our most serious constraint to hunger elimination in this century.

To solve the problem, government action established the Soil Erosion Service as part of the Department of the Interior. The service was later transferred to the Department of Agriculture and now exists as the U.S. Soil Conservation Service. Its functions have been broadened, but still major among its activities is the control of erosion. The possibility of predicting erosion losses in conservation planning was presented in detail in the last chapter. In this chapter you will become acquainted with practices for *controlling* erosion, with emphasis on erosion caused by water. Wind erosion is considered briefly but is discussed at greater length in the Great Plains section of Chapter 16. There the farmer must constantly be ready to guard against losses caused by strong winds.

WATER CONSERVATION

Soil Management Principle

Soil conservation and management include water conservation.

Water conservation, in the broad sense, is the most important practice, or group of practices, in the field of soil management. This is true because the practices that conserve water are also those that conserve soil and promote a high state of productivity, the real goal in soil management.

Water is any nation's cherished resource. A civilized community may do without other commodities for a longer period than it can do without water. Industry locates where water is available. Native vegetation is sparse or dense in accordance with water supply, and the species of plants reveal the climate largely a function of water supply.

In humid areas, where drought periods are severe but of short duration, deep sands and gravelly soils are low in crop productivity because they have low water-holding capacity in the region of root growth. The water from rain and snow sinks to such deep levels that it is lost to use by growing crops. Water conservation on such soils becomes a matter of holding more of it within the reach of plant roots. Also, the selection of crops as to depth of rooting, water requirement, and growing season, is a matter for careful consideration.

Over still larger acreages of agricultural lands, water conservation is a matter of getting water to penetrate the soil **where it falls.** The first step in farm planning (farm and home development, land-use planning, or **soil management,** it makes little difference what term is used) on many, perhaps most, farms is the drawing of a complete, all farm water management plan. The objective of such a plan is to provide for the greatest possible use of the water that falls on the land. This means as much penetration as possible where it falls, controlled removal of excesses, and protection of the soil against damage from falling and running water. Such protection is accomplished through the correct cultural practices and certain engineering structures and by maintenance of adequate cover. We must bear in mind that the farmer wishes at the same time to use the soil to the extent of its capability. This is necessary if the farmer is to obtain an adequate living from the land.

Infiltration

As discussed in this chapter, water conservation is largely a matter of infiltration and how it is connected with cultural practices and erosion-control measures. Broadly speaking, soil-forming processes, including soil organism activities, tend to increase the rate of infiltration. As soon as we started cultivating land, the infiltration rate decreased and erosion increased.

We have seen in Chapter 4 that the state of aggregation is correlated positively with the content of easily decomposed organic matter in the soil. Well-aggregated soils are high in volume of large pores through which water can readily enter. It follows then that the infiltration rate may be directly related to organic-matter management. Growing crops are always adding organic matter to soil. Even while plants are growing, roots and leaves are sloughing off to start decomposition and increase aggregation. Dead plants leave open root channels through which water may enter. Furthermore, vegetation prevents rain from striking the soil, thereby preventing destruction of soil aggregates. On bare soil, a single beating rain may move, by splashing, more than 100 tons of soil per acre. At such a time, suspended soil quickly fills soil pores and the infiltration rate falls. Runoff increases, and the amount of soil it carries depends largely on the slope.

But runoff can be heavy. One study where a 4.4-in. rain in 2-1/2 hours penetrated bare granitic soil to a depth of only 7 cm found about 80 percent of the storm total lost as runoff.

Infiltration is slower under row crops than under grasses and is intermediate under small grains. One of the chief values of a crop residue mulch is the protection against raindrop splash and the resulting decreased infiltration. Of course, the mulch material eventually adds to the soil organic matter.

Musgrave presented the curves shown in Fig. 7-4 as evidence of the effect of crop on rate of infiltration. Under meadow, the rate at the end of 1 hour was 0.8 in/hour, whereas under corn the rate at that time had fallen to less than 0.5 in./hour. At the end of 5 hours, infiltration had leveled off under both crops but was still much faster under the meadow than under the corn.

Always plan for a drought. In some areas (Michigan, for example), crop yields are lessened by lack of water **every year.** The seriousness and extent of the drought varies locally and with seasons. The odds are so great, however, that moisture shortage will occur sometime during the year, that every farmer should assume the drought is "just around the corner." Practices that may avoid drought injury involve hastened infiltration and efficient use of water after it enters the soil.

1. Install tile. It seems a paradox that water removal may lessen drought injury. Effective tile drainage allows plants to root deeply. Thus they

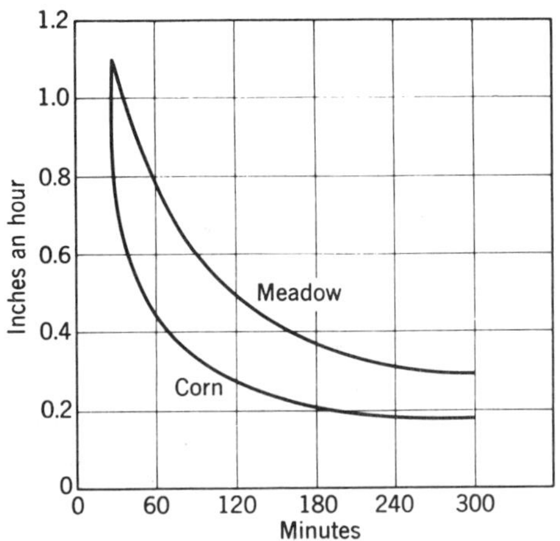

FIGURE 7-4 A comparison of infiltration under bluegrass pasture and under corn, showing the more rapid decline in rate for the row crop. (From G. W. Musgrave. How much of the rain enters the soil. In *Water, The Yearbook of Agriculture.* U.S. Department of Agriculture, Washington, D.C., 1955, p 154.)

have more soil volume to draw on and available soil moisture supplies are usually greater at the lower depths.

2. Destroy hard pans that may limit root penetration.

3. Avoid excessive transpiration losses. Green manures are not practical in desert areas unless irrigation water is plentiful and cheap. Where temporary drought threatens in humid areas, green manures should be plowed under or otherwise destroyed while soil moisture is still plentiful.

4. Keep soils loose. This hastens infiltration, increases pore space for greater storage, and favors the quick formation of a soil mulch.

5. Increase soil organic matter (see Chapter 4).

6. Grow adapted crops. Consider season of growth and rooting depth.

7. Eradicate weeds.

8. Apply needed plant nutrients so crop plants may grow fast enough to compete with weeds. Also, well-fed plants develop root systems that are more efficient at gathering moisture. Over the years, many researchers have shown that plants on good soil or those on poor soil to which needed fertilizers (N, P, or K) were applied used less water per unit of growth than did those poorly supplied with nutrients. Kozakiewicz and Ellis (1967) showed that beans and corn (*Phaseolus vulgaris* and *Zea mays*) made much more efficient use of soil moisture as a result of adequate fertilization with manganese. The soil was Houghton muck (Typic Medisaprists).

9. Irrigate where practical (see Chapter 3).

10. Make use of practices that hold the water on the surface or slow up its movement to allow infiltration (contouring, strip-cropping, plant-residue mulches, and terracing). These practices are so effective that they are considered separately in the remaining pages of this chapter.

Soil Management Principle

The best management practices must be used to hold water on the surface of a soil to allow for maximum infiltration.

CONTOURING

Water loss and erosion are lessened where contouring is practiced. This is farming on the level at right angles to the direction of slope (see Fig. 7-5). Effectiveness depends on the terraces and ridges left by tillage implements and on the method of tillage. If soils are worked down to a level condition before planting, contouring may be quite ineffective in controlling runoff. On the other hand, planting on the furrow, or wheel-track planting of corn, may result in contouring being very effective as an erosion-control practice.

FIGURE 7-5 Contour strip-cropping can offer excellent erosion control.

Contouring has an advantage in addition to that of slowing up and reducing runoff and reducing erosion; it actually reduces power requirements. This will partly compensate for the inconvenience of working irregular-shaped areas.

Slope, Permissible Percentage, and Length

Contouring is most effective on slopes ranging from 3 to 8 percent. It is quite ineffective on slopes more gentle than 3 percent, and its effectiveness decreases as slopes increase above 8 percent until 25 percent is reached. In the most effective range, contouring may reduce soil losses to as little as 50 percent of the loss when up-and-down tillage is used.

Length of slope[1] must be considered. Several practices may have an effect on allowable lengths. Soil erodibility varies with soil type. See Table 6-3 for K values used in the water-erosion equation. Adjustments may be made for land use (rotation) and method of tillage.

In general, slope length should be limited to 300 ft when degree of slope is 3 to 6 percent. Slopes in this percentage range that are longer than 300 ft should be strip-cropped or terraced for effective erosion control. Where slopes are greater than 8 percent, 200 ft is the greatest length that may be

---

[1]**Length of slope** may be defined as "the distance that water will travel at right angles to a contour line until it reaches a defined water channel which is or should be protected." G. M. Browning, C. L. Parish, and John Glass. A method for determining the use and limitation of rotations and conservation practices in control of soil erosion in Iowa. *J. Am. Soc, Agron.* 39:65–73, 1947.

safeguarded by contouring. On longer slopes, other water control methods should be used.

Contouring will be much more effective under soil management practices that maintain aggregation and volume of noncapillary pore space, because of their effect on rate of water intake.

Inaccurately or inadequately designed erosion-control measures may actually increase erosion in certain locations because of runoff concentration. On real long slopes, contouring may direct so much water into a natural channel or depression that a gully may develop. Such a disaster may sometimes be avoided by use of a grassed waterway or other form of diversion channel. It may be wise to consult a specialist from the local Cooperative Extension Service or the U.S. Soil Conservation Service for assistance if it seems that a diversion channel or grassed waterway is needed. Soils will produce to the extent of their capability only when water is properly managed and erosion is controlled.

STRIP-CROPPING

Strip-cropping is the practice of growing row and/or grain crops in alternate strips with meadow. Strip-cropping may include winter grains and row crops. The dense cover provided by the meadow (or winter grain) stops runoff and eroded soil from reaching the adjoining clean-tilled strips. The width of the meadow strips (or winter grain strips) should vary with degree of slope. As degree of slope increases, meadow strips width should be increased. In a strip-cropped area, meadows are usually cut for hay or are used for green manure production.

Field Strip-Cropping

The term **field strip-cropping** is applied to the practice of using uniform-width strips placed across the general slope but with no attempt at exact contouring. This arrangement is used in areas where the topography is too irregular for contour strip-cropping.

In some areas, field strip-cropping and wind strip-cropping can be successfully combined. Such a combination will only be successful, however, where the direction of the prevailing winds is roughly parallel with the main slopes. Of course, in some areas there are no main slopes, so the water erosion control strips can be in any direction.

Contour Strip-Cropping

The practice of growing row or grain crops in alternate strips with meadow is termed **contour strip-cropping.** The strips are arranged on the contour. The dense cover provided by the meadow stops runoff and erosion from adjoining clean-tilled strips. Experimental results show that soil losses from contour strip-cropped areas are one-fourth of those from up-and-down hill cropped areas and one-half of those from contoured areas. To obtain such maximum benefit from contour strip-cropping, 50 percent of the area must be in meadow.

Slope, Degree, and Length

Other conservation measures, especially contour tillage and proper rotation of crops, must be supplementary to strip-cropping. Grassed waterways must be used for disposal of concentrated water. On long slopes, diversion terraces are needed to carry away the water that collects from the tilled strips. On slopes up to 12 percent, strip-cropping alone can be depended on to prevent erosion if the length of the slope is less than 400 ft. On slopes longer than 400 ft, terraces must also be used. Where percentage of slope ranges between 12 and 20 percent, strip-cropping alone will be sufficient only on slopes up to 300 ft in length.

Strip Width

For effective runoff and erosion control, strip width should vary inversely with percentage of slope. Other factors, however, must be taken into consideration. For example, strips may be wider on permeable than on impermeable soils.

Type of rotation should also be considered. The width of a row-crop strip can be greater when the crop follows a meadow crop than when it follows a small grain or other row crop. Method of tillage also has an effect. Row-crop strips can be wider when soils are left loose as in minimum-tillage methods. The more compaction, the narrower the permissible widths.

The soil factors worked out by the Soil Conservation Service, Experiment Stations, and the Agricultural Research Service may be used as a relative measure of allowable strip width. Strip widths for row crops should range with slope as follows.[2]

2.1 – 7.0 percent slope	100 – 88 ft width
7.1 – 12.0 percent slope	88 – 74 ft width
12.1 – 18.0 percent slope	74 – 60 ft width
18.1 – 24.0 percent slope	60 – 50 ft width

Wind Strip-Cropping

The administration of wind strip-cropping varies from that of strip-cropping for water-erosion control in that wind direction must be considered (see Fig. 7-6). Strips should be at right angles to the direction of the prevailing winds.

[2]These widths were developed by the U.S. Soil Conservation Service and the U.S. Agricultural Research Service.

FIGURE 7-6 Wind erosion can be damaging to the field being eroded and to fields receiving the wind-blown particles and also be a hazard to people.

On sloping soils it is sometimes necessary to consider both water and wind erosion.

The width of cultivated strips should be kept at a minimum, usually not more than 100 ft for row crops or small grains.[3]

TERRACING

Cropland that cannot be effectively protected by other conservation practices (rotations, contour tillage, strip-cropping) may be terraced. After lands have been terraced, other conservation practices should also be followed because terracing alone is not an effective means of maintaining satisfactory yields.

Terracing is expensive. On some soils, in some geographic locations, the practice is too expensive to be practical. Proximity to markets, for instance, may have considerable bearing on the value of farm products. Other factors, as well, may help farmers to decide whether they can afford to terrace and follow a less-protective cropping system or whether it might be more profitable to depend on strip-cropping and grow more protective crops. For instance, one may terrace a Miami loam soil and follow the same rotation (clover, row crop, oats) on 2 to 12 percent slopes as on the level areas. Without terracing, adequate protection of 400-ft-long slopes would require 3 years of

[3]Consult *Strip Cropping for Conservation and Profit.* U.S. Department of Agriculture Farmers Bulletin, 1981.

alfalfa in rotation with a row crop and small grain. In other words, terracing makes it possible to crop land with slopes to 12 percent as if it had less than a 2 percent slope.

Runoff loosens and transports topsoil when it attains a velocity of 2 to 3 ft/sec. The finer the soil, the lower the velocity necessary to move a certain amount of soil. Less velocity and less runoff near the top of a slope results in little soil loss. Volume of runoff and the velocity of the flowing water increases with distance down the slope. Terraces, to be effective, must intercept the water before it gains sufficient velocity to pick up appreciable amounts of soil. The percentage of slope, the texture and structure of the soil, the nature of the cover, and maximum rainfall intensity are factors that largely determine the advisable spacing and size of terraces.

Types of Terraces

A **terrace** may be defined as an embankment of earth constructed across a slope in such a way as to control water runoff and minimize erosion. There are two general types, absorption and diversion.

Ridge Terraces

Absorption terraces are built in the form of a broad, low ridge constructed on the absolute contour. The channel should be wide and shallow, with no depressions for impounding water that might destroy crops. This type of terrace is used on permeable soils where water conservation is of primary importance, as on the Great Plains and the sandy coastal plains.[4]

Depending on the importance of total water conservation, ridge terraces may be constructed with closed ends or with the ends built somewhat lower than the ridge, so excess water may spill over there rather than over the ridge.

In many areas where ridge terraces are used, wind erosion must also be controlled. This can be accomplished because the ridges are low enough for machinery to cross. Crop strips may even cross the ridges where wind stripping is of great importance. Under such management, however, it is very difficult to maintain the ridges.

Broad-Channel Terraces

Where water conservation is of lesser importance, broad-channel terraces are used to conduct excess water from the fields at nonerosive velocities. The channel should be wide, relatively shallow, and have gently sloping sides. The height of the ridge determines the capacity of the channel. An attempt at too much capacity results in the movement of excessive amounts of topsoil. The topsoil in the channel may thus become too thin. Real shallow soils cannot be successfully terraced.

[4]Please refer to page 85 of *Manual of Soil and Water Conservation Practices*. U.S. Soil Conservation Agricultural Handbook 61, June 1954.

The maximum allowable gradient of a broad-channel terrace depends mostly on the depth of the channel. The deeper the water, the greater its scouring effect at a constant gradient. The aim then should be to construct broad, shallow channels with a gradient just less than that which will result in scouring. The cross section of the channel should be uniform over its entire length. The gradient of the channel should increase gradually from the upper to the lower end to avoid deepening water toward the lower end.

The gently sloping sides of the channel and of the lower side of the ridge make it possible to use machinery freely over the terraces.[5]

Nichols terraces are simply broad-channel terraces, so named after the person who widely recommended their use. The Mangum terrace, first constructed in North Carolina by P. H. Mangum in 1885, is different in that soil is moved toward the ridge from the lower as well as the upper side. The Mangum terrace is somewhat easier to construct than is the Nichols terrace.

Bench Terraces

Very steep land, up to 50 percent slopes, may be cultivated by means of bench terraces. The slope is cut and filled to make strips or steps with a slight gradient across the slopes and with a slight slope from the front to the back of the step. The risers between steps are almost vertical when made of rock and are as steep as possible when made of earth. Earthen risers must be heavily vegetated. Where very intense cropping is desirable, as in the *Oriental* countries, the risers are used for productive crops such as strawberries, or when grassed, the forage is gathered for feed.

The California or Puerto Rico terrace may be constructed gradually by allowing natural erosion and plowing to move the soil. The first step is to plant narrow grass barriers across the slope at distances apart that must vary with the slope and therefore with the needed width of terrace. The grass barriers should be on the absolute contour. After the barriers are established, the land between them is plowed and planted. For each crop, the furrows should be turned toward the barriers. Thus, after several plowings, a satisfactory bench will result. Risers must be covered with the grass used for the barrier, and careful erosion-control practices should be used on the benches.

Terraces in the Orient

Because terraces control erosion, they make it possible to raise crops on very steep land. This is of great importance in Asia where land for food-crop production is very scarce. Much of the area is mountainous, so it is very difficult to cultivate. Rice is a main food crop throughout South Asia, Taiwan, the Philippines, Indonesia, and Malaysia. The best rice grows where fresh water continually comes into the paddy. This can occur most readily in side-hill terraces such as those shown in Figs. 7-7, 7-8, and 7-9. Wherever

---

[5]Please refer to page 94 of *Manual of Soil and Water Conservation Practices.* U.S. Soil Conservation Agricultural Handbook 61, June 1954.

FIGURE 7-7 In these rice terraces in central Taiwan, wherever a flow of water comes down from the mountains, the slopes are terraced. The best rice grows on such terraces because the water is always fresh.

FIGURE 7-8 A beautifully engineered set of bench terraces for rice production in Taiwan. A small terraced tea garden is in front of the house.

FIGURE 7-9 This farm in Taiwan has rice terraces (bench) in front of the house and a tea planting in back. Tea bushes live many years, so it is very essential that the land be properly terraced (this time diversion terraces) to carry the water away but leave the soil.

water flows down a mountain, farmers build walls to form level bench terraces about 12 in. (30 cm) deep, so arranged that water may run from the upper one to the next one below and so on down. Note that in Fig. 7-7 a channel protected by grass is left to carry excess water. A typhoon will wash out such construction, but if the farmers are persistent, they simply carry the soil and rocks back up the mountain and rebuild the terraces.

Figure 7-8 shows a beautifully constructed set of bench rice terraces. Notice the farmer's house at one side of the picture. In front of the house is a small terraced tea garden. Tea is usually grown on steep slopes, as shown in the background of Fig. 7-9, with rice paddies in the foreground. Again note the family buildings on the side hill.

Planning the System

Terrace systems should be carefully designed and constructed (Fig. 7-10). Fundamental engineering principles are involved, but erosion-control experience is also essential. Soil characteristics, crops to be grown, and rainfall intensity must be considered. The entire farm water system, including drainage outlets from the terraces, must be accurately constructed to avoid terrace wall breaks as occurred in the terrace system shown in Fig. 7-11. Also, one must be careful to avoid infringing on the rights of neighboring property owners, sometimes difficult in an area of small farms as the region of Guate-

FIGURE 7-10 Terraces for sugarcane on the slopes, with a rice paddy in the valley. Note the two people working in the rice, probably weeding. Incidentally, they usually carry the weeds home to feed the water buffalo and/or pigs.

FIGURE 7-11 These bench terraces on a Guatemala farm were not accurately engineered. Water had broken the terrace walls in several places.

mala shown in Fig. 7-11. You are urged to seek the advice of local water-management specialists. The U.S. Soil Conservation Service Agricultural Handbook 61 offers excellent information on the general subject of water conservation. Local personnel from the Service may have the Handbook on file and are ready to offer advice.

Forests Control Erosion

Perhaps the ultimate in erosion control is the maintenance or establishment of forests. One of these writers was once asked to assist the government of Thailand in a forest conservation survey that was to be used in an effort to lessen the destruction of forests in the highlands of that country. It was largely designed as a planning program to protect the cropland from injury caused by eroding sediment from the hills. More about the role of forests in erosion control in Chapter 20.

REFERENCES

Kozakiewicz, A., and B. G. Ellis. (1967). Water consumption by *Phaseolus vulgaris* and *Zea Mays* as influenced by manganese fertilization. *Soil Sci. Soc. Am. Proc.* 31:123–125.

Musgrave, G. W. (1955). How much of the rain enters the soil? In *Water, The Yearbook of Agriculture*, U.S. Department of Agriculture, Washington, D.C.

Stallings, J. H. (1951). *Soil Conservation*. Prentice–Hall, Englewood Cliffs, N.J. pp 151–159.

CHAPTER 8

Land Capability and Soil Management

Soil Management Principle

Soils should be managed according to their inherent capacity for sustained production.

Over fertilizing a soil that has an inherent low productive capacity is poor management. Limitations due to climate and local hazards must be considered. Farmers should also consider their personal likes and dislikes, desires, managerial abilities, markets, and the type of farming in the locality in which they operate. It is necessary for them to make a living from their land. The important considerations then are crop species grown, their volume, and unit cost of production. Volume of production depends on acreage and yield per acre. Acreage may be fixed by farm boundaries, contracts, or allotments, but yields are flexible.

Unit cost of production for a given crop depends on yields per acre and fixed and operating costs; in other words, management. Where business is concerned, the management program should be one that will result in a maximum profit from the crops produced over a long period, preferably an indefinite period. In preparing the plan, the natural capabilities of the soils must be considered.

THE SOIL SURVEY, A SOIL INVENTORY

Soil surveys are basic to any work in which the kind of soil must be considered when making plans or recommendations. The information obtained by the survey and the soil maps prepared from that information are used for making recommendations to land owners and for many purposes other than in agriculture. When making a soils map, the surveyor studies the soils in detail, including the lower layers of the profile as well as the surface layer (see Fig. 8-1). The soil characteristics that are found indicate the soil series[1] and phases. The surveyor also records the degree of slope and erosion and any other important features. He or she records on the map the boundary lines between different kinds of soil (including slope and erosion), using symbols to designate each. The surveyor also indicates other features that may aid in making management recommendations. These may be water courses, wet spots, deep gullies, and roads or lanes.

In field mapping, the surveyor usually uses a three-part symbol to designate the kind of soil. The first part indicates the series and phase, the second part the slope, and the third part the degree of erosion. In some cases, the surveyor differentiates between wind and water erosion. Soil type and degree of erosion are usually designated by a numeral and the slope by a letter or a letter and numeral. In each state, the soil scientists assign their own numbers

[1]The **soil series** is the lowest category in the soil taxonomy system. The series name will appear on current maps where the "mappable" unit is predominately one series. For further discussion, see *Soil Taxonomy*, Soil Conservation Service, USDA Agricultural Handbook 436, pp. 81–82. 1975.

FIGURE 8-1 These Michigan State University students, under the direction of Professor I. F. Schneider, are studying a soil profile. They are preparing for a career in soil survey.

to designate the series and phase. Although slope designations do vary throughout the United States, within the North Central states, slope and erosion are commonly designated as shown following.

Slope is indicated by capital letters A through F (percent of slope shows the number of feet-fall in 100 ft horizontal).

A	0–2 percent	D	12–18 percent
B	2–6 percent	E	18–25 percent
C	6–12 percent	F	25+ percent

The degree of erosion is shown by the arabic numbers 0 to 5 placed after the slope letters A to F.

As an illustration, the number 450B2 (sometimes written 450/B2) means Miami loam (Typic Hapludolf) with 2 to 6 percent slope and slight erosion. A number 5 in place of 2 in the symbol would mean that the land is severely gullied. This complete information means much more to the farm planner than simply the fact that the soil is Miami loam. In fact, there would be more difference in the management plans for Miami loams varying through slopes B, C, and D than between similar soils with the same slope in three different series, for instance, Miami, Onaway (Alfic Haplorthods), and Guelph (Glossoboric Hapludolf).

In the past 20 years, soil profiles have been studied more thoroughly than had been previously done. More laboratory studies have measured chemical properties such as pH, cation exchange capacity (CEC), and free iron oxides in addition to the physical properties of color and texture. Likewise, drainage and conditions brought about by variations in drainage have been more carefully considered. These refinements have been built into the new soil taxonomy system and have been desirable from a soil-management viewpoint because it makes it possible to explain differences in productive capacity that heretofore remained mysteries.

As a result of the Soil Conservation District program, a part of the detailed soil mapping to designate series and phase[2] is done on individual small tracts (farms). Some is done on a township basis and some on a county basis. At the present time, many of the soil survey reports are available for distribution in published form. Also, the soil maps are available for study in the local offices of the U.S. Soil Conservation Service and in some states at the Experiment Station and local offices of the Cooperative Extension Service. The Cooperative Extension Service and the Soil Conservation Service frequently prepare progress maps showing the area covered by detailed soil surveys. A detailed soil survey map is illustrated by Fig. 8-2. Inquiry should be made at the state offices of the U.S. Soil Conservation Service.

---

[2]The **soil phase** is the taxonomic unit (mapping unit) that gives the most information regarding the nature of the soil and how it should be managed. In the new concept of the term it is a subdivision of a series based on the texture of the surface soil (or degree of decomposition in organic soils) topography, degree of erosion, depth of solum, depth to layers of unconformable material, and salinity of the soil.

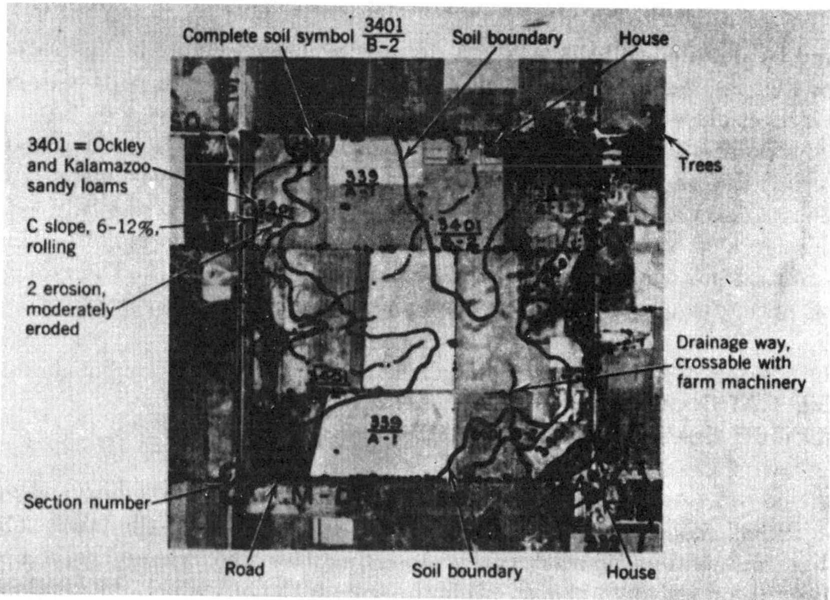

FIGURE 8-2 A detailed soil survey map of 160 acreas in Calhoun County, Michigan. A typical soil profile description follows:

As pointed out earlier, the number of soil phases has been increased during the past few years. The many uses of soil survey maps and reports have indicated a need for arranging similar phases into soil-management groups. The Soil Conservation Service recognizes this need and has arranged what it calls Land Capability Units, the grouping actually used by district personnel in planning individual farms.

THE NATIONAL SOIL CONSERVATION SERVICE LAND CAPABILITY SYSTEM

The soil survey, as already indicated, results in a detailed inventory of all soil resources. The **mapping unit** is a portion of the landscape that has similar characteristics and qualities and that has limits fixed by precise definitions. It is the soil unit about which the greatest number of precise facts and predictions can be made. The soil mapping units provide the detailed soils information needed for arranging capability units, forest-site groupings, crop suitability groupings, range-site groupings, engineering groupings, and, finally, soil-management groupings. Yield predictions are made on a mapping unit area. Mapping units are the **soil phases** mentioned previously.

Soil Management Principle

Soil surveys can define the capability of a farmer's soil resources.

Although it is recognized that detailed mapping units and closely associated mapping units are the only reliable bases for farm planning, the Soil Conservation Service and others recognize a need for some broader soil groupings, some that indicate general land use capabilities to be used in broad program planning and general conservation needs studies. Accordingly, the **Capability Classes, Subclasses,** and **Units** were established.

Land Capability Classes

The U.S. Soil Conservation Service recognizes eight Land Capability Classes distinguished according to permanent characteristics and qualities of soil that limit land use or impose hazards to the soil when used. The Classes show the location, amount, and general suitability of the soils for agricultural use. Capability Classes are so established that soils having the greatest alternative uses are in Class I and those with the fewest alternative uses are in Class VIII, and the risks or limitations become progressively greater from Class I through Class VIII.

There are four Classes of land generally considered suitable for cultivation and four best-suited for permanent vegetation. Under special conditions of treatment and management, the latter Classes may be cultivated. The first four classes are also suited to use for permanent vegetation. The eight classes are indicated on individual farm maps by Roman numerals and/or by color.

Some Characteristics of the Capability Classes[3]

Class I (Light Green). Soils in Class I have few or no limitations or hazards (see Fig. 8-3).

Class II (Yellow). Soils in Class II have few limitations and hazards. Simple conservation practices are needed.

Class III (Red). Soils in Class III have more limitations and hazards than do those in Class II. They require intensive conservation practices.

Class IV (Blue). Soils in Class IV have more limitations and hazards than do those in Class III. Very intensive conservation practices are needed when these soils are cultivated.

Class V (Dark Green). Soils in Class V have little or no erosion hazards but have other limitations that prevent normal tillage for cultivated crops.

Class VI (Orange). Soils in Class VI have severe limitations and hazards that make them generally unsuited for cultivation.

Class VII (Brown). Soils in Class VII have very severe limitations and hazards that make them generally unsuited for cultivation.

---

[3]The following material is a summary taken from the report of the Committee on Capability Grouping of Soils, which met in connection with the National Technical Work-Planning Conference of the Cooperative Soil Survey at St. Louis, Missouri, March 11–16, 1957.

FIGURE 8-3 This Class I land in Michigan is suitable for intensive production of row crops. Class I land in the North Central states is ideal for corn, soybeans, and grains.

Class VIII (Purple). Soils in Class VIII have hazards that limit their use to recreation and wildlife (see Fig. 8-4).

Land Capability Subclasses

The Land Capability Classes are subdivided to indicate the kind of limitation encountered in use and management. Subclasses are groups of **Capability Units** that have the same major conservation problem. The subclass designation thus indicates the particular nature of the hazard involved. Thus the Capability Class and Subclass together indicate both extent and kind of hazard.

Soil Management Principle

The Subclass designation indicates the particular nature of the hazard involved.

---

e. RUNOFF AND EROSION. All lands with sufficient slope to require water erosion control measures are placed in this Subclass. In general, it includes soils with more than 2 percent slope.

w. WETNESS AND DRAINAGE OR EXCESS WATER. In general, all

FIGURE 8-4 A typical view of Class VIII land in a Western area. The soil is shallow, with a high percentage of the surface occupied by rock outcrop. Vegetation is mostly brush, and the land is not suitable for grazing.

soils with an excess water hazard or limitation are placed in this Subclass. They may need drainage or protection from flooding. Water erosion is not a problem, although some very gently sloping soils are included. After drainage, there may be danger of wind erosion. Textures vary from fine to coarse. Organic soils are included as also are stream-bottom mineral soils.

s. ROOT ZONE OR SOIL PROBLEMS. Lands where soil limitations such as shallowness, stoniness, drought, low fertility, and salinity are dominant are included in this Subclass. Erosion, wind, and/or water is a problem on most of the soils.

c. CLIMATIC HAZARD. Lands where temperature and/or rainfall are the dominant limitations in use. Soils in the far north may be nonproductive entirely because of low temperatures whereas in the desert areas a shortage of water and high temperatures may interact to prevent profitable land use. In semiarid regions, irrigation may remove this restriction.

INTERNATIONAL SOIL CLASSIFICATION

Since the middle of this century, soil scientists in the United States have been attempting to develop a system of nomenclature that could be used throughout the world. The first introduction of their efforts to the soils community was entitled *The Seventh Approximation*, released about 1960. A 1975 version of their efforts entitled *Soil Taxonomy, A Basic System of Soil Classification for*

Making and Interpreting Soil Surveys is by the Soil Survey staff of the USDA Soil Conservation Service. A look at that publication will show why it is not within the scope of these pages to present more than a brief introduction to the system. Features of diagnostic horizons are shown in Table 8-1.

TABLE 8-1 Derivation and Major Features of Diagnostic Horizons

Horizon	Derivation	Major features
Surface Diagnostic Horizons-Epipedons		
Mollic	L. *mollis*, soft	Thick, dark-colored, high base saturation, and strong structure so that the soil is not massive or hard when dry.
Umbric	L. *umbra*, shade	Same as mollic but highly H saturated and may be hard or massive when dry.
Ochric	Gr. *ochros*, pale	Thin, light-colored, and low in organic matter.
Histic	Gk. *histos*, tissue	Very high organic matter content and saturated with water at some time during the year unless artificially drained.
Anthropic	Gr. *anthropos*, man	Mollicklike horizon that has a very high phosphate content resulting from long-time cultivation and fertilization.
Plaggen	Ger. *plaggen*, sod	Very thick, over 20 in., produced by long-continued manuring.
Subsurface Diagnostic Horizons		
Argillic	L. *argilla*, white clay	Illuvial horizon of silicate clay accumulation.
Natric	*Natrium*, sodium	Illuvial horizon of silicate clay accumulation, over 15% exchangeable sodium and columnar or prismatic structure.
Spodic	Gk. *spodos*, wood ash	Illuvial accumulation of free iron and aluminum oxides and organic matter.
Oxic	L. *oxide*, oxide	Altered subsurface horizon consisting of a mixture of hydrated oxides of iron, aluminum and 1 : 1 clays.
Cambic	L. *cambiare*, to change	An altered horizon due to movement of soil particles by frost, roots, and animals to such an extent to destroy original rock structure or aggregation into peds or both.
Agric	L. *ager*, field	An illuvial horizon of clay and organic matter accumulation just under the plow layer due to long-continued cultivation.

Source: USDA Handbook 436, 1975 as adapted by H. D. Foth. *Fundamentals of Soil Sciences.* 7 ed. Wiley, New York, 1984.

The highest categories in the system are called "orders" with two or four "suborders" in each. Perhaps it is correct to say that climate has the greatest influence in bringing about the differences between orders. The list of 10 orders, as shown in Table 8-2 is taken from the USDA Handbook 436. Included is information as to the derivation of the names. It seems advisable to include here a brief explanation of each order. Note that each name includes the syllable "sol," an abbreviation of solum meaning soil. In Table 8-2 we present the list of 10 orders with the formative syllable, derivation, and meaning of each. Included also are the approximate equivalents of the orders in the old classification.

Land Capability Units

Let us now go back to the U.S. Soil Conservation Service Land Capability System. As we know, the smallest mapping unit is the soil type,[4] made even smaller when slope and degree of erosion are mapped. On most farms, two or more soil types must be farmed together if soil management is to be practical. Actually, a group of mapping units that are nearly alike in suitability for plant growth and response to management becomes a "Land Capability Unit," a division of a Land Capability Subclass. Such a unit is large enough to be used in detailed farm planning, whereas the area of a single mapping unit is generally too small, especially in a glaciated region such as Michigan. For instance, several mapping units would be included under the Land Capability Unit Iw2a. The soils would all be Class I land with a drainage hazard (w). They would vary somewhat in profile and in texture from loam to silt loam. Slope would not be sufficient to make erosion a hazard.

Soils in the same Capability Unit should be sufficiently similar in characteristics and qualities to have similar cropping potentials and hazards. They should be sufficiently uniform to (1) produce similar kinds of plants with similar management practices; (2) require similar conservation treatment and management under the same kind and condition of vegetative cover; and (3) have comparable potential productivity.

Land Capability Units for Genesee County, Michigan

Genesse County is located in East Central Michigan. It is a farming area quite typical of much of the North Central region of the United States, excluding perhaps the Central Prairie subregion. The county contains soils in six orders, 45 series, 85 mapping units, and 32 Capability Units. As one would suspect by studying the meaning of the order names (Table 8-2), four orders are missing — Aridisols, Oxisols, Ultisols, and Vertisols. This emphasizes the role that climate plays in establishing soil characteristics.

---

[4]The term "type" is not used in the new system of nomenclature, but since these pages do not fully explain the new system, we take the liberty hereafter of continuing to use the word as a part of the lowest category name. In other words, "Miami loam" is a soil type (soil phase).

TABLE 8-2 New Soil Orders and Approximate Equivalents in Great Soil Groups of 1949 System

Order	Formative syllable	Derivation	Meaning	Approximate equivalents
1. Entisol	ent	Coined syllable	Recent soil	Azonal soils and some Low Humic Gley soils
2. Vertisol	ert	L. *verto*, turn	Inverted soil	Grumusols
3. Inceptisol	ept	L. *inceptum*, beginning	Inception, or young soil	Ando, Sol Brun Acide, some Brown Forest, Low Humic Gley, and Humic Gley soils
4. Aridisol	id	L. *aridus*, dry	Arid soil	Desert, Reddish Desert, Sierozem, Solonchak, some Brown and Reddish Brown Soils, and associated Solonetz
5. Mollisol	oll	L. *mollis*, soft	Soft soil	Chestnut, Chernozem, Brunizem (Prairie), Rendzinas, some Brown, Brown Forest, and associated Solonetz and Humic Gley soils
6. Spodosol	od	Gk. *spodos*, wood ash	Ashy (Podzol) soil	Podzols, Brown Podzolic soils and Ground-Water Podzols
7. Alfisol	alf	Coined syllable	Pedalfer (Al-Fe) soil	Gray-Brown Podzolic, Gray Wooded, Noncalcic Brown, Degraded Chernozem, and associated Planosols and Half-Bog soils
8. Ultisol	ult	L. *ultimus*, last	Ultimate (of leaching)	Red-Yellow Podzolic, Reddish-Brown Lateritic (of United States), and associated Planosols and Half-Bog soils
9. Oxisol	ox	F. *oxide*, oxide	Oxide soils	Laterite soils, Latosols
10. Histosol	ist	G. *histos*, tissue	Tissue (organic) soils	Bog soils

Source: USDA Handbook 436, 1975, as adapted by H. D. Foth. *Fundamentals of Soil Science,* 7th ed. Wiley, New York, 1984.

TABLE 8-3 Soil Management Groups used in Michigan

Texture of upper 3 ft of soil profile	Mineral soils			Organic soils (M)	
	Natural drainage and surface color				
	Well-drained, light-colored (*a*)	Imperfectly drained; moderately dark-colored (*b*)	Poorly drained, dark-colored (*c*)	Very poorly drained, very dark-colored shallow 12 to 42 in. thick	deep over 42 in. deep
0 Clays (over 55%)			0*c*	*M*/1*c* over clays	
1 Clays and silty clays	1*a*	1*b*	1*c* 1*c-c*		
2 Clay loams, silt loams, and loams	2*a* 2*a-a* 2*a-af* 2*a-c*	2*b*	2*c* 2*c-c* 2*c-l*	*M*/3*c* over loams	*Mc* *Mc-L*
3 Sandy loams	3*a* 3*a-a* 3*a-af* 3*a-L* 3/2*a* 3/*Ra*	3*b* 3*b-a* 3*b-c* 3/1*b* 3/2*b*	3*c* 3*c-c* 3*c-L* 3/1*c* 3/2*c* 3/*Rc*		
4 Loamy sands or sands with some finer textured layers	4*a* 4*a-a* 4*a-L* 4/2*a* 4/*Ra*	4*b* 4/1*b* 4/2*b* 4/*Rb*	4*c* 4*c-L* 4/1*c* 4/2*c* 4/2*c-c* 4/*Rc*	*M*/4*c* over sands	
5 Sands	5*a* 5*a-h* 5/2*a*	5*b* 5*b-h* 5/2*b*	5*c* 5*c-c*		*Mc-a*
G Gravelly or stony	*Ga* *Ga-c*		*Gc* *Gc-c*	*M*/*mc* over marls	
R Bedrock at less than 42 inches	*Ra* *Ra-c*		*Rc* *Rc-c*		

The numbers indicate the relative coarseness of the mineral materials from which the soils were formed — from 0, the finest textured clays, to 5, the coarsest textured sand. The small letters immediately following the numbers or capital letters indicate the natural drainage under which the soil developed — *a* for well drained, *b* for imperfectly drained, and *c* for poorly drained conditions.

Where capital letters are the first part of the symbol, they represent important soil characteristics as follows: *G* for gravelly or stony soils; *M* for mucks and peats; and *R* for rocky soils where bedrock is close to the surface.

Where soils are formed from one kind of material on top of another kind of material, a fractional symbol is used. The capital letters or number above the line refers to the upper layer, and the capital letter or

Now let us take a look at the Capability Units mapped in Genesee county, 32 in all, much easier to manage than would be 85 mapping units handled separately. In each unit, the Roman numeral indicates the National Capability Class, the small letter the subclass (kind of hazard — e, w, s, or c), and an arabic numeral to identify the capability unit within each subclass. The symbols within the parentheses in each case (arabic numerals and small or capital letters) identify the soil management groups used specifically in Michigan. See Table 8-4.

1. Capability Unit I-1 (2.5a, 2.5b)
   Celina-Conover, Tuscola series (Alfisols)

2. Capability Unit IIe-I (1.5a)
   Morley silt loam (Alfisols)

3. Capability Unit IIe-2 (2.5a, 2.5b, 3/2a, 4/2a)
   Celina, Conover, Metea, Miami, Owosso, Sisson, Tuscola (Alfisols)

4. Capability Unit IIe-3 (3a)
   Fox sandy loam (Alfisols)

5. Capability Unit IIw-2 (1.5b, 1.5c)
   Del Rey (Alfisols)
   Lenawee (Inceptisols)

6. Capability Unit IIw-3 (1.5b)
   Del Rey silt loam (Alfisols)

7. Capability Unit IIw-4 (2.5b, 2.5c, 3/2b)
   Brookston, Conover, Metamora (Alfisols)
   Sebewa (Mollisols)

8. Capability Unit IIw-5 (2.5b, 3b, 3/2b)
   Conover, Kibbie, Metamora (Alfisols)

9. Capability Unit IIw-6 (2.5c, 3b)
   Colwood (Mollisols)
   Kibbie (Alfisols)

10. Capability Unit IIw-8 (3/2b, 3/2c)
    Breckenridge (Inceptisols)
    Brevort (Entisols)
    Metamora (Alfisols)

---

number below the line refers to the lower layer. For example, 3/1 represents sandy loam 18 to 42 in. thick over clays; 5/2 represents sand 42 to 66 in. thick over loams to clays; and *M*/4 is for muck or peat 12 to 42 in. thick over sands.

When a letter follows the small letter that indicates the natural drainage, it indicates other characteristics of the soil as follows.

a. Very strongly acid soils
c. Free lime in surface layers
f. Fragipan — compact, impermeable layer in soil profile
h. Hardened and cemented subsoils
L. Lowland soils subject to seasonal overflow

11. Capability Unit IIw-9 (L-2a)
    Landes (Mollisols)

12. Capability Unit IIw-10 (M/3c)
    Linwood muck (Histosols)

13. Capability Unit IIs-2 (3a)
    Fox, Miami (Alfisols)

14. Capability Unit IIIe-4 (1.5a)
    Morley silt loam (Alfisols)

15. Capability Unit IIIe-5 (2.5a, 4/2a)
    Celina, Metea, Miami, Sisson (Alfisols)

16. Capability Unit IIIe-9 (4a, 4/2a)
    Arkport, Boyer, Metea, Spinks (Alfisols)
    Oakville (Entisols)

17. Capability Unit IIIw-5 (4b)
    Minoa (Inceptisol)
    Wasepi, Spinks (Alfisols)

18. Capability Unit IIIw-6 (4c)
    Gilford (Mollisols)
    Lamson (Inceptisols)

19. Capability Unit IIIw-9 (4/2b, 4/1b, 4/1c)
    Allendale (Spodosols)
    Pinconning (Inceptisols)
    Selfridge (Alfisols)

20. Capability unit IIIw-10 (4/2c)
    Brevort loamy sand (Entisols)

21. Capability Unit IIIw-11 (5c)
    Granby loamy sand (Mollisols)

22. Capability Unit IIIw-12 (L-2c)
    Ceresco, Cohoctah, Sloan (Mollisols)

23. Capability Unit IIIw-15 (Mc, M/3c)
    Carlisle, Linwood, Lupton, Rifle (Histosols)
    Wallkill (Inceptisols)

24. [5]Capability Unit IIIs-3 (4a, 4/2a)
    Oakville (Entisols)
    Arkport, Boyer, Metea, Perrin, Spinks (Alfisols)

25. [5]Capability Unit IIIs-4 (4a, 4/2a)
    Oakville (Entisols)
    Arkport, Boyer, Metea, Perrin, Spinks (Alfisols)

26. Capability Unit IVe-4 (2.5a, 4/2a)
    Miami, Metea (Alfisols)

---

[5]You will notice that these two Capability Units include the same six series, but soils in Unit IIIs-4 are slightly more droughty because they lie at a somewhat higher elevation.

27. Capability Unit IVe-9 (4a)
    Oakville (Entisols)
    Boyer, Spinks (Alfisols)
28. Capability Unit IVw-2 (5b)
    Au Gres (Spodosols)
29. Capability Unit IVw-5 (M/4c)
    Markey (Histosols)
30. Capability Unit IVw-6 (M/mc)
    Edwards (Histosols)
31. Capability Unit IVs-2 (5a, 5/2a)
    Oakville (Entisols)
    Croswell (Spodosols)
32. Capability Unit VIe-2 (2.5a)
    Miami (Alfisols)

Note that in 12 instances two soil orders are included in the Capability Unit. Two of the units include three soil orders. The value is having both systems included in the list of Capability Units is illustrated by the last unit in the list. The Michigan symbol does not show that the slope is too steep for cultivation as does the national symbol VIe-2.

With 32 Capability Units in one Michigan county (there are 83 counties in the state). You will see how we cannot show them all. Those interested in a specific county, province, or country may contact local agriculture extension personnel for a list of Capability Units (Management Groups) in their area. Now we finish this chapter with a look at Michigan's Management Groups.

SOIL MANAGEMENT GROUPS USED IN MICHIGAN

The present Michigan system of grouping soil types into management groups will serve, for the purposes of this text, as an example of a local effort to group soils into usable units. The system, worked out cooperatively by Michigan Agricultural Experiment Station, Michigan Cooperative Extension, and U.S. Soil Conservation Service personnel, has already been referred to in several publications and is outlined in Table 8-3. It is logically arranged to group soils on the basis of texture, natural drainage, and certain other characteristics. The latter refer to organic versus mineral, double story, rockiness, cementation, differentiation of B horizon, and lime content. An explanation of the way these characteristics are designated is given in the footnote for Table 8-3.

Farm Planning

In planning for the actual use of the land in a certain field, it is necessary to know more about the soil than the Soil Management Group, or, in other

TABLE 8-4 Crop Rotation Suggestions for Soil Management Groups 2a, 2b, and 2c

	Slope		Practices				Approximate
Class	Percent	Length, in feet	None	Contouring	Strip-cropping	Terracing	Land capability class
A	0–2	All	CRO				I
B	2–6	100	AARW	CRO	AAARRO	CRO	II
		200	AARW	AROW	AAARRO	CRO	
		300	AAARW	AARW	AAARRO	CRO	
		400	AAWO	AAWO	AAARRO	CRO	
C	6–12	100	AAAARW	AAARO	AAARRO	CRO	III
		200	AAWW	AAWO	AARO	CRO	
		300	AAWW	AAWW	AARW	CRO	
		400	AAW	AAW	AAARO	CRO	
D	12–18	50	AAWO	AARW	—	Not recommended	IV
		100	AAWW	AAWO	AARW		
		200	AAW	AAW	AAAARW		
		300	AAAW	AAAW	AAWO		
E	18+	All	Permanent vegetation—grass or trees				VI
F, G	18+	All	Permanent vegetation—grass or trees				VII

A, alfalfa; C, clover; R, row crop; W, winter grain; O, spring grain.

words, more information is needed than simply where the soil falls in Table 8-3. That is only what might be termed "basic information" about the soil. It is also necessary to know the percentage and length of slope and the amount of erosion that has already occurred. With that additional information, from an actual cropping table as illustrated by Table 8-4, taken from the Odessa Township (Michigan) report, the cropping system may be found that may be used with the slope and erosion-control practices to be followed. Note that percentage of slope, length of slope, and erosion-control practices are all important in determining the most intensive, or least protective, cropping system that may be used. Terracing makes it possible to use the same rotaton on any percentage of slope up to 12 and any length up to 400 ft. It is assumed that the soils are to be limed and fertilized as needed.

The Soil Management Groups shown in Table 8-3 become U.S. Soil Conservation Service Land Capability Units by indicating percentage of slope, degree of erosion, and Land Capability Subclass. Class, of course, is also given. For example, the Land Capability Unit IIIE-2a/(C-1) may be explained as follows:

III = Land Capability Class
IIIE = Land Capability Subclass
2a = Soil Management Group
C = Slope (6 – 12%)
1 = over slight erosion

The old group designation for IIIE – 2a/(C-1) was IIIE1 (see Table 8-5). Other Land Capability Units under El and for soils falling in the same Soil Management Group 2a are IE – 2a/(B-1), IIE – 2a/(B-3), IVE – 2a/(B-4), IIIE – 2a/(C-1), IVE – 2a/(C-3), and IVE – 2a/(D-1). Note that these units are all in the same Soil Management Group (2a), but they vary with respect to percentage of slope and extent of erosion. The latter two variables determine the Capability Class, I to IV in these illustrations. As slope percentages and extent of erosion increase, Capability Class also increases.

Table 8-5 shows how the Soil Conservation Service Land Capability Units as currently used in the North Central region compare with the designations now being used in Michigan.

Local Adaptations

Readers and teachers using this text are urged to contact the local offices of the U.S. Soil Conservation Service or the State Cooperative Extension Service or local government agencies for a listing of the Soil Management groups or Land Capability Units used in their areas. It may also be possible to obtain

TABLE 8-5 A Comparison of Land Capability Unit Designations used in the North Central Region with Those used by the Michigan Soil Conservation Service[a]

North central region	Michigan
IW2a	IW2c
IIW1c	IIW3c(m)
IIW1b	IIW4/2c
IIW4a	IIW3/2b
IIIW4b	IIIW1b
IIIW6a	IIIW4b(m)
IVW3	IVW0c
VIW16	VIWGbc
IIIE1	$IIIE-\dfrac{2a}{C-1}$
IIIE2	$IIIE-\dfrac{2/Ra}{A-1}$
IIIE6	$IIIE-\dfrac{1a}{C-1}$
VIIE4	$VIIS-\dfrac{Ra}{A-1}$
IIS1a	$IIS-\dfrac{3a}{B-1}$
IIS1b	$IIS-\dfrac{3/1a}{A-1}$
IIIS2	$IIIS-\dfrac{4a(m)}{B-1}$
IVS3	$IVS-\dfrac{5a(m)}{A-1}$
VIIS5	$VIIS-\dfrac{5.3a}{A-1}$

[a]Assisted by Clarence Engberg, Michigan State Soil Scientist, U. S. Soil Conservation Service.

from those offices a list of local soil types (phases) included in each Management Unit. Such information is necessary for detailed land use planning.

REFERENCES

Mokma, D.L. (1982). *Soil Management Units and Land Use Planning.* Michigan Cooperative Extension Bulletin E-1261.

Robertson, L. S., D. L. Mokma, D. L. Quisenberry, et al. (1976). *No Till Corn: 3. Soils.* Bulletin E-906. Cooperative Extension Service, Michigan State University.

U.S. Department of Agriculture Handbook 436, (1975). U.S. Government Printing Office, Washington, D.C.

Warncke, D. D., D. R. Christenson, and M. L. Vitosh (1985). *Fertilizer Recommendations for Vegetable and Field Crops in Michigan.* Bulletin E-550. Cooperative Extension Service, Michigan State University.

CHAPTER 9

Tillage and Planting Methods

If soils were animate, if they had personalities, we would say they are happiest when they are working hard, certainly not when they are idle. Unless water is extremely limited, as in desert areas, nature provides some kind of vegetative cover to keep them working, to protect them, and gradually to build or replenish their supplies of organic matter. As already mentioned, the rate of organic-matter accumulation depends on the species of plants providing the cover, temperature, water conditions, and the nature of the soil. Soils are different! The materials that weathered into fine-textured soils contained more mineral nutrients than did those that gave rise to sandy soils, which helps produce more organic matter per hectare for fine-textured soils. Also, organic-matter decomposition was slower in the fine-textured materials. Thus rate of organic-matter accumulation varied because plant growth varied and the activity of decomposing organisms varied. One thing is sure, however; nature intended soils to produce plant growth to the limit of their possibilities (production capacity), and without tillage.

Nature did not provide means for stirring, or **tilling** soil, except through action of micro- and macro-organisms. Major tillage operations are inovations introduced by farmers. We surely have evidence that, on the whole, these inovations resulted in detriment rather than good. Unless they have been very well managed, cropped soils are less fertile, less productive, than are virgin, untilled soils. This is undoubtedly true of grasslands in a temperate climate. Virgin grassland soils are often considered ideal. They are as we find them without ''benefit'' of forces that do to the soil what farmers do with the moldboard plow, disk, packer, harrow, roller, or cultivator. If nature did not need all these tools to build a good productive soil, why do people need them? The answer is, they **do not!**

Humans developed tillage practices because they wished to grow a different species of plant than was dictated by nature. The advisability of this is freely admitted. A buffalo may live on grass or a deer on cedar browse, but people need grain and vegetables and their cattle need forage and grain. As a result, we must destroy the native vegetation in order to establish plants of our choice or need.

Soil Management Principal

The best management practice will be the least amount of tillage necessary to grow the desired crop.

Admitting then that virgin soils are ideal, or nearly so, without having been plowed, disked or harrowed, why do we not take the "tip" and work our soils as little as possible? Under some conditions, a good job of moldboard plowing and immediate planting is all that is necessary for satisfactory results. And in many cases, zero tillage may be even better. With few exceptions, tillage increases the chance of soil erosion. Within certain cropping systems, moldboard plowing may not be necessary nor desirable. Conservation tillage has been introduced and developed to control sheet erosion while still producing satisfactory yields. Hilner says, "Conservation tillage appears to have a long run solution to the sheet erosion problem. But it is not a panacea for all erosion problems, it is another tool, a powerful tool."[1] Other tillage tools are replacing the moldboard plow. These will be discussed later under the subject of conservation tillage and zero tillage. It will suffice to say here that we are in the midst of a dramatic change in the United States. Twenty-nine million acres were conservation tilled in 1972, and 100 million were conservation tilled in 1982, just 10 years later. In the midst of such change, we must constantly examine why we till—to prepare a seedbed, to loosen compacted soil, and to control weeds. Any additional tillage is for appearance or the convenience of the farmer, and it may actually be to the detriment of the soils as well as the crop.

Soil Management Principal

Tillage is to prepare a seedbed and/or to control weeds.

OBJECTIVES IN SEEDBED PREPARATION

The first objective of tillage, in the minds of many, perhaps most *humid* region farmers, is the coverage of existing vegetation, manures, and crop residues. In most cases, the primary tillage tool is the moldboard plow, which

---

[1] H. R. Hilner. Leadership in conservation tillage. Michigan State University, Michigan Department of Agriculture Soil Conservation Service, Michigan Department of National Resources.

is the most effective and efficient of all implements for burying trash. Incidentally, **the moldboard plow is also the best implement for loosening and crumbling soil,** for temporarily increasing the pore space of the fraction of the soil that feeds the plants, and furnishes most of their water. But is must be remembered that the objective of preventing sheet erosion is met by leaving residue on the surface. And a tool other than a moldboard plow (e.g., a chisel plow) more nearly meets this objective.

The corresponding objective of the farmer who inhabits the semiarid regions is to destroy existing vegetation without burying it and in such a way as to furnish the least-possible hindrance to planting and cultivation machinery.

In any region, an extended period of seedbed preparation is an aid to weed control. The period must be of several weeks duration, however, to have an appreciable effect. Weed seeds do not germinate until they are brought near the surface. After the seedlings emerge, they may be uprooted or buried by subsequent tillage. It is true, of course, that summer fallow is very effective in the control of such perennial weeds as quackgrass, Johnson grass, Bermuda grass, thistles, wild morning glory, and plants of the mustard family, especially in regions having low summer rainfall. Farmers in humid areas, however, generally overestimate the effectiveness of seedbed tillage as a weed-control measure. Sometimes all spring crops are planted soon after the soil is first dry enough to till. At the longest, planting must follow within 2 or 3 weeks of the date of plowing. Soil moisture is usually high at that time, so many of the uprooted weeds quickly reestablish themselves, and many weed seeds are germinated, but the emerging plant has not reached the surface. Moving such seeds with tillage implements does not destroy them.

Soil Management Principal

Plant immediately after plowing to give crops an equal chance to compete with weeds.

It is better to plant immediately after plowing and give the crop plants an equal chance to compete with the weed plants. Chemical weed control can effectively substitute for tillage in weed control.

Soil Management Principal

Chemical weed control can and should substitute for tillage.

Farmers have long recognized the need for loosening the soil so that planters might operate and cover the crop seeds properly with pulverized, moist soil. They have long recognized the need for close contact between soil and seed so water may move into the seed and start the germination process. That is why planters have press wheels and why writers have urged the preparation of a "fine, firm seedbed." They failed, however, to recognize the fact that only the soil directly below and above and an inch or two away affects this movement of soil water to the seed. The soil **between** the rows is not a part of the seedbed but rather is a part of the root bed and as such should be **loose** to promote aeration as well as root penetration and water storage.

Plant roots and soil organisms must have oxygen. Farmers recognize this as a reason for seedbed preparation. Penetrometer measurements (see Fig. 9-1) have shown, though, that the conventional amount of tillage after plowing often packs the soil as firmly as it was before plowing. When clayey soils are a little too moist, several workings with a disk may pack them even more firmly than they were previous to plowing. Thus aeration may be reduced or slowed rather than increased or hastened, and future tillage becomes more difficult.

Infiltration rate is increased by moldboard plowing, especially if the direction of travel is across the slopes, and if subsequent tillage is held to a minimum. The stresses caused by the moldboard crumble the furrow so completely that large pores allow ready entry of water (see Fig. 9-2). Gravity eventually settles the furrow, but soils not tilled after plowing remain loose long enough for the next crop to provide a protective cover.

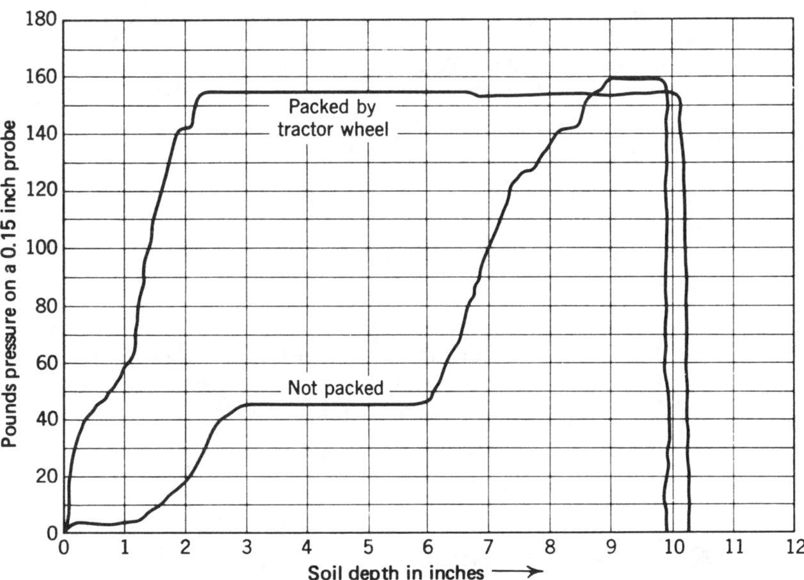

FIGURE 9-1 Curves made with a self-recording penetrometer show that three passes with a medium-size tractor packed the plowed layer as firmly as the undisturbed subsoil. To obtain the upper curve, the penetrometer probe was forced into the soil between corn rows where the cultivating tractor wheel had passed three times. The lower curve is the result of a reading in the "middle," where tractor wheels had not passed since the land was plowed. The curves show that the tractor wheel packed the soil to a density equal to that of the soil below the plow depth.

FIGURE 9-2 A properly adjusted moldboard plow, in the absence of subsequent tillage ("minimum tillage"), leaves the soil easily receptive of water. This field was the site of a minimum-tillage demonstration at Berne, Indiana. The left side of the field was fitted and planted in the old conventional manner. The right side was simply plowed and planted. The soil is Blount-Pewamo association ranging from silt loam to clay loam. Planting was completed at 2:00 P.M. on May 16. One inch of rain fell between 4:30 and 6:00 P.M., and the picture was taken one-half hour later. Tillage like that conducted on the left side of the field results in serious erosion on sloping fields. This demonstration was conducted by Leo N. Seltenright, County Agricultural Agent, Adams County, Indiana. (Photograph courtesy of Cliff Spies, Extension Agronomist, Purdue University, and Charles Bruffy, Federal Chemical Company.)

METHODS OF PREPARING SEEDBEDS

Mulch Tillage

The practice of leaving crop residues on the soil surface covered only sufficiently to anchor them against blowing is referred to as **mulch tillage.** The practice reduces soil-water evaporation and is commonly recommended over much of the Great Plains and in other low water areas. (See Chapter 16 for a full discussion of this important tillage method for dry areas.)

Mixing of Vegetation with Shallow Surface Soil

Any one of several machines may be used to mix residues and vegetation with the surface soil. Chief among them is the rotary tiller, a power take off or motor-driven implement that pulverizes the soil by means of tines or knives rotating at high speed around a horizontal shaft. Certain Michigan experiments showed rotary tillage to be rather unsatisfactory. Soils are left fluffy and powdery, subsequent weed infestation was great, and power requirement was high. More time was required for removing weeds after the rotary tiller than after any other implement. Furthermore, rotary tillage required more horsepower hours per acre than any other tillage method included in the experiments. Rotary tillage did not result in high yields. In fact, in most instances, yields after the rotary tiller were lowest, largely as a result of greater weed competition.

The small tiller commonly used in gardens is quite a different machine. The tines rotate slowly and do not leave the soil fluffy and powdery as do the large, high-speed machines. They are quite satisfactory for small areas. Even though they may be rather hard on the gardener, they are usually favored over the spade and hoe.

The auger plow (a horizontal auger) does a thorough job of mixing vegetation and residues with the top soil to a depth of about 4-1/2 in. To attain this depth, it is usually necessary to cover the land a second time. As the cut is only 20 in. wide, with a full two-plow tractor, power requirement is high. Also, lack of coverage of weeds and grass results in a high cost for weed control.

The vertical-disk plow (wheatland disk-tiller) is satisfactory as far as power and labor requirements are concerned, but the method used in non-arid areas may partially fail because of poor weed control. As mentioned in Chapter 16, this implement is excellent in the drier regions where residues should be buried only sufficiently to anchor them against movement by wind. In humid areas, however, where grasses are not easily subdued, planting cannot immediately follow disk tilling. Intermediate tillage to eradicate grasses and certain other weeds is necessary and, of course, expensive. In other words, this type of tillage in humid areas should be confined to summer fallow or possibly to fitting stubble land for fall-planted grains.

Subsoiling or Chiseling

The subsoiler or chisel has long been used as an implement for breaking tillage pans or naturally tight subsoils. During the early days when horses supplied the power, the work was done in the spring when the pans were softened by plenty of moisture. The soft, high clay content soil did not shatter, so the effect of the chisel was confined to a narrow band. The pans regained their former consistency within a few months, so the effort was not worthwhile.

Soil Management Principal

Deep tillage should be used only when a soil-compaction problem exist.

Chiseling during the dry season has been more successful. Relatively dry, tight subsoils may shatter for a distance of several feet from the chisel. Followed by a minimum of surface tillage, the effect may then last for a considerable period. In dry soil, however, power requirement is high.

Chiseling or even deep plowing with a moldboard plow may be hazardous for deep clay soils where the tight material extends well below the loosened soil, unless the chisel slits are on a slope with an outlet provided for excess water. Otherwise, such a soil may absorb so much water that the time required for evaporation will delay further tillage until planting may be dangerously delayed.

Shallow pan soils underlaid by more pervious material are quite likely to be improved by some kind of deep tillage. It is advisable to study soils carefully before spending money for chiseling. You cannot gain by going through the motion of breaking a pan that does not exist. Local Extension Service and U.S. Soil Conservation Service personnel have information regarding the soil types that may be improved by chiseling.

Moldboard Plowing — Minimum Tillage

The moldboard plow is the best implement ever devised for loosening and crumbling clay soils. Those low in organic matter that have been excessively tilled, perhaps when moisture was high, are especially in need of such loosening. The lifting and turning action of the moldboard shatters the soil in several directions, which greatly increases pore space. Measurements have shown that a freshly turned furrow in a clay loam soil may occupy 50 percent more volume than it did prior to plowing. If the unplowed soil has 50 percent pore space, the plowed soil then contains twice as much pore space as before plowing (see Fig. 9-3). Thus the possiblity of water storage is greatly increased and aeration is improved.

Three types of moldboard plows are available.

1. **Stubble plows** have short moldboards with abrupt curves to give maximum shearing effect. The furrows are thoroughly crumbled as they are turned.

2. **Breaking or sod plows** have long, gently sloping moldboards for use where the furrow should be turned without breaking. Grassland soils already possess good structure and are loose and friable. Subsequent planting is easier if the furrows are turned smoothly.

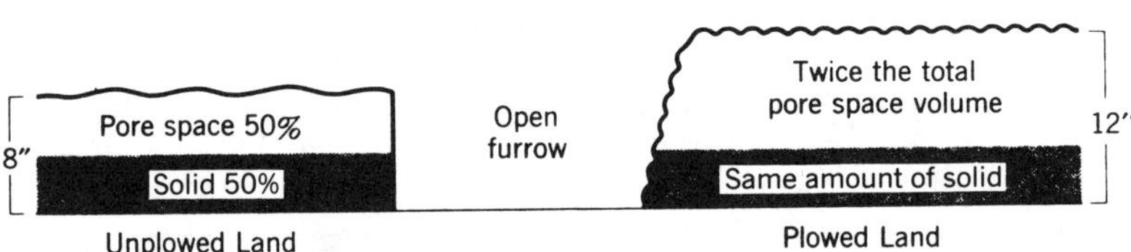

FIGURE 9-3 Freshly plowed clay loam soil may contain twice the pore space of the same soil before plowing. To obtain this effect, a properly adjusted, short moldboard plow must be used and soil moisture must be optimum.

3. **General purpose plows** are intermediate between the stubble and breaking plows. They are favored by farmers who prefer to invest in only one
   plow. At low speeds these plows turn sod quite satisfactorily, and at high
   speeds they crumble the soil fairly well.

Time of Plowing

Generally, plowing should immediately precede planting. Vegetative cover,
where one exists, is destroyed by plowing. Since the aim of the conservation-
minded farmer is to keep the soil covered as much of the time as possible, it
follows that the farmer should plant a new crop as soon as he destroys the old
one.

Fall plowing with a moldboard plow has both advantages and disadvantages. It results in a more efficient distribution of labor and more efficient use
of machinery. In regions where winter temperatures drop below freezing,
soils that turn over in a lumpy condition may be broken down by frost action.
It should be remembered, however, that the same forces that crumble large
lumps may also reduce the size of desirable smaller aggregates, thus tending
to destroy favorable soil structure. Land that turns over in a cloddy condition
has probably been worked or pastured when too wet.

In the Northern states, certain insects, for example, the grasshopper, are
partially controlled by fall plowing.

Compacted soils without vegetative cover may be loosened by fall plowing so infiltration rate is increased and erosion is thereby lessened. Contour
plowing with the furrows turned uphill makes the practice particularly effective. But it must be remembered that vegetative cover and/or residues on the
surface are much more effective at reducing erosion.

Spring plowing has much in its favor. In a good soil-management program, vegetation and/or crop residues are always to be turned under if
erosion is not a problem. Plowing thus exposes the soil to losses by erosion,
and leaching losses are always greater while soils are bare. When soils are
plowed just before the next crop is planted, a new cover is soon established
and losses from these two destructive forces are held to a minimum.

Less tillage is required after spring plowing. In fact, only moldboard
plowing is sufficient tillage for most crops to be planted immediately with
wheel-track planting or other suitable planter.

Weed control in subsequently planted crops is easy where planting is
done the same day the land is plowed. This is largely because weed seeds are
not given opportunity to germinate nor stolons to start before the crop seeds
are planted. Thus the crop and the weeds have an even start.

Spring-plowed soils are loose enough to allow rapid infiltration, so losses
by erosion are much less than from fall plowed land, even **after** crops are
planted. Minimum tillage sets the stage for the maximum rate of infiltration
on spring-plowed land. Hays (see Table 9-1) determined the effect of minimum tillage as compared with conventional tillage on runoff and erosion.

TABLE 9-1 The Effect of Minimum and Conventional Tillage on Runoff and Erosion from Fayette Silt Loam on 15 Percent Slope[a]

| | Runoff water (inches) | | Erosion soil (tons) | |
Kind of tillage	1959	1955–59	1959	1955–59
Conventional	2.50	0.81	7.6	2.9
Minimum	1.13	0.35	1.5	0.7

[a]Rotation is corn, grain, hay. In both treatments, the land is spring-plowed. On the conventionally tilled plots, the soil is double-disked and worked with a field cultivator prior to seeding. There is no seedbed preparation after plowing on the minimum-tilled plots. Corn is wheel-track planted.

These data were obtained from plots at the Upper Mississippi Soil and Water Conservation Station, LaCrosse, Wisconsin. They are included here as a result of the courtesy of Orville E. Hays, Superintendent of the Station.

The soil was Fayette silt loam (Typic Hapludalfs) on 15 percent slope at LaCrosse, Wisconsin.

During the period 1955 to 1959, there was more than twice the water loss and four times the soil loss from the conventionally tilled as from the minimum-tilled wheel-track planted plots. Recent observations suggest that 30 years later, similar results can be expected. This effect of lesser erosion from minimum-tilled plots has been observed at other locations. Thus erosion control is effected by the physical condition of the soil during the early crop season as well as by the protective cover during the late fall, winter, and early spring.

Spring-plowed land contains more pore space than does fall-plowed land during the growing season. Thus it can hold more total moisture during the critical period when the need of the crop is great. The slow germination sometimes caused by a shortage of moisture in the surface of spring-plowed land is the result of improper planters or planter adjustment. Planters should be adjusted to press the seed firmly into the soil, and some provision for firming the soil under and around the seed is desirable. Corn (maize) growers have done this by planting in tractor wheel tracks or by use of a press wheel ahead of or behind the planter shoe. Soybeans are being planted in the same manner.

Water loss is slow from spring-plowed land, except from the very surface. The top inch dries quickly to form a mulch that protects remaining soil water against evaporation; provided, of course, the plowed surface is not packed by excessive tillage. The quick drying also reduces weed growth.

Sand and loamy sand soils are seldom too wet to plow, but care must be exercised with sandy loams, loams, and soils high in clay content. Plowing of very dry soils is expensive because of the power required, the wear on equipment, and the fact that clay soils break up into clods. However, the first rainy season thereafter corrects such damage.

Depth of Plowing

The plowed surface is the most effective part of a soil. Early experiments failed to show advantage (based on yields) of plowing deeper than 6 to 7 in. probably owing to faulty methods. Fertilizer applications were inadequate, and plant populations were usually too low. Also, tillage after plowing was excessive, according to present-day standards. The value of plowing, beyond the coverage of residues and vegetation, as has been pointed out, is through loosening the soil (see Fig. 9-3). It is surely true that any value that might result from loosening the soil will be nil if the plowed layer is immediately repacked.

Deeper plowing may be beneficial on some soils if accompanied by improved fertilizer, organic matter, and tillage practices. Experiments at the Michigan Experiment Station have proved this to be true. Rates of fertilizer application should be in proportion to the volume of soil being fertilized and the yield expected. Also, more organic matter must be incorporated with a deeper soil, and the plowed layer must be kept loose so that the organic matter can decompose. In other words, soil-formation conditions must be maintained to the entire depth of the deeper surface soil. Otherwise, the value of deeper plowing is never attained. Figure 9-4 shows the deeper spread of corn roots where Kalamazoo sandy loam soil (fine-loamy, mixed, mesic, Typic Hapludalf) had been plowed to 18 in. in depth.

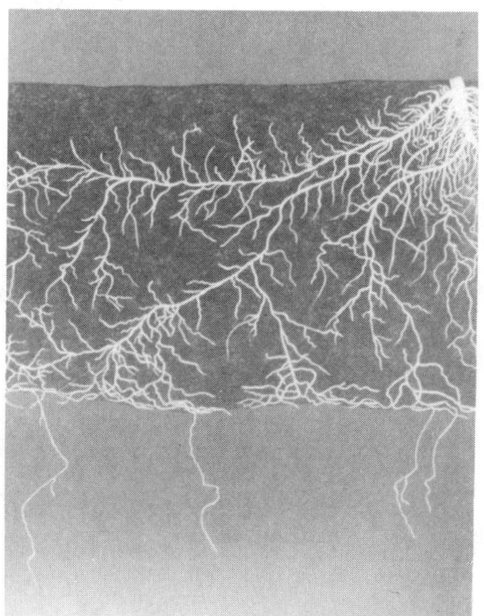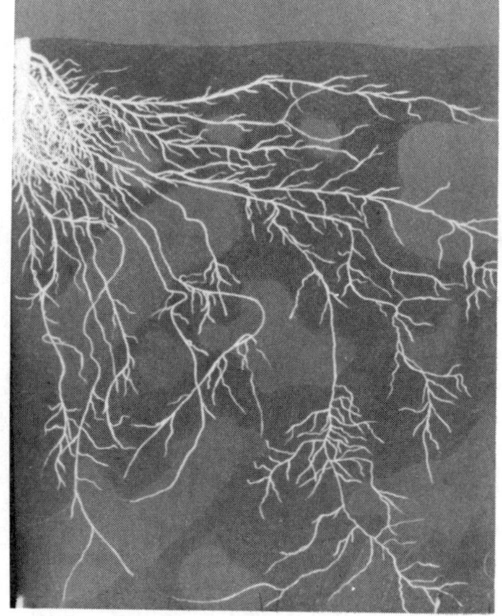

FIGURE 9-4 Penetration of corn roots in Kalamazoo sandy loam soil, plowed to 9 in. on the left and 18 in. on the right. Roots were actually exposed by preparing monoliths.

How Much Tillage After Moldboard Plowing?

"Minimum tillage" may be defined as the least amount needed for quick germination, a good stand, and satisfactory yields. Experiments conducted in Michigan, Wisconsin, and New York show that top yields of corn, oats, sugar beets, and beans are possible where the only tillage is done with the moldboard plow. Under many conditions, top yields are possible with no tillage. Thousands of farmers have proved the same thing on their own farms. Iraqi farmers have worked their soils only with the moldboard plow for 4000 years.

Planting and the first cultivation (perhaps the only cultivation) is facilitated on some soils by using a smoothing implement attached to the plow. Some of the more common ones are described here.

Plow Packer. The plow packer (Fig. 9-5) is not a new tool; several companies have manufactured it for many years. It was one of the first implements used in a series of Michigan State University tillage experiments started in 1946. Draft requirement on a level clay loam soil was only 100 lb. As shown in Fig. 9-6, the plow packer did only a small amount of tillage. Yields, however, from plots where the plow packer was used have been just as high as where any other implement was used.

The plow packer works best on loam and lighter soils having some cover other than a heavy sod. It does very little smoothing, but it crushes soft lumps near the surface.

Mulcher. The mulcher—a double cultipacker with spring teeth between the packers—does much more fitting than does the plow packer.

FIGURE 9-5 The plow packer does as little tillage as any of the once-over implements. Experiments, however, have shown that it is sufficient as long as adequate crop stands are obtained. This picture was taken in 1947, a year after the Michigan minimum-tillage experiments were started on the University farm.

FIGURE 9-6 Many farmers like the revolving bladed tiller. Note the desirable condition of the soil after its use. Note also that the farmer is wheel-track planting corn as fast as he plows. Even with rain that night, he did not need to refit his soil. Furthermore, the corn germinated ahead of the weeds.

It smooths the soil and is especially good where an ideal job of plowing is impossible. Do not set the spring teeth deep enough to pull up sod or trash. On a level clay-loam soil, the draft is 250 lb. This can be varied with the setting of the spring teeth.

Yields where this implement was used on experimental plots were not greater than where the plow packer was used. In many cases, however, first cultivation of row crops was easier on plots that had been worked with a mulcher.

Revolving Bladed Tiller. Very fine-textured soils, soils of poor structure, and very heavy sods on any soil, can be effectively tilled with the revolving bladed tiller (Fig. 9-6). The blades press the furrows and exert some firming below the surface. At the same time the frame, hinged in the middle, does some smoothing. This action does more to correct plowing defects than does the plow packer. Draft is about twice that needed for the plow packer.

Spike Harrow. Under most soil conditions, one section of a spike harrow is very satisfactory as a minimum-tillage tool. Used with a two-bottom 14-in. plow, it covers the soil twice. With larger tractors and plows, more sections of a spike harrow would, of coarse, need to be used. A rather poor job of plowing can look quite presentable when a spike harrow is attached. This implement was used in the field shown in Fig. 9-7.

FIGURE 9-7 This corn was wheel-track planted on May 14, the day the land was plowed. A spike harrow attached to the plow did all the tillage. The first cultivation was on June 18; the second on June 25. The row width is 56 in. The picture was taken on July 9. (Note the absence of weeds.) This corn yielded 85 bu of dry, shelled corn.

FIGURE 9-8 The long-tined weeder, designed for weeding row crops, is a good minimum-tillage tool, especially on sandy soils.

Spike harrows collect trash. This usually happens on the ends of the field but is not so much of a problem if a boom is arranged to lift the harrow when the plow is lifted. This setup also automatically cleans the harrow at each end of the field.

Weeder. The weeder shown in Fig. 9-8 has very little draft. It seldom clogs with trash and does a very good job of breaking up soft clods. It does very little, however, in smoothing out defects caused by poorly adjusted plows. It is especially well adapted for use on loam and sandy soils. The tines are too flexible for clay or clay-loam soils except for those very high in organic matter.

Conservation Tillage

The term **conservation tillage** describes a number of tillage operations or combination of operations with a common objective — to prepare a seedbed suitable for high productivity with a minimum loss of soil by erosion. The amount of erosion that occurs following tillage is directly related to the amount of crop residue left on the surface. Thus, tillage with a chisel plow, sweeps, or a heavy tillage disk may be effective in leaving residue on the surface and hence in reducing erosion.

Soil Management Principal

No-till is the most effective management practice in reducing erosion in row-crop production on soils with good structure and drainage.

The reduction in soil loss from no-till (corn or soybeans) has been dramatic. In general, soil losses under no-till are neglible. Other forms of conservation tillage (i.e., chisel plow) show some reduction in erosion as compared to plowing.

Although no-till systems are excellent from the standpoint of controlling erosion, a number of soil-management factors are changed, and these may require considerable skill in developing procedures for maximum production. First, since all residues are returned to the surface, the soil profile will become considerably different from that which results from conventional tillage (i.e., moldboard plow). Organic matter in the surface 0 to 5 cm was nearly double (4.82 versus 2.40%) for no-till corn compared to conventional tillage in Kentucky. But less organic matter was contained in the 15 to 30 cm depth. The long-term effects of accumulating organic matter on the surface of the soil are as yet unknown, but it may improve infiltration of water, make the soil slower to warm in the spring, and alter the chemistry of the surface layer.

Second, tillage is no longer used for weed control, hence chemical weed control must be practiced. In fact, modern herbicides have really made no-till a reality.

Third, fertilizers and lime can not be incorporated to a considerable depth in no-till operations. Hence they must be surface applied or band placed. This is leading to nonuniform soils from a pH and nutrient standpoint. Obtaining a good soil sample for soil-testing purposes is then very difficult.

When no-till is properly practiced and effective weed control is obtained, yields are equivalent to those obtained with other management systems. But yields may be less under a number of conditions. The reason for no-till is to reduce erosion, and it is on erosive soils that no-till has the greatest benefit.

Cultivation

One cultivation is usually enough to control the weeds in row crops. When planting is done on the day of plowing, crop plants usually emerge before the weeds. Adequate application of starter fertilizer, in bands close to the seed, results in rapid early growth. Proper timing of the cultivation then makes it possible to cover the weeds in the row without covering the much larger crop plants. Corn cultivation may be 4 to 5 weeks after planting. Extra nitrogen may be applied at the same time. A second cultivation, if necessary, may be accompanied by the seeding of a cover crop.

Chemical weed control may, of course, entirely eliminate the need for the one or two cultivations, but sometimes weather conditions (too dry or too wet) may make the chemicals (herbicides) less effective. So many farmers like to have a cultivator available in case it is needed.

PLANTING METHODS

Firm Seedbed

Seeds placed in close contact with firm, moist soil will germinate quickly. They grow more rapidly, however, on loose soil, as shown by the alfalfa in Fig. 9-9. The wise home gardener spades the soil, rakes it lightly, makes a shallow trench, and drops the seeds. After covering the seed, the home gardener presses the soil and seed by walking directly **on the row.** The modern planter is equipped with press wheels to bring about the same close contact between seed and moist soil. Over short distances, water moves readily through firm soil but slowly or in the gaseous phase through loose soil.

As already pointed out, minimum tillage is most easily practiced by planting immediately after a good job of moldboard plowing. A properly adjusted plow leaves the soil very loose, too loose for quick seed germination. The planting operation must provide the needed firming in the row. This may be illustrated by the soil condition that exists where corn has been wheel-track planted in freshly plowed soil (see Figs. 9-10 and 9-11).

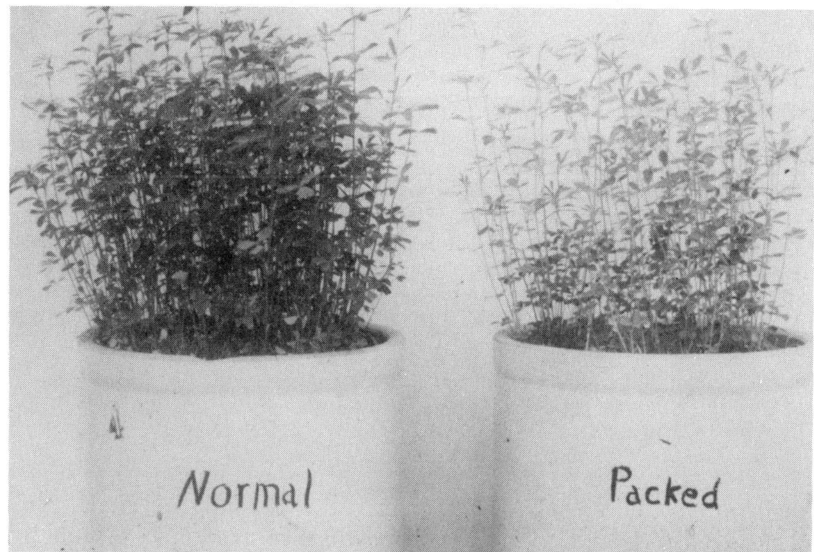

FIGURE 9-9 Alfalfa was planted in fertile loam soil. The soil was poured loosely into the left pot and was packed in the right one, the same quantity of soil in each pot. Lack of oxygen prevented normal growth in the packed soil.

FIGURE 9-10 Tractor wheels leave an ideally firm soil for corn planting. If the soil is a little dry, the wheels crush lumps and promote capillary water movement. This tractor, owned by Orla, Harold, and Darwin Sheathelm of Dansville, Michigan, was modified for four-row wheel-track planting. The front wheels are 40 in. apart and the rear wheels are 120 in. apart.

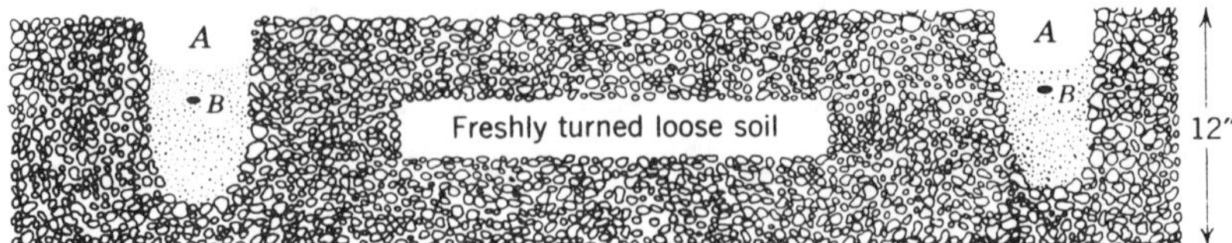

FIGURE 9-11 Wheel-track planting in freshly plowed soil makes an ideal seedbed but leaves loose soil between the rows, an ideal rootbed. This diagram illustrates the mechanical condition of soil immediately after plowing and at planting. The plow was set to cut 8 in. Note that soil compaction under the tractor wheels extends almost to plow depth and that the 12 in. of loosely turned soil between the rows of seed will allow quick penetration of rainwater. (Diagram from *Hoard's Dairyman*, May 10, 1956.)

Fertilizers are Important

The success of the minimum-tillage principle and weed control with little or no cultivation is tied closely to fertilizer use. Adequate nutrients must be placed close to the seeds and in a location where roots will reach them immediately after germination starts. Care must be taken, however, to avoid contact between seed and an amount of fertilizer that may delay germination. One role of the fertilizer is to hasten growth of seedlings so the crop may satisfactorily compete with weeds (see Fig. 9-12) and so that a canopy is

FIGURE 9-12 This corn on Miami loam soil was wheel-track planted on May 30, the day of plowing. Six rows at the left were not fertilized. The picture was taken on July 4, immediately after the second and last cultivation. Weeds in the unfertilized rows were not well-controlled because the corn was too small at the first cultivation.

rapidly developed to combat water erosion. These objectives fail where fertilizer contact delays germination. For a complete discussion on fertilizer placement see Chapter 14.

Planting Row Crops

Spring moldboard plowing and immediate planting (plow-plant) has worked well for all row crops planted in early spring, a period when soil moisture is adequate or excessive. Wheel-track planting, pictured in Fig. 9-10, has given the best results for corn and soybeans in the humid areas. In drier regions, the lister planter (Fig. 9-13) is widely used. Plow-plant has been successful for sugar beets in the Eastern area (Michigan, Wisconsin, Ohio, Ontario) without resorting to wheel-track planting. The fact that beets are planted earlier in the season than corn, at a time when soil moisture is high, has had a bearing on the success with which growers have accepted the practice. As early as 1953, 66 percent of Michigan's top beet growers were using minimum tillage. By 1987, practically all are working their soil less than they did before minimum tillage was recommended.

Planters to be operated in very loose soil (not in tractor wheel tracks) should be provided with a special means for firming the soil in the row. Press wheels ahead of the seed furrow opener and more pressure on the seed press wheels have helped to solve the problem. Such a planter unit is on the Michigan experimental plow-plant unit shown planting soybeans in Fig. 9-14.

Where crops are to be planted during a season when soil moisture is likely to be low, it may not be safe to delay plowing until the calendar says it is time to plant. At that time the soil may be too dry to plow, especially if the soil is protected by a green cover crop as it should be. Disking to destroy the cover

FIGURE 9-13 The lister planter is widely used in the drier regions. (Photo by Wesley Chaffin; from Oklahoma Agricultural Experiment Station Circular 635.)

FIGURE 9-14 Plowing, smoothing, and planting soybeans in one operation. Fertilizer was being dropped on top of the soil to be plowed under and placed in a band beside the seed for starter effect. Although it was not being used here, a mechanism was also available to apply a pre-emergence weed spray. Use of the spray for corn has entirely eliminated the need for row cultivation for weed control. (Courtesy C. M. Hansen and L. S. Robertson.)

crop and form a mulch to lessen evaporation is one alternative. Another is to plow early, while moisture is plentiful, let the land lie until time to plant, then destroy growing weeds with as little tillage as possible, and plant. Under such management, use of carefully selected and properly applied herbicide may be advisable. Advice from local extension personnel is suggested. In such a program, a smoothing implement should trail the plow. Late spring planted crops such as navy beans and late potatoes in Michigan, and possibly soybeans in other parts of the Corn Belt, should be handled in this manner.

No-till and Ridge-till Planting of Corn

The term "no-till" is a misnomer. Actually, tillage is performed in a narrow strip for each row, enough so that the planter units may open the soil, drop the fertilizer and seed, and cover both. It is simply another method of minimum tillage that works best on sand and sandy loam soils. A similar statement may be made about till planting on ridges. The finer textured loams and clay loams are usually in need of being loosened with a moldboard plow or chisel plow. A compact subsoil like the one shown in Fig. 9-15 will result in shallow rooting and damage from summer drought.

FIGURE 9-15 No-till corn grown on soil with a plow pan that interferes with both water movement and root penetration. Such zones of compacted soil should be destroyed with tillage, either deep moldboard plowing or plowing with a chisel plow. (Taken with permission from W. J. Cook and L. S. Robertson. Michigan State University Extension Bulletin E-1354, November 1979.)

Soil Management Principal

Successful no-till and ridge-tilled corn are always associated with good drainage and good soil structure.

A special planter is necessary for successful no-till or ridge-till planting. Such a planter is shown in Fig. 9-16.

FIGURE 9-16 A modern no-till planter used for strip-tillage and ridge-till planting. (Photograph furnished by William A. Hayes, Fleischer Manufacturing, Inc., Columbus, Nebraska, March 1983.)

A farmer who contemplates the purchase of a no-till planter would be wise to examine the profile of his soil, determine its texture and structure, and perhaps consult local soils extension personnel.

Planting Small Grains

Very little tillage is necessary for spring-planted small grains. Use of the plow-press drill shown in Figs. 9-17 and 9-18 is ideal. The soil is put into a condition very similar to that resulting from wheel-track planting corn. The narrow metal press wheels, carrying the weight of the drill, press the seed firmly into otherwise loose soil. Emergence is rapid, and yields are very satisfactory. There is only one disadvantage; dead furrows remain open and are somewhat of a handicap to harvesting operations. Combining in the same direction as planting does much to take care of the difficulty, however.

In lieu of a narrow plow-press drill, farmers have adapted other equipment for plow-planting spring grains. A harrow between the tractor and the drill can be designed to throw soil into the tractor wheel marks. Without the harrow, the tracks are so deep that seeds are not evenly covered. A trailing cultipacker will press soil on the seeds but is less desirable than press wheels because we prefer to have the soil pressed only in the row, not between the rows. Weeds come faster and offer more competition with the grain where the cultipacker rather than the press wheels is used.

FIGURE 9-17 The plow packer and press drill combination has been a very good method for planting oats and legumes. Note the good soil condition following the drill. Seeding is through five straight disk openers. A packer wheel follows each opener, and soil is pressed on the seed to aid in quick germination.

FIGURE 9-18 This picture was taken on the Northern Plains, where press drills have been used for many years. Here two plows, two packers, and two drills are being hauled by one tractor. In this semiarid region, it is essential that spring wheat and barley seeds be planted immediately after plowing. Then germination occurs before surface soil has time to dry out. A minimum of labor is involved. (Photograph by Leo E. Orth, Sinclair Petrochemicals, Inc.)

One farmer's answer to the urge to plow and plant oats in one operation is illustrated by Fig. 9-19. He saved several dollars per acre and harvested a good crop. Conventional tillage and planting is too costly for this low-value crop.

Planting Legumes

The results of experiments conducted in Michigan have shown that the minimum-tillage principle works for legumes as well as for small grains. The same precautions should be taken, however, regarding the season of the year and its effect on soil moisture. When legumes are to be seeded during a period when the odds are great that soil moisture will be plentiful (April in Michigan and Wisconsin), plowing and immediate planting are recommended. The practice works especially well for alfalfa, because the seeds germinate quickly and taproots grow rapidly. A band of high phosphorus fertilizer placed directly below the seed causes rapid early growth, so the crop may compete successfully with weeds. Under Wisconsin conditions, it was

FIGURE 9-19 Don Pace, Calhoun County, Michigan, has figured out how to plow and plant oats in one operation. The rear plow has three 14-in. bottoms and a tandem packer, with the 11-hole (7-in. spacing) grain drill attached to the outer end of the packer frame. Ahead is a two-bottom 16-in. plow with packer. Oats are fertilized and legumes are seeded in one operation, leaving no tractor wheel marks. Press wheels would hasten germination.

found that old pastures may be plowed and immediately reseeded to alfalfa in this manner. Even on steep slopes, the old sod held the soil against erosion until the new crop offered the protection needed. In fact, it was found that a crop of wheel-track planted corn with alfalfa seeded between the rows resulted in good alfalfa coverage without appreciable erosion. The previous crop had been bluegrass for 12 years.

Plow-plant is not recommended for seeding alfalfa during a dry season (the summer months in Michigan). There is always a possibility that such a plan will be successful, but the odds are too great that rainfall will not be sufficient for a good job of plowing and for germination of the seed. It is better to plow early, harrow enough to eliminate weeds, and plant when season and weather are favorable. Certain herbicides may be used to avoid the necessity of excessive tillage between plowing and planting. Herbicides should be used discreetly, however, or injury to the alfalfa seedlings may result.

Band seeding with the use of press wheels is recommended by Michigan Cooperative Extension personnel whether the crop is seeded alone or with a small grain companion crop and whether in the spring with minimum tillage or in the summer with more tillage. Firm soil in the row is always desired.

REFERENCES

Aldrich, S. (1956). New styles in seedbeds. *What's New In Crops and Soils.* 9(2): November.

Bevins, R. L., G. W. Thomas, M. S. Smith, et al. (1983). Changes in soil properties after 10 years non-tilled and conventionally tilled corn. *Soil Tillage Res.* 3:135–146.

Cook R. L., H. F. McColly, L. S. Robertson, and C. M. Hansen (1958). *Save Money-Water-Soil with Minimum Tillage.* Bulletin 352. Cooperative Extension Service, Michigan State University.

Cook, R. L., L. M. Turk, and H. F. McColly (1953). Tillage methods influence crop yields. *Soil Sci. Soc. Am. Proc.* 17:410–414.

Gebhardt, M. R., T. C. Daniel, E. E. Schweizer, and R. R. Allmaras, (1985). Conservation tillage. *Science* 230:625–630.

King, A. D. (1983). Progress in no-till. *J Soil Water Conserv.* 38:160–161.

Laflen, J. M., W. C. Moldenhauer, and T. S. Colvin (1980). Conservation tillage and soil erosion on continously row cropped land. In *Crop Production with Conservation in the 80's.* American Society Agricultural Engineers, St. Joseph, Mo.

Moldenhauer, W. C. (1984). A comparison of conservation tillage systems for reducing soil erosion. In *Preliminary Proceedings for a System Approach to Conservation Tillage.* Kellogg Biology Station, Hickory Corners, MI.

Mulvaney D. L., and L. Paul (1984). Rotating crops and tillage—both sometimes better than just one. *Crops and Soils.* 36:18–19.

Peterson, A. E. (1958). Wheel track planting corn and interseeding alfalfa. Joint Meeting Agronomists with the Fertilizer Industry, National Plant Food Institute, February 13–14.

Peterson, A. E., O. I. Berge, J. T. Murdock, and D. R. Peterson (1958). *Wheel Track Corn Planting.* Wisconsin Extension Circular 559.

Robertson, L. S., D. L. Mokma, D. L. Quisenberry, et al. (1976). *No Till Corn: 3. Soils.* Bulletin E-906. Cooperative Extension Service, Michigan State University.

Robertson, L. S., and A. E. Erickson, (1983). *Till Planting on Ridges.* Bulletin E-1683. Cooperative Extension Service, Michigan State University.

Tesar, M. B. (1976). Good stands for top alfalfa production. Bulletin E-1017. Cooperative Extension Service, Michigan State University.

Wendt, R. C., and R. E. Burwell (1985). Runoff and soil losses for conventional, reduced, and no-till corn. *J. Soil Water Conserv.* 40:450–454.

Soil Testing

Soil testing makes complete nutrient control a possibility. Fertilizer experiments are being patterned to determine economically optimum rates of nutrient application. High yields with low production costs per unit are a **must** in modern farming. In the old days, farmers were largely selling their own labor! They invested little cash, so they could get by with low yields. During unusually poor years, they got along on less or went into debt, hoping the next year would be a "good" year so they might pay last year's debts. Often this was the case, and many continued thus through their lives. Others, less fortunate or with less ambition, lost or sold their lands and drifted to the cities.

Farmers of today are different in that failure is more certain and sooner unless they are obtaining reasonably high yields. Improved drainage, many improved cultural practices, better varieties, and control of insects and diseases have helped to set the stage for high yields. As a result, the demand on the soil has gradually increased. Soil testing lets farmers know how much and what kind of fertilizer they must apply to be sure of returns from their investments in other improved practices.

Soil Management Principle

Soil-fertility control is possible where soil testing is practiced.

BIOLOGICAL TESTS

Some kind of biological test is necessary for the development, calibration, and interpretation of a chemical test. The statement that a soil contains 175 lb of potassium per acre means nothing to the chemist without agronomic background. To the agronomist from the North Central region of the United States, however, who was told also that the extracting solution was 1 *N* ammonium acetate pH 7.0, the figure would mean that the soil contained a medium level of this essential nutrient and that alfalfa would respond to an application of soluble potash fertilizer. On the other hand, if the test result was 250 lb/acre, that same agronomist would expect very little response to potash fertilizer.

A biological soil-test method was used to establish the pounds of potassium range over which alfalfa may be expected to respond to applications of potash to Michigan soils. In effect, it was a series of field experiments on 105 farms scattered over the state. Phosphate and potash fertilizers were applied singly and in combination. Soils were sampled and tested for extractable potassium, and yield increases were recorded. It was then possible to divide low-response from high-response soils and correlate yield response with soil-test results. The dividing line between low and high response was set at 150 lb extractable potassium per acre. You are urged to remember that some correlation of this kind is necessary if chemical tests are to have real meaning.

Soil-test research in Ohio has recently concluded that "the Bray-1 'available P' is a good index of the available P to which crops are responding both in P uptake and yield." Furthermore, in discussing the basis for "improved K recommendations," it was pointed out that three parameters are needed for calculating the amount of fertilizer K needed for obtaining maximum crop yields. They are (1) The initial (existing) level of exchangeable (and soluble) K in the soil, (2) the sufficiency level of exchangeable K for maximum yield (KSL), and (3) the K fixation factor which indicates what the difference in No. (1) and No. (2) must be multiplied by, to obtain the fertilizer K required. Number 1 has been measured in a soil-testing laboratory, and number 2 has been the result of fertility studies in the field, greenhouse, or growth chamber. This, in effect, is similar to the Michigan study on alfalfa just mentioned.

Greenhouse pot tests may be as good as field plots for the solving of certain soil-fertility problems, among them the nutrient status of the soil relative to the production of a certain crop.

Some investigators have used indicator crops, for example, romaine lettuce, as an indicator of the nutrient status of soils.

For many years, the fungus, *Aspergillus niger* has been used to evaluate the nutrient status of agricultural soils.

Recently, de Vries (1980) has listed several of the environmental conditions and management techniques that must be controlled if greenhouse pot experiments are to be reliable in nutrition studies. One of his suggestions, which we have found to be particularly important, pertains to the decision as to when to terminate the experiment. Size of plants, size of pots, kind of soil, and frequency and method of watering are some of the factors that determine the effective termination date.

A biological test known as the **Neubauer test** has been widely used in Europe, but only to a limited extent in the United States. One-hundred grams of soil is diluted with quartz sand and placed in a small container. One-hundred carefully selected rye seeds are planted and allowed to grow for a predetermined number of days. The plants are then analyzed for the nutrients under study. Neubauer tests require a great deal of time and are too expensive for the routine testing of farm soils.

As already mentioned, perhaps the greatest possibilities in the use of biological soil tests lie in the development of chemical tests. The latter are so much faster and cheaper that they can be of much greater use in evaluating soil nutrient status for fertilizer recommendations. However, in the last analysis, the chemical test must reveal the availability of a certain nutrient to a specific plant or its value is limited or nil. Much of the fertility research conducted at experiment stations during the last 20 years involves this need for soil-test interpretation. The answers must come through correlation of chemical soil test results with data obtained in tests with plants; in other words, biological tests.

Soil Management Principle

A soil chemical test must be correlated with field studies to be of value.

CHEMICAL TESTS

Chemical soil tests may be divided into two broad groups, tests for exchangeable bases and tests for soluble or extractable nutrients. The tests are made in laboratories located in every state and in almost every county of some states. They have as their primary objective the determination of the available nutrient status of the farm, garden, greenhouse, and lawn soils. Data so obtained are used in selecting the fertilizer and/or soil amendments needed for the production desired. An alternative to the field experiment is the pot culture experiment in greenhouses, like the one shown in Fig. 10-1.

Test for Exchangeable Cations

Agronomists quite generally agree that plants have the ability to use nutrient cations attached to the soil colloids. Fine clay particles and decomposed organic matter (organic colloids) possess this important property known as **cation exchange.** The phenomenon is often called **base exchange,** but the term is incorrect because hydrogen ions react in this respect as do the nutrient bases. There is some lack of agreement as to just how a nutrient cation moves from the soil colloids to the root, whether directly or in solution, but it finally reaches its goal, the inside of the root. Furthermore, the ease with which the ion reaches its goal depends on the total amount of the nutrient attached to the soil colloidal fraction and secondly on the ratio that exists between that nutrient and the sum of all the cations so attached. This latter phenomenon is known as degree of **ionic saturation.**

FIGURE 10-1 A typical greenhouse where variations in plant growth help to establish minimum nutrient levels for specific crops. Such pot culture experiments may partially substitute for field trials.

Graham (1950) describes in detail in Missouri Agricultural Experiment Station Circular 345 how the tests for exchangeable cations are used in studying the nutrient status of soils. It is interesting that on the cover of the Circular, entitled, *Testing Missouri Soils,* is an air photograph of a large experimental field with an inset showing a farm advisor and a farmer in a soil-testing laboratory. Dr. Graham has pictorially shown the relationship that must exist between an experimental field (a biological test) and the chemical tests being run in the laboratory.

The Soil Test

For the three exchangeable nutrient cations, potassium, magnesium, and calcium, the soil test is based on the fact that they can be *exchanged* from the soil colloidal complex by introducing a large quantity of a known but different cation. In most of the states in the North Central region (in a total of 19 states in the United States), the extracting reagent for these three cations is neutral/normal ammonium acetate. In nine Southeastern states, the extracting reagent is 0.05 normal hydrochloric acid (HC1) plus 0.025 normal sulfuric acid (H_2SO_4) whereas the Morgan solution (1.4 normal sodium acetate-acetic acid buffered at pH 4.8) is used in five other states. The particular extracting reagent and procedure used is not so important as long as the test results have been correlated with field plot or greenhouse pot experimental results.

All soils below 7.5 pH contain enough hydrogen, according to Graham, to make necessary a determination of that ion in total. Otherwise, total exchange capacity cannot be calculated. (It is assumed that total exchange capacity of humid region soils is essentially the sum of the quantities of calcium, magnesium, potassium, and hydrogen. In arid regions, it is necessary to include sodium and in some cases it is possible to omit hydrogen.)

After the four determinations are complete, total exchange capacity is simply the sum of the exchangeable ions of calcium, magnesium, and potassium plus the total of hydrogen. The percentage saturation of each ion can then be calculated by simple arithmetic.

Soil Management Principle

For best interpretation, soil test must be related to soil series.

To use **degree of saturation** as a guide to fertilizer application, it is necessary to know something about the nature of the clay minerals and/or to correlate with crop response to variable potassium levels. The total exchange capacity of kaolinitic clay is much lower than that of montmorillonitic clay. It is necessary to have a rather high percentage of potassium saturation in soil with largely kaolinitic colloids, perhaps as high as 9 percent in soils having an exchange capacity of 4 meq/100 g.

Montmorillonitic clay colloids (smectites) have a high exchange capacity, and potassium attached to these colloids is easily replaced. As a result, the degree of potassium saturation need not be high (2 percent potassium may be satisfactory in soils having a total exchange capacity of 18 meq/100 g. For exchange capacities between 4 and 18 meq, potassium saturations should range from 9 to 2 percent.

The range of desirable saturation with potassium varies with the percentage and nature of the soil organic matter. It should be verified by field or greenhouse experiments in each state and perhaps in each management group of soils.

Likewise, favorable ranges of calcium and magnesium saturation may be established. A degree of magnesium saturation as high as 10 percent is usually considered satisfactory for plant growth. The presence of 2 : 1 clay minerals necessitates a higher degree of saturation because of the possibilities of entrapment of the magnesium ions between the expanding and contracting sheets.

Calcium saturation should be 70 percent or higher in soils high in 2 : 1 expanding minerals, whereas in 1 : 1 clays (kaolinitic) 40 to 50 percent saturation is high enough for plants to obtain the calcium they need.

There is considerable indication of a need for balance between the calcium and magnesium available to plants. Consequently, as acid soil is limed and the degree of calcium saturation is increased, there may be a need to increase the degree of magnesium saturation to avoid magnesium deficiency. The use of dolomitic limestone accomplishes this end.

High calcium saturation is important from the standpoint of potassium relations. Potassium is readily replaced by calcium when the latter is added as a neutral salt to an acid soil. Thus an application of calcium chloride may increase losses of potassium by leaching. Such a loss does not happen, however, when calcium is applied as a liming material such as calcium hydroxide, because the calcium replaces hydrogen ions to form water and the colloidal complex becomes more highly calcium saturated. Very little potassium is replaced because calcium more readily replaces the hydrogen.

When potassium fertilizer is then applied, the potassium ions replace the calcium more easily than they do hydrogen, so liming, in effect, has facilitated the change from soluble to exchangeable potassium. There is less chance then of potassium loss by leaching.

Another reason for maintaining a high degree of calcium saturation is that calcium in the soil solution forces the plants to use more. The tendency for total cation concentration to remain constant then results in their taking in less potassium. This may have an economic advantage because calcium is cheaper than potassium. It may result, however, in potassium deficiency. These relationships furnish the arguments in favor of testing for exchangeable nutrient cations and total exchange capacity.

Correct fertilizer management is easier, perhaps only possible, when the cation exchange capacity (CEC) of a soil is known. Hue and Evans (1983) have published two computer-assisted methods of estimating CEC. Results compared favorably with results obtained by the standard $1\ N\ NH_4OAc$ method.

Tests for Soluble Nutrients

The tests for soluble nutrients are commonly called **quick tests.** A sample of the soil is shaken with a certain amount of solvent for a definite length of time before filtering. A measured quantity of the filtrate is then tested for the nutrients under study. The tests are based on the idea that the fraction extracted is a measure of the fraction available to the plant during its period of growth. The fallacy of such reasoning is at once apparent. Variations in the chemical nature of the soil cause marked variations in the solution effect of the extracting reagent. Also, plants vary in their ability to take nutrients from a given soil. The only way then for such tests to be of agronomic value is to standardize them by means of a biological test in which the plant in question has been used as an indicator. In other words, quick-test results on an unknown soil mean nothing as far as availability of nutrients to an unknown plant is concerned.

Phosphorus

Dilute acetic acid, dilute sulfuric acid, dilute hydrochloric acid, strong hydrochloric acid, dilute hydrochloric acid with ammonium fluoride, sodium acetate, sodium bicarbonate, and water saturated with carbon dioxide are all being used as extracting reagents for the determination of available phosphorus. All give different answers in terms of parts per million or pounds per acre. When a soil test is interpreted, it is always necessary to know what extraction reagent was used.

Generally, dilute acids have rated better than have strong acids when compared against biological test results. The soil testing services in several states have changed in recent times from strong to dilute acid methods. This has, for instance, been true in Nebraska, Ohio, and Michigan. It was noticed, for instance, in Michigan, that $0.135 \, N$ HCl extractions consistently showed about 30 lbs more available phosphorous per acre where 100 lb of rock phosphate had been applied than where the soil had not received the rock. According to the calibrations of the test, that amount was enough to class the soils as high rather than low in phosphorus. Plants growing on the soils were not able to pick up the phosphorus from the rock, so the strong acid was producing results way out of line with those obtained when plants were used as indicators. Dilute acids, either acetic or hydrochloric with ammonium fluoride (Bray method), were not indicating the presence of the rock phosphate in the soil. Since plant yields, appearance, and tests did not indicate the presence of the rock, it seems necessary to rate the weak-acid methods above the strong acids as a measure of nutrient availability to the plants being grown.

The criticism that a strong acid extraction removed too much phosphorus from applied rock phosphate is in agreement with the statement commonly made that strong acids cannot be used for measuring phosphorus availability in alkaline soils because they dissolve tricalcium phosphate and apatite not available to the crops. Thus the results are misleading in that they lead farmers to believe that their soils will furnish more phosphorus than is actually the case.

The National Soil Test Work Group found the Bray P_1 test ($0.025 \, N$ HCl and $0.03 \, N$ NH_4F) to produce results that correlated better with crop responses than did the other methods considered. Many states have adopted this method, now used in 12 states of the North Central region. North Dakota uses $NaHCO_3$ because so many soils are alkaline.

McLean, Arscott, and Hannan (1983) showed that soil characteristics must be considered when interpreting the results of soil tests. They worked with two soils from Ohio and two from Bangladesh. The soils were treated with five rates of phosphorus applied as triple superphosphate and cropped with wheat in a growth chamber. Tests for Bray P_1 phosphorus, among other analyses, were run on all soils. Wheat yields were higher on the Ohio soils than on those from Bangladesh. Of special interest as far as soil testing is concerned is the fact that 95% of maximum yields (substituting in the Mitscherlich-Bray equation) were attained on the Ohio soils at one-half the level of Bray P_1 available phosphorus as was required for the Bangladesh soils. The writers believe the difference lies in the fact that the Bangledesh soils are deficient in zinc and they became progressively more deficient as rates of phosphorus were increased.

Saturated Media Extract

Warncke has developed the saturated media extract method specifically for the testing of greenhouse growth media that are not natural soils. It is simply soaking and leaching a large volume of the growth medium with distilled water, then testing the leachate by standard methods.

Potassium

Many soil-testing laboratories provide tests only for reaction, phosphorus, and potassium. Where percentage saturation is not considered, a number of extractants are in common use (weak acetic acid, 0.1 N HCl, water, etc.). The exact extractant is not important as long as there is calibration with field plot or other biological test results.

A picture of the forms of potassium in soil makes clear the fraction being removed in a 1-min. extraction with such an extractant as 0.1 N HCl.

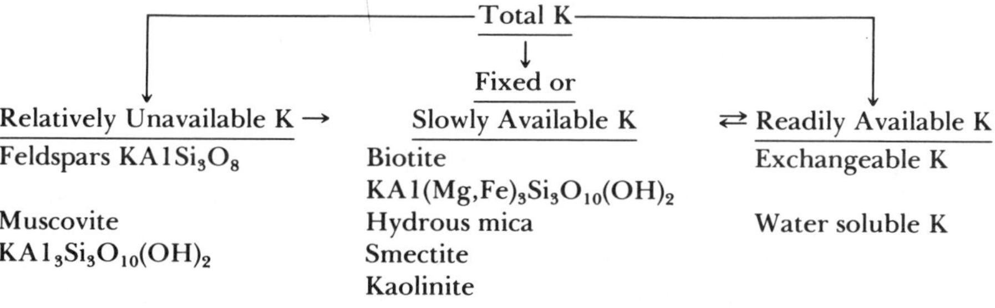

We would expect an extraction with 0.1 N HCl to remove the water-soluble potassium and a portion of the exchangeable potassium. Just how much of the total exchangeable potassium is removed depends on dilution, extraction time, and the nature of the soil, particularly as to the saturation of the soil with other bases. The considerations for soil type enters, of course, into field or greenhouse calibration studies. The work of the National Soil Test Work Group already referred to shows that soil-testing laboratories across the country are doing a good job of testing for potassium.

Calcium and Magnesium

Most of the tests for calcium and magnesium are made by methods similar to those described earlier in this chapter under the heading "Test for Exchangeable Cations." The acetic acid method of Spurway and Lawton, (1949) is an exception. The test works well and is rather widely used in the testing of greenhouse, garden, and lawn soils. The extractant is 0.018 N acetic acid. The method has recently been compared with two others in experiments with cyclamen and poinsettia and found to be satisfactory.

Chlorides and Sulfates

Tests for these two anions are important where fertilizer rates are very high, as in greenhouses, and/or where irrigation water contains soluble salts. Dilute acetic acid is a very good extractant because the results seem to correlate well with tolerance figures for most of the crops grown in greenhouses and gardens.

Nitrogen

Nitrogen is the most important of all plant nutrients because it is used in large quantities by all plants (it is part of protein and of chlorophyl) and further because its form in the soil changes rapidly as a result of microorganism activity. A soil may furnish plenty of nitrogen for a corn crop during one season but fall far short the second season, either for a repeat of the same crop or a different one.

Farmers have become conscious of the need for controlling available nitrogen levels, so they are asking for available nitrogen soil tests. They wish to know how much nitrogen is needed at planting time, and how much may be needed later as topdressings or sidedressings.

Almost all the nitrogen in soil is in the form of organic matter. It must be mineralized by microorganism activity before plants can use it. It has often been said that a test for the end product (NO_3) means nothing as a measure of the ability of a soil to furnish sufficient nitrogen for a crop. **This is not true!** It depends entirely on when and how often the test is made. Greenhouse managers have used the test to advantage for years. They make the test at frequent intervals during the vegetative growth period. A nitrate extract below 20 ppm (0.018 N acetic acid extraction method as devised by Spurway) shows a need for more nitrogen fertilizer, whereas a test higher than 50 to 100 ppm, depending on the plant species, indicates an excess of available nitrogen, and perhaps a need for leaching.

Corn growers may successfully use the same test, and in pretty nearly the same manner. Sample the soil carefully when the corn is 6 to 10 in. high. Twenty cores with a sampling tube are not too many for one sample. Extract immediately or dry immediately for later extraction. If extraction is to be delayed for a few hours, add 3 ml of toluene per kilogram of soil. Seal the container and store in a refrigerator. This will greatly lessen but not entirely stop ammonification and nitrification. If the samples are to be held for more than 2 days, dry immediately in a ventilated oven at 55°C. Then store in sealed jars until time is available for testing.

Repeat weekly for 4 to 6 weeks. A blank test shows the crop is starving or getting *just enough* nitrogen. It would be wise to apply nitrogen fertilizer and be on the safe side. More research is necessary to establish the minimum level. It is probably lower than the 20 ppm used for greenhouse soils. A test as high as 20 ppm would surely indicate a sufficiency at that particular time, but it would still be wise to repeat the test in another week. It is not wise to draw conclusions during very dry weather. A good rain may greatly change nitrogen status within a few hours.

In drier regions where nitrate leaching is not a factor, measurement of total nitrate in the profile prior to planting a crop has been a useful guide to nitrogen fertilization.

The incubation test developed at the Iowa Experiment Station is designed to estimate the rate at which soil organic nitrogen is converted to the nitrate (NO_3) form.

Troug and others at the Wisconsin Station developed an ammonia-distillation method for determining nitrogen availability. The soil is digested for 5 min. with alkaline permanganate, which oxidizes a certain fraction of the organic matter. The strength of the permanganate has a bearing, of course, on the portion of organic matter oxidized.

Organic matter or total nitrogen determinations are indicative of the ability of a soil to furnish sufficient nitrogen for a crop, and, with reasonable attention paid to cropping history, the data are quite helpful in planning the nitrogen fertilizer program.

Easily oxidizable organic matter may be determined by use of potassium dicromate and sulfuric acid. The colors are read on a photoelectric colorimeter, with the use of a standard curve established by testing a set of soils of known organic matter. The results obtained by such organic-matter determinations must be calibrated against data obtained from field-plot soils where nitrogen status has been studied.

Organic-matter tests do not separate forms of organic matter to the extent that changes brought about by the growth of a single legume crop are detected. In this respect, they fall short of what is desired because the production of 1 year of a legume, such as sweetclover or alfalfa, makes a marked difference in the amount of nitrogen needed for top production of the following crop. This is expected because such a crop may add as much as 100 lb of nitrogen per acre and leguminous organic matter decomposes rapidly. Thus, it may be seen why it is so important to consider cropping history along with the organic-matter tests. Carson has favorably reported the results of nitrate nitrogen testing as a means of determining the need of a crop for fertilizer nitrogen. Prasad, Spiers, and Ravenwood (1981) obtained a very good quadratic relationship between soil values (R^2 nearly always greater than 0.9) and nitrogen uptake by chrysanthemum and verbena. The testing procedure evaluated for nitrogen was 1 : 1.5 water extraction procedure. Again we see that soil testing can make nutrient control possible if all procedures are proper. In the work of Prasad and colleagues it was the method of testing that was critical.

Soluble Salts

Expected yields and rates of fertilizer application are continually climbing. In the event of unbalanced applications relative to the demands of the crop, toxic soluble residues may occur, especially in regions of lesser rainfall. Also, growers of high-value crops, such as in vegetable gardens, and under glass often apply too much of all fertilizers. A soluble salt test will reveal such a condition. Donahue, Miller, and Shickluna (1983, p. 317) give good suggestions for determining soluble salts.

LIMITATIONS AND INTERPRETATIONS

Soil testing has brought nutrient control within our grasp. Improvements are yet needed in some of the methods, especially those for available nitrogen. Much can be gained from more uniformity in methods across the country, as interchange of samples and experiences would then be possible. The Soil Test Work Group of the National Soil Research Committee is endeavoring to bring this about.

Soil tests should be considered as tools for the evaluation of soil fertility. The immediate objective of most soil tests is to obtain information to make possible better selection of fertilizers and other soil amendments. Field plots are the most dependable soil tests. We cannot, however, conduct a plot experiment on every field; we cannot even conduct plot tests on all soil series. The next best thing then is to establish criteria whereby we may compare unknown soils with those on which we have conducted field trails. Chemical soil tests serve as a tool for the establishment of some of the criteria. The determination of texture is another tool. In fact, soil series identification is the first step in such a process and should be ahead of soil testing. It is our opinion that soil-test results mean very little unless soil series, or at least soil management group, can be identified.

Furthermore, soil-test results mean very little unless field-plot results, or at least greenhouse pot culture results, have preceded the soil testing. In some countries, (e.g., Thailand, Taiwan, and Iran) field plots have been established entirely for the purpose of calibrating soil tests.

Test Results are Relative

As already mentioned, soil testing is a "must" in deciding on grade and quantity of fertilizer or the need for such soil amendments as lime or sulfur. The pH test is a reliable guide for the use of certain nutrient elements. Zinc, boron, manganese, iron, and copper are likely to be needed on high lime soils, whereas calcium and molybdenum are more often needed on acid soils.

The actual level of a nutrient as shown by a soil test cannot, however, be directly used to calculate the pounds of that nutrient needed to grow a certain expected yield, and yet, too many people are attempting such calculations. For instance, suppose we aim for 600 bushels of potatoes per acre, a high but not unusual yield in many areas. Crop composition tables tell us the crop will take 180 lb of K_2O from the soil. If the soil test shows 180 lb, we might assume that fertilizer potash is not needed. This is absurd. We know from experience and experiments that the soil test would need to be at least twice 180 before we would omit potash from the fertilizer. In the first place, the test may not be an actual measure of what is available to the plant. Second, the plant roots do not contact every single particle of soil as does the soil-test extracting reagent.

Likewise, it is absurd to figure that if the crop needs 180 lb and the soil test indicates the presence of 80 lbs, we must supply 100 lb of fertilizer, especially if the fertilizer is applied broadcast.

How then should soil-test results be used? First, we should assume that soil testing is for the purpose of comparing the present nutrient status of similar soils where field experiments **have** been conducted and therefore where crop response to fertilizer is known. Starting with such an assumption, we then may say that a certain soil-test result with respect to potassium, say again the foregoing test of 80 lbs, indicates that this particular **series and phase** of soil will require perhaps 160 lb of K_2O as fertilizer to expect a yield of 600 bushels of potatoes an acre. The figure 160 is taken from some field experiments on that soil series where the soil contained 80 lbs potassium and required that amount of potash fertilizer to produce the desired yield.

Experiments have shown that a certain soil-test result does not call for the same ratio of fertilizer on all soils. In Michigan, a $1:4:2$ ratio fertilizer would be recommended for wheat on most clay loam soils testing low in both phosphorus and potassium, but a $1:4:4$ ratio would be recommended for wheat on sandy loams having identical test.

Nutrient Goals?

The question is commonly asked, "Should we attempt to apply enough fertilizer to raise the nutrient status of a soil to the levels that research workers have indicated as high or at least adequate for satisfactory production?" The answer is, "No, such applications are seldom practical." Rather, we should apply the quantity of fertilizer that experimental results show is advisable for soils in that series having the nutrient status indicated by the test results at hand. When such a course is followed, the nutrient levels in the soil will gradually rise, because it is impossible for the crops being grown to pick up all the nutrients in applied fertilizers. Residues build up and are detected in later tests. The build-up was very striking in the soil of the Ferden Experimental Field in Saginaw County, Michigan.

A crop sequence experiment was in progress on this Sims clay loam (Mollic Haplaquepts) for 16 years. The soils on all plots were carefully sampled when the experiment started in 1940. They were sampled again in 1956. At the time of the 1956 sampling, all plots had received 3200 lbs of 4-16-8 fertilizer per acre.

Soil-test results showed that even at these relatively low rates of fertilizer application over the 16-year period, both phosphorus and potassium levels did build up but at an uneven rate. It was time to change the fertilizer ratio. Soil testing is a sure guide to effective nutrient control.

TAKING SOIL SAMPLES

Soil-test results are worthless, may even be misleading, unless the samples tested actually represent the area to be treated. Too many soil samples are taken carelessly from two or three places in a field, perhaps without regard to soil type and previous soil use. An investment of several hundred dollars in fertilizer may later be based on the laboratory results.

The pint of soil sent to the laboratory is only about one 2-millionth part of an acre, and the teaspoonful actually used in the test is 1 percent of the pint. Obviously, the pint must be carefully selected from the acre furrow depth and the pint must be well mixed before the spoonful is taken.

When to Test

Soil Management Principle

Soil test should be repeated every 2 to 5 years.

Once every 2 to 5 years is often enough for reaction tests and tests for the mineral nutrients where hay, grains, and row crops are grown in sequence. Annual or biennial testing is recommended where high-value row crops such as beets, corn, cotton, potatoes, and tobacco are alternated with cover crops. Nitrate tests must be made several times during the season of rapid vegetative growth if they are to be used as a guide to nitrogen fertilizer use. Generally, very heavy rates of fertilizer application should be accompanied by more frequent soil testing.

Soils may be sampled at any season when moisture and temperature are favorable. Late summer and early fall is a good time in the Northern states, whereas in warmer regions, late fall and early winter is just as favorable. Too many farmers wait until early spring when they and the testing laboratories are rushed and soils are too wet for satisfactory sampling and mixing.

Equipment Needed

The soil probe or sampling tube is a very convenient sampling tool. It not only allows speed but it makes more accurate composite samples than any other tool on the market, as it may always be inserted to a marked depth and it removes the same amount of soil at each insertion.

Other sampling tools and a graphic story of how to use them are shown in Fig. 10-2.

Sizing up the Field

Before sampling a field, size it up for differences in soil characteristics. Consider its productivity, topography, texture, drainage, color of topsoil, and past management. If these features are uniform throughout the field, each composite sample of the topsoil can represent as much as 10 acres. However, if there is considerable variation in these features, and this is usually the case,

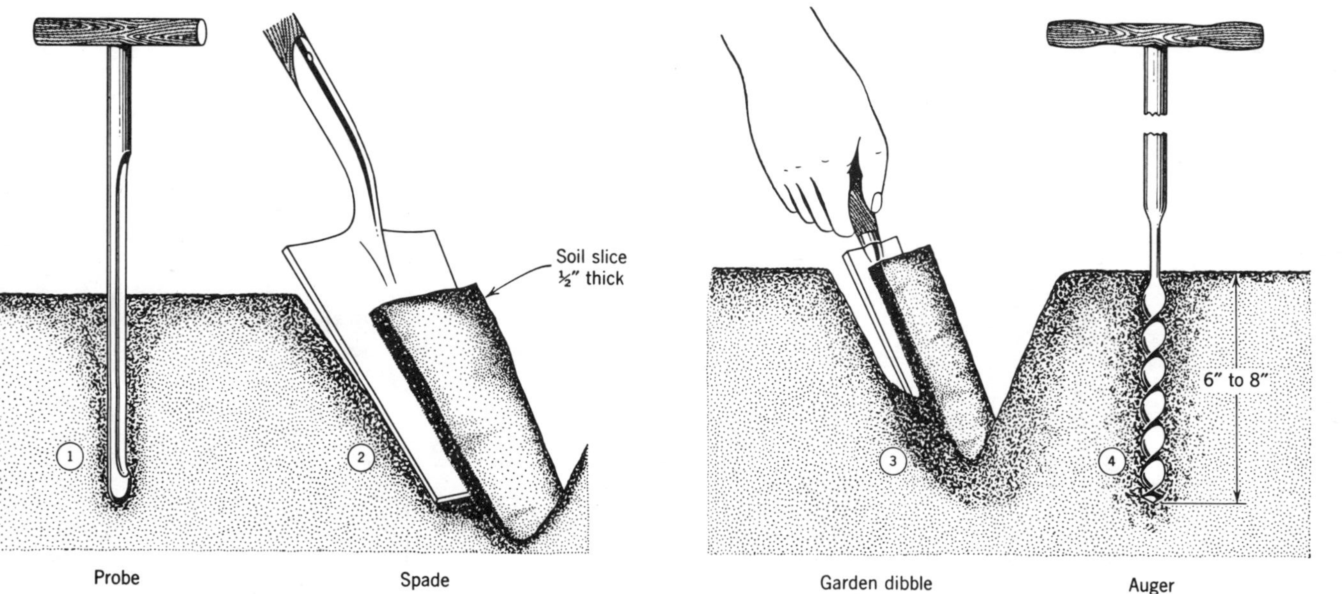

1. Sampling probe provides uniform soil cores—easy to use —saves time
2. A spade or shovel can do the job.
3. Use a narrow (1½-inch) garden dibble to take a slice of soil ½-inch thick.
4. A satisfactory soil auger may be made by welding a 1¼-inch or 1½ inch wood bit into a ½-inch pipe equipped with T-handle.

FIGURE 10-2 How to use sampling tools.

divide the field accordingly, perhaps as suggested in Fig. 10-3. Then take a composite sample from each predetermined area. A composite sample made up of samplings from two distinctly different areas is not representative of either area. In fact, one core of organic soil would spoil, for testing purposes, 19 cores of sandy soil.

Making the Composite Sample

From each predetermined soil area, prepare a composite sample by taking 20 samplings consisting of vertical columns of soil approximately 1 in. in diameter. The length of each column should correspond to plowing depth.

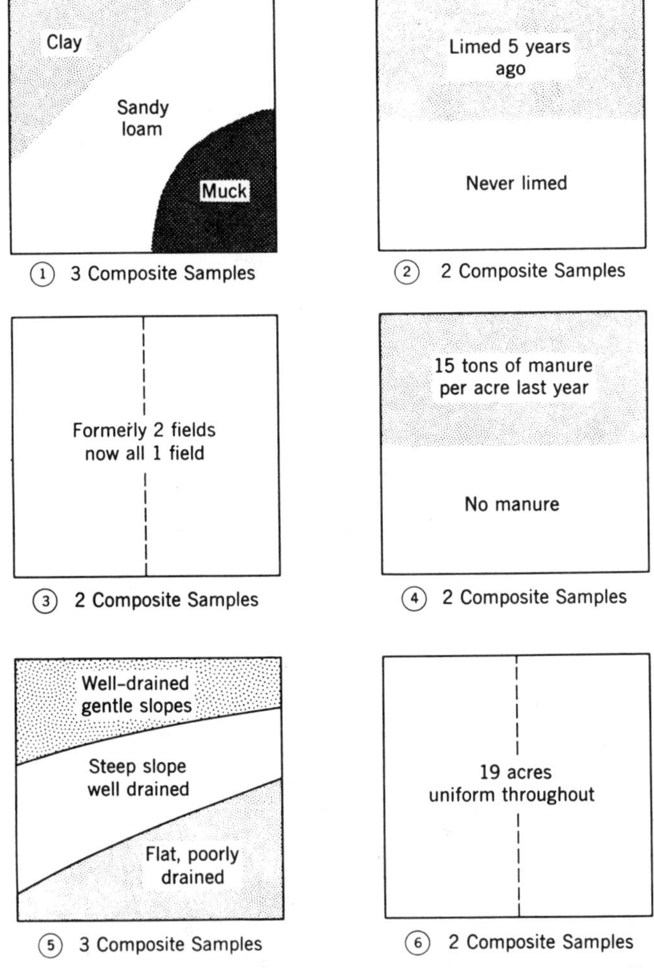

FIGURE 10-3 Take a number of composite samples corresponding to the distinctly different soil areas. Differences may be the result of variations in soil type, texture, topography, erosion, or past management. In uniform areas, limit a composite to 10 acres.

Avoid sampling unusual areas unless such locations are sampled and packaged separately: those close to gravel roads, in dead furrows, previous locations of lime or manure piles, burned muck areas, or from where timber piles have been burned or allowed to rot.

Subsoil samples taken at a depth of 18 to 24 in. will often aid in making recommendations. This is especially important with muck or peat soils. Such samples need not be a composite but should be taken from a location that appears to be average for the area.

Course to Follow

The samplings in each predetermined area may be taken in a course somewhat as suggested in Fig. 10-4. The exact course would vary with the shape of the area to make up the composite sample.

Mixing the Samplings

If the individual samplings are taken with a spade or dibble, reduce and place them in a clean pail until 20 are collected from the area. Then mix the soil throughly by hand and by rotating the pail at an angle of 45 degrees as shown in Fig. 10-5. Fill a clean pint container with the well-mixed soil and take or send it to the soil-testing laboratory. Be sure to label it with your name and field number.

Other Information Needed

As the medical diagnostician is aided by the patient's history of health, diseases, operations, and previous treatments, so is the agronomist, who is to make a soil management recommendation aided by the liming, fertilizing, and cropping history of the field where a soil sample was taken. Some of the information that may be helpful is as follows.

1. Legal description of the farm or exact distance from a well-known landmark such as town, church, or school.

2. Soil series name or soil-management group designation. Soil Conservation District Cooperators have this information in their farm plans.

3. Crop or crops on the land 2 years ago, last year, and this year. Next crop to be grown.

4. When the field was last limed, rate of application, and kind of liming material.

5. When the field was last manured and at what rate.

6. When was fertilizer last applied, what was the analysis, and at what rate was the material applied?

7. Is the land artificially drained?

8. Are there special problems specific to the area?

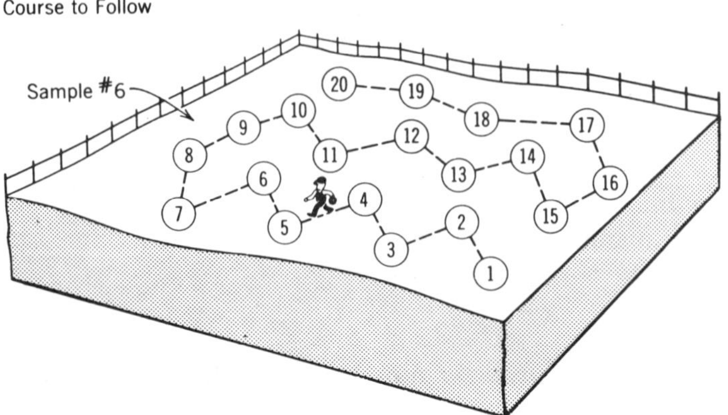

FIGURE 10-4 Course to follow in taking the samplings that make up a composite sample.

TESTING SERVICES

Testing services of some nature are available in all states. In most states, the Cooperative Extension Service is in some way tied in with soil testing. Michigan, for instance, has a central laboratory at Michigan State University. Farmers take samples to the County Extension office to be mailed to the laboratory. Results go back to the farmer through the county office.

Most soil samples have, in the past, been taken by the farmers. Lack of

FIGURE 10-5 A plastic pail should be used to collect, mix, and reduce the cores to make a dependable composite sample. Reducing is not necessary if a soil probe is used.

information about how to take samples and the human tendency to "put off" has been a bottleneck in the entire program. Many farmers would like to hire this service. To a limited extent, the Agricultural Stabilization Committees (ASC) have made this possible. Some state and county laboratories are starting to offer sampling services on the bases of a certain charge per acre with a minimum charge per call. In some areas such services include written reports, follow-up calls, and general soil-management assistance.

Commercial soil sampling and testing services are available in some states. The services vary in extent as do those offered by the Extension-supported laboratories. The more complete programs include soil mapping, sampling, testing, and recommendations, with follow-up calls to check results. The farmer should be sure that recommendations received from such laboratories are backed by adequate research performed on the soil in question.

U.S. readers are urged to consult local Extension or U.S. Soil Conservation Service Personnel for information regarding soil testing in the home county.

REFERENCES

Carson, P. L. (1980). *Recommended Potassium Test.* North Dakota Agricultural Experiment Station, North Dakota State University, Fargo. No. 499 (rev.).

Cook, R. L., and J. F. Davis (1957). Residual effect of fertilizer. *Advances in Agronomy* 9:206–216.

Cook, R. L., and C. E. Millar (1946). Some techniques which help to make greenhouse investigations comparable with field plot experiments. *Soil Sci. Soc. Am. Proc.* 11:298–304.

De Vries, M. P. C. (1980). How reliable are results of pot experiments? *Commun. Soil Sci. Plant Anal.* 11:895–902.

Donahue, R. L., R. W. Miller, and J. C. Shickluna (1983). *Soils, an Introduction to Soils and Plant Growth,* 5th ed. Prentice–Hall, Englewood Cliffs, NJ.

Graham, E. R. (1950). *Testing Missouri Soils.* University of Missouri Agricultural Experiment Station Circular 345.

Hue, N. V., and C. E. Evans (1983). A computer-assisted method for CEC estimation and quality control in a routine soil-test operation. *Commun Soil Sci. Plant Anal.* 14:655–667.

Jenny, H., J. Vlamis, and W. E. Martin (1950). Greenhouse assay of fertility of California soils. *Hilgardia* 20:1–8.

McLean, E. O. (1982). Chemical equilibrations with soil buffer systems as bases for future soil testing programs. *Commun. Soil Sci. Plant Anal.* 13:411–433.

McLean, E. O., T. G. Arscott, and M. A. Hannan (1983). An evaluation of soil solution and chemical extraction methods for assessing phosphorous availability to wheat plants grown in Ohio and Bangladesh soils. *Commun. Soil Sci. Plant Anal.* 14:1–13.

Mehlich, A., E. Truog, and E. B. Fred (1933). The Aspergillus niger method of measuring available potassium in soils. *Soil Sci.* 35:259–278.

Munson, R. D., and G. Stanford (1955). Predicting nitrogen fertilizer needs of Iowa soils: IV. Evaluation of nitrate production as a criterion of nitrogen availability. *Soil Sci. Am. Proc.* 19:464–468.

National Soil Test Work Group (1956). *North Carolina Agricultural Experiment Station Technical Bulletin* 121.

Prasad, M., T. M. Spiers, and I. C. Ravenwood (1981). Soil testing of horticultural substrates. *Commun. Soil Plant Anal.* 12:811–823.

Prasad, M., R. E. Widmer, and R. R. Marshall (1983). Soil testing of horticultural substrate for cyclamen and poinsettia. *Commun. Soil Sci. Plant Anal.* 14:553–573.

Sonneveld, A. J., J. van den Evde, and P. A. van Dijk (1974). Analysis of growing media by means of 1:1.5 volume extract. *Commun. Soil Sci. Plant Anal.* 5:183–202.

Spurway, C. H., and K. Lawton (1949). *Soil Testing, a Practical System of Soil Fertility Diagnosis.* Michigan State University Agricultural Experiment Technical Bulletin 132 (4th rev.).

Warncke, D. D., D. R. Christenson, and M. L. Vitosh, (1985). *Fertilizer Recommendations for Vegetable and Field Crops.* Bulletin E-550. Cooperative Extension Service, Michigan State University.

CHAPTER 11

Tissue Testing

Tissue testing is another diagnostic tool. As such, it is not new. Director A. D. Hall, Rothamsted Experiment Station, published a paper in volume 1 (1905) of the *Journal of Agricultural Science* entitled "The Analysis of the Soil by Means of the Plant." Similar work had in fact been going on since the days of Theodore de Saussure who, in 1804, showed that the composition of plant ash varied with the nature of the soil and with the age of the plants and that it consisted mostly of alkalis and phosphates. He showed further that plants grown from seed in water did not gain in ash, thus that ash constituents came from the soil.

J. B. Boussingault, starting field experiments on his farm in Alsace in 1834, weighed and analyzed manures and crops and drew up balance sheets to show which ingredients come from the soil and which from other sources such as manures or air and water. This work was soon followed by that of Justus von Liebig, father of agricultural chemistry, who advanced his "mineral theory," that plants grow by taking minerals from the soil or from manures and, therefore to maintain productivity, minerals must be replaced in the soil. He believed that from plant analysis it should be possible to decide what should be applied as fertilizer to bring about satisfactory growth. His "law of the minimum," still quoted by agriculturists, is that the growth of plants is limited by that plant nutrient element present in the smallest quantity, all others being present in adequate amounts.

Lawes and Gilbert, in their work at Rothamsted, spent much of their time attempting to prove or disprove the findings and ideas of Liebig. By 1855, they had published the idea that the compositon of the ash of plants is not proof of the **quantities** of the constituents required by the plants.

Director Hall, by studying plant composition data obtained at Rothamsted, attempted to show that a deficiency or excess of a particular constituent is indicated by the variation in the amount present in the ash from that in the ash of normal plants. He finally reached the conclusion that results of plant-ash analyses were not indicative of soil conditions and could not be used instead of soil analyses. He concluded further that such use of plant tests might be possible after "constants" had been determined for the quantities of nutrients in test plants occurring naturally on unmanured land. We are still attempting to establish such "constants," actually percentages of the nutrients, that may be considered normal.

TESTING THE ENTIRE DRIED PLANT

Plants within a species tend to have a definite chemical (mineral and organic) composition. If it were not so, tables like those published in Morrison's *Feeds and Feeding* could not be assembled. The figures in such tables are, of course, based on averages. Many factors, of which soil nutrient status is one, affect the composition of plants within a species. Other factors are soil moisture, light intensity and duration, temperature, variety of plant, plant organ or tissue analyzed, and stage of maturity. Since there are so many factors influencing the uptake of nutrients by a plant, we quickly see the fallacy of using plant composition as a measure of soil nutrient status unless careful steps are taken to control or measure the influencing factors other than those being studied.

Balance versus Intensity of Nutrients

Balance rather than intensity of nutrients is important as far as mineral composition of plants is concerned. Composition is most readily changed by eliminating or inducing deficiencies or excesses. Conversely, raising or lowering nutrient intensity will have very little effect on composition. An illustration of this came from the data in a dairy nutrition experiment conducted at the Michigan Agricultural Experiment Station.

A part of a depleted Michigan farm was limed and fertilized, and the remainder was left in a depleted condition. The crops grown on these two areas over a 10-year period were fed to two comparable herds of dairy cattle. Samples of the crops were analyzed for the mineral and certain organic constituents. The analytic results and the performance of the diary cows showed that mineral composition and the nutritive value of the feeds produced were not affected by the soil treatments.

A word about the nutrient status of the two areas explains the results and furnishes an excellent example of the importance of **balance** of nutrients. The nutrients in the depleted soil were at very low levels (intensity) but were in good balance. This was evidenced by the fact that crops grown on the area appeared normal in every way except in yield. Deficiency symptoms did not appear, and stands were always adequate. Yields were very low.

Fertilizers, including trace elements, and lime were applied to the other area of accordance with need as indicated by very carefully conducted soil tests. Again on this part of the farm, crops were always normal with no signs of deficiencies or excesses. Intensity was increased, but nutrient balance had been maintained. As a result, yields were greatly increased but plant composition remained essentially the same.

How Does Nutrient Balance Effect Composition?

Plants possess the ability of selective absorption to only a limited degree. Generally, the amount of any one nutrient in a plant is a measure of the amount in the soil, and of the effect of other factors including levels of other nutrients, rather than of the amount needed by the plant.

Soil Management Principle

A deficiency of one nutrient in a plant results in accumulation in the tissue of some or all of the other nutrients.

Starvation for a nutrient results in accumulation in the tissue of some or all of the other nutrients. The relative concentration of excess nutrients will be in the order of their concentrations in the soil in forms available to the plant. Then the application of the single deficient nutrient will result in increased growth and a decrease in the plant content of some or all of the other nutrients, both essential and nonessential, unless they were present in the soil in extremely large amounts. You may readily see then that if a plant starving, let us say, for nitrogen is analyzed for mineral content, all nutrients might be higher than in normal plants, even though the concentrations of those same nutrients in the soil were low.

Another soil might be very low in available phosphorus, low in easily available nitrogen, and medium in potassium, calcium, and magnesium. Corn on such soil might test just slightly below normal for phosphorus, probably normal for nitrogen, and very high for the other three mineral nutrients. It might be assumed then from plant composition that all nutrients are in sufficient supply, whereas if phosphorus had been applied as a fertilizer, perhaps the extra growth would have depleted the soil enough to lower the plant test below normal for **all** other nutrients.

Composition versus Yield

When a crop yield is seriously limited by a deficiency of a single nutrient in the soil and when that nutrient is applied in regularly increasing increments, yield and percentage of the nutrient in the plant follow a rather definite pattern with respect to each other. Up to the point where rate of nutrient application and yield have a straight-line relationship, percentage of the nutrient in the plant will remain essentially constant. Sometimes it actually drops slightly. This percentage has been called the "minimum percentage." The theoretical curves in Fig. 11-1 show this percentage to be 1.0, unchanged

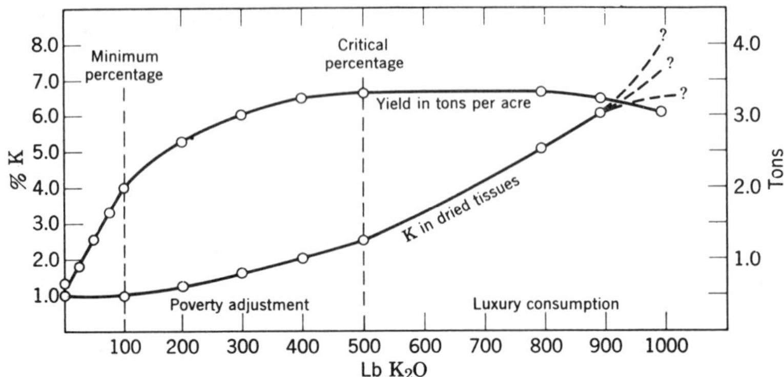

FIGURE 11-1 The effect of potash (K_2O) fertilizer on yield and potassium content of alfalfa (a hypothetical case).

up to the break in the yield curve where rate of yield increase falls. Percentage composition then starts to rise, gradually accelerating as rate of yield increase declines until the "critical percentage" is reached. The latter is defined as the percentage composition where yield does not increase further.

Still further applications of fertilizer cause a greater acceleration in the nutrient percentage rise, continuing at least to the nutrient rate that becomes toxic and where yield declines. After that rate is reached, percentage composition may continue upward at a greater or lesser rate, depending on the nature of the toxic effect. A content of the nutrient higher than the critical percentage is called **luxury consumption.**

Some agronomists claim that if you wish always to have top production it is necessary to maintain soil nutrient levels high enough to bring about luxury consumption. This is probably not true because if the nutrient balance is maintained, plants, owing to their tendency to maintain a constant level of cations and anions, cannot take in enough total nutrients to bring all levels above the critical percentages.

Cation and Anion Constancy

Many researchers have found that plants under uniform environmental conditions tend to take in a constant amount, on an equivalent basis, of nutrient cations including ammonium. Similarly, the sum of all anions tends to remain constant. Chief among the environmental conditions that affect the constant is climate.

As would be expected, the sum of the cations divided by the anions is more definitely a constant than either taken separately. You may readily see then how ammonium fertilizers may bring about a greater intake of anions. As a plant absorbs NH_4+, there will be a reduction in uptake of other cations, for instance $Mg^{2}+$. Total cations may, however, increase to some extent. In keeping with even a slight increase in cations, there must be an increase in anions. Whichever anion is most plentiful will be taken to keep the balance.

Experiments have shown that phosphate ion uptake may be increased in this manner. That the effect is not simply the result of nitrogen fertilization was proved by using nitrate rather than ammonium as a source of nitrogen.

Plant Parts and Organs and Stage of Maturity versus Composition

Plant parts vary greatly in chemical composition. Many data on stems versus leaves, grain versus straw, young leaves versus old leaves, tubers versus roots, and young versus mature plants are available in the literature. All such data emphasize the fallacy of attempting to analyze entire plants. It is impossible to collect physiologically similar plants from different soils or even from different plots on the same soil. Uncontrolled variables would cause more variations in composition than one might expect to result from the treatments or variables under investigation. Should we then throw away the plant test as a diagnostic tool? No, the thing to do is select physiologically similar tissues from the plants to be compared and be sure the parts selected are those that research has shown may give a good measure of the nutrient relationship between plant and soil. It is therefore necessary to determine the "critical percentages" for the particular tissues involved. This has been done for a number of crops, and because leaves have usually been selected, the term foliar analysis has commonly been used for such investigations.

FOLIAR ANALYSIS

The early work on the analysis of plants as a means of testing the soil has served one important function, that of pointing to some of the reasons why plants vary so much in chemical composition and why it is so difficult to accurately analyze entire plants.

Knowledge regarding "why" certain desired results are not obtained always leads to intensified efforts to refine the methods so the desired goals may be reached. This is exactly what has happened in plant analysis. Numerous investigators have worked long and hard during the past few decades on better ways of using plant testing as a diagnostic tool. Enough favorable results were obtained by the early analyzers of mature plants that researchers were sure the tool had possibilities.

It has been known for a long time that morphologically different tissues vary greatly in composition. Consider corn (maize), for instance. The difficulty involved in preparing an entire plant for testing is very great, especially after maturity is reached. Leaf blades, sheaths, stalks, kernels, cobs, and roots are all greatly different in mineral and organic composition. Even after thorough grinding, specific gravity differences between tissues from the different parts make it almost impossible to do a good job of mixing and sampling. Furthermore, uncontrolled environmental factors, such as those that affect pollination, may cause variations in grain percentage in the entire plant. This can have a marked effect on total composition. Likewise, soil aeration may affect root-top ratios. Total nitrogen in roots is low compared to that in the tops, thus the percentage in the total plant will be reduced by an increase in the percentage of root tissue.

The use of entire plants involves so much material that we are limited to a small number of plants to make a sample. Individual plant variations (genetic) are thus more important.

Age of tissue affects plant composition. Kwong and Boynton, (1959), for instance, tagged strawberry leaves on June 5. They selected only "youngest mature leaves" at the time of tagging. Samples of the tagged leaves were picked and analyzed at intervals of 9, 27, 45, 57, and 73 days after June 5. During that time, the potassium content of the petioles dropped from 3.5 percent to about 1.1 percent. The low point was reached on the forty-fifth day. During the entire 73 days, potassium in the leaflets dropped regularly from 1.5 percent to about 0.8 percent. You are reminded that these leaves were becoming physiologically older at each sampling.

During the same period, and at each sampling, Kwong and Boynton also selected samples of leaves for analysis that were the "youngest mature leaves," in other words, they were the same physiologic age as the tagged leaves were on June 5. Tests showed that the potassium content of "youngest mature leaves" did not vary significantly during the June 5 to October 8 period. This has tremendous significance to persons interested in using strawberry leaf analysis as a means of testing the soil.

Incidentally, these investigators found strawberry leaflets and petioles to be quite different in mineral composition. They found further that results from leaflet tests were more consistent than were those from petiole tests and that lateral and terminal leaflets were almost identical to each other — either could be used satisfactorily. They interestingly compared their strawberry results with results obtained from work on the grape, where it was shown that petioles were more satisfactory than leaves as tissue for diagnostic purposes. This is explained on the basis that grape leaves are so large that a sufficient number for a representative sample make a cumbersome amount for drying and grinding.

These workers discovered one other very important point, that "youngest mature leaves" were constant in composition over the season (June 5 to October 8) **only if they were physiologically active.** Nutrient deficiency and/or drought have an upsetting effect of causing rather marked changes. In the selection of leaves for diagnostic purposes, such influencing conditions should be avoided.

Testing Fruit Plants

Trees are so large that it is difficult to take soil samples that represent the area from which a tree gathers its mineral nutrients. An apple tree may spread over an area great enough to represent two or more distinct soil series. Roots may be deep in one soil, shallow in another. It is possible, but surely very difficult, to take samples to represent the entire volume of root-filled soil.

How can you test a plant the size of an apple tree? Certainly the tree cannot be ground and sampled. Even if it were possible, the fruitgrower would not like the tree destroyed. Certainly the way to proceed is to test small, representative portions without injuring the tree.

From the works of several investigators it may be concluded that fruit plants vary with respect to the tissue best suited for diagnostic work and that with some plants a certain tissue should be tested for one nutrient but a different tissue for other nutrients. Some of the information is shown in Table 11-1. Some tissues have been found to be less satisfactory for diagnostic

TABLE 11-1 Comparative Evaluation of Various Tissues for Diagnostic Work with Deciduous Fruit Plants

Element	Plant	Most desirable tissue	Other tissues studied
Nitrogen	Raspberry	Leaf blade	Leaf petiole
	Persimmon	Leaves, fine roots	Shoot, trunk, large roots
	Grape	Leaves	Shoot, trunk, large roots, fine roots
Nitrate (NO$_3$)	Grape	Petiole	Leaf blade
Phosphorus	Raspberry	Petiole	Leaf blade
	Persimmon	Fine roots	Leaves, shoot, trunk, large roots
	Grape	Petiole	Leaf blade, trunk, large roots, fine roots
Potassium	Apple	Leaf blade	Petiole
	Peach	Leaf blade	One-yr wood, two-yr wood, petiole, fruit minus pit, fruit pit
	Apple, plum, gooseberry, red current	Leaves	Bark, wood, fruit pulp
	Raspberry	Petiole	Leaf blade
	Persimmon	Leaves, fine roots	Shoots, trunk, large roots
	Gooseberry	Leaves, stems	Berries
	Grape	Petioles	Leaf blade, trunk, large roots, fine roots
Calcium	Raspberry	Petiole	Leaf blade
Magnesium	Raspberry	Leaf blade	Petiole
	Apple	Leaves	Shoot
Boron	Plum	Fruit flesh	Leaves

Source: From F. H. Emmert. Chemical analysis of tissue as a means of determining nutrient requirements of deciduous fruit plants. *Proc. Am. Soc. Hort. Sci.* 73:521–547, 1959.

work. In most instances, leaves, leaf blades, or petioles are the most satisfactory. It is generally conceded that for diagnostic work with tree fruits, the most satisfactory results may be obtained by using whole leaves from the middle portions of new shoots. Some researchers think it is better to select leaves only from spurs.

Time of Sampling — How May Leaves?

Diurnal changes in the composition of leaves are relatively small, so it may make little difference when they are removed from plants. But slight errors due to this effect may be minimized by always taking the samples in the morning. This, of course, gives the workers the advantage of having time to process the samples. Bowen (1978) recommends that samples should be taken within 3 hours after sunrise and always at the same hour. He seems not to be sure that diurnal changes are great enough to worry about, but it is well to be on the safe side. Furthermore, the moisture content of plant tissue decreases during the day. This may affect nutrient levels, so an early morning sampling is probably wise.

Many investigators have recommended the midportion of the growing season as the best time for taking fruit-tree samples. During early season, cell proliferation and development are progressing at a rapid rate, so leaf composition changes from day to day. Again, toward the end of the season, as leaf senescence approaches, mobile elements are moving back into the wood, so leaf composition again changes rapidly. The composition is the most stable between these two periods.

In making a survey of 47 Stayman Winesap and Delicious orchards in Ohio, Beattie and Ellenwood (1950) collected 25 leaves from each of four trees to make a sample. All leaves were from the middle of new shoots and were taken between July 12 and August 5. They were wiped with damp cheesecloth to remove spray residues and were dried at once in a forced draft oven at 70°C, then were ground to 40-mesh and stored for analysis.

Kenworthy (1953) made a similar survey of Michigan orchards. He collected for each sample 100 leaves from "middle of the current season's growth," taking two leaves from each shoot, from a total of four trees. Orchards in good vigor, without deficiency symptoms were selected. The four trees sampled in each orchard were selected as the average for the orchards. Leaves selected were free of visible insect, disease, or mechanical injury. The leaves were wiped with damp cheesecloth to remove spray residues, dried at 100°C, and ground for analysis.

In both these investigations, nitrogen was determined by the Kjeldahl method and other constitutents by conventional methods. Kenworthy used a spectrograph whereas Beattie and Ellenwood used a flame photometer and the semimicro methods of Peech. They determined boron by the quinalizarin-sulphuric acid method of Berger and Truog.

The procedures used by Kenworthy and Beattie and Ellenwood are quite respresentative of those followed by many other horticulturists. We feel safe in recommending them. However, a more modern method of analysis, use of the plasma emission spectrograph, makes it possible to measure many trace nutrients of low concentration.

Testing Herbaceous Crops

Again in testing herbaceous crops the problem of obtaining morphologically homologous tissue in all samples to be compared must be considered. The investigator should always keep in mind that the results from tissue tests should be considered as comparative rather than as absolute, whenever standards for comparison are available. Also, tissue tests should be used routinely with soil analysis to monitor the soil-plant environment.

Tests should be made as early in the season as possible, because results are generally better at that time and there is opportunity for later tests to be made for purposes of confirmation. In fact, a well-planned diagnostic program will include several tests on the same crop. An additional advantage of early testing is that corrective treatments, particularily for nitrogen, may yet benefit the crop if applied by sidedressings of fertilizer.

During very early growth, it is sometimes wise to select entire plants, but only when a dozen or more plants may be used for a single sample. Otherwise, most investigators recommend using the oldest leaf that will remain functional throughout the season. Where deficiencies are likely to appear, lower leaves are liable to become chlorotic or necrotic. These should be avoided. In work with corn, Tyner (1946) "used the sixth leaf from the ground, taken at the time of full silk and tassel with pollen shedding." He selected this time because then all corn varieties are morphologically and physiologically comparable. From this time on, all corn varieties mature in about the same number of days. Furthermore, this is a stage when nutrient use has been high for some time with demand on the soil probably at a maximum for the season. Bowen (1978) is in fairly good agreement with Tyner. He suggests "the entire leaf at the ear node (or immediately above or below it)." Leaves from 15 to 25 plants should constitute a sample. Incidently, Bowen includes directions for sampling a large number of field crops, fruits and nuts, vegetables, and ornamentals. For alfalfa, it has been found that it is better to test new growth for determining the boron status of a soil. The top 2 in. of the alfalfa plant should be used. Results are much better than when entire plants or lower portions of the plants are tested. Furthermore, plant tests are more valuable than soil tests as a tool for evaluating the need of alfalfa for boron applied as a fertilizer.

Godfrey, Aggrey, and Garber (1979), working with cassava in Sierra Leone, found that *bark* nutrient concentrations reflected soil fertility levels. They found that bark analyses proved to be a useful tool for predicting fertilizer recommendations in the management of soil for cassava production. The research on this very important food crop was done jointly by personnel of the Department of Agronomy, Njala University College, Njala,

Sierra Leone, and the Department of Statistics, University of California, Riverside, California.

Nutrients not readily translocated from old to new tissue (boron, iron, calcium, and manganese) must be available directly from the soil to the meristematic areas. It seems logical then that tests for these nutrients should be made on relatively young leaves or, as with alfalfa for boron, on the apical portions.

Critical Nutrient Level or Nutrient Balance

After tissue tests are made, the problem of interpretation arises. What do they mean? There are two ways in which they may be used. First, does the amount found in the sample equal the amount experiments have shown to be necessary for the desired growth of the plant in question? These so-called "critical nutrient levels" must be determined by controlled pot or field experiments or they may be established by surveys of broad areas including tests of plants from large numbers of fields on many types of soils. Combinations of the two methods have been used. This was true in the boron work conducted by Baker and Cook (1950), who found 20 ppm boron in the apical 2-in. portions of alfalfa to be the dividing line between deficiency and sufficiency. Perhaps the critical level for boron in alfalfa is not too different from that for the same nutrient in apples. Dowd (1949) tested leaves from three orchards in Michigan where he had applied borax to certain trees. Leaves from untreated trees in the three orchards contained 20.5, 21.3, and 23.4 ppm boron, respectively. Adjoining treated trees in the same orchards contained, respectively, 31.5, 36.8, and 30.3 ppm. Dowd knew that boron was needed in each orchard because of the increase in rate of terminal growth that resulted from borax treatment. In Ohio, boron tests as low as 16 ppm in a number of Stayman Winesap and Delicious orchards have been reported. Dowd noted deficiency symptoms on Delicious, Jonathan, McIntosh, Steeles Red, Baldwin, Red Astrachan, and Yellow Tranparent.

Tyner found the critical levels for N, P, and K in the sixth leaves of corn in West Virginia to be 2.90 percent, 0.295 percent, and 1.30 percent, respectively. For each change of 0.1 percent N, P or K from those values he obtained differences in yield amounting to 4.43 ± 1.1, 25.3 ± 0.67, and 2.05 ± 0.93 bushels per acre, respectively.

In a review entitled "Potassium in Plant Nutrition," Lawton and Cook (1954) reported critical potassium levels in the dried tissue of several crops as follows.

Sugar beets	1.0 −1.5 percent
Barley	0.5 −1.0 percent
Truck crops	0.75−1.5 percent
Apples	0.75−1.0 percent
Peaches	1.0 −1.5 percent
Citrus	0.35−1.0+ percent

As reported, these ranges are the result of variations obtained by different workers. Some of the variations are probably due to varying environmental conditions and are proof of the need for more work to establish narrow ranges or to furnish information regarding environmental conditions that change the critical levels.

The **survey method** has been used extensively by horticulturists in determining ranges in tissue composition within a certain geographic area. Averages of the results obtained from all samples have been considered as "normals" to be used in the same manner as critical levels determined by pot culture or field plot experiments. When tissue tests are used to supplement the use of other diagnostic tools, "normals" established in this manner may be very useful.

Beattie and Ellenwood (1950) tested leaves from 47 commercial orchards in Ohio. They determined averages and assumed that test results below average should be a warning to a grower that nutrient status is not quite what it should be and perhaps remedial measures should be considered. At least the grower should investigate further.

Kenworthy (1953) tested leaves from 124 commercial orchards, consisting of 52 apple, 26 peach, 36 cherry, and 10 pear orchards. His results, shown in Table 11-2, are in fairly good agreement with those obtained by Beattie and Ellenwood. Considered together, the results from the two investigations cover a large geographic area containing a wide range of soils.

The Michigan data (Table 11-2) reveal some very interesting facts. High nitrogen levels, especially in peach leaves, reflect liberal nitrogen applications. Only orchards in good vigor were selected. Potassium levels were essentially the same in the four kinds of fruits, whereas certain other constituents varied considerably. Magnesium, iron, copper, and boron were the most variable. These facts might be taken as an indication that more research is needed on the role of these nutrients in tree fruit production.

Kenworthy stresses, and rightly so, the importance of nutrient balance, already mentioned in these pages. He designed a very clever arrangement, called "Wheels of Nutrition," as a means of explaining nutrient balance to students and laypersons. As Kenworthy points out, it is difficult for such persons to visualize balance between such figures as nitrogen 2.41 percent, phosphorus 0.266 percent, potassium 1.58 percent, calcium 1.48 percent, magnesium 0.435 percent, manganese 110 ppm, iron 249 ppm, copper 18 ppm, and boron 55 ppm. (see figures for apples in Table 11-2) because of the wide variations, particularly when percentages and parts per million are cited together. It is somewhat like the necessity of a new bride learning that her larder is balanced when it contains 25 pounds of flour, 10 pounds of sugar, 1 pound of coffee, 4 ounces of tea, and perhaps only 1 ounce of pepper. Absolute amounts are not equal, but comparatively they are all 100 percent of need.

Kenworthy used the wagon wheel as an analogy. When all spokes are of the same length the wagon rolls easily, the tractor can pull a heavier load with less work, and production can be at a maximum. If one spoke is short, how-

TABLE 11-2 Composition of Leaves from Michigan Apple, Cherry, Peach, and Pear Orchards

	Apple	Cherry	Peach	Pear
Nitrogen (%)	2.41 ± 0.05	2.83 ± 0.07	3.98 ± 0.14	2.50 ± 0.08
Phosphorus (%)	0.266 ± 0.034	0.267 ± 0.034	0.238 ± 0.046	0.135 ± 0.052
Potassium (%)	1.58 ± 0.12	1.54 ± 0.12	1.55 ± 0.14	1.45 ± 0.16
Calcium (%)	1.48 ± 0.12	1.91 ± 0.13	1.95 ± 0.14	1.90 ± 0.13
Magnesium (%)	0.435 ± 0.022	0.740 ± 0.044	0.672 ± 0.61	0.397 ± 0.054
Manganese (ppm)	110 ± 11	114 ± 13	113 ± 40	133 ± 13
Iron (ppm)	249 ± 44	280 ± 52	191 ± 44	140 ± 56
Copper (ppm)	18 ± 4	55 ± 9	12 ± 3	54 ± 5
Boron (ppm)	55 ± 7	67 ± 9	53 ± 12	23 ± 7

Source: A. L. Kenworthy. *Nutritional Condition of Michigan Orchards: A Survey of Soil Analysis and Leaf Composition.* Michigan State University Agricultural Extension Station Technical Bulletin 237, 1953.

ever, the load will stall when it comes to rest with that spoke bearing the load, or more energy will be required to keep the wheel rolling.

Figure 11-2 shows the nutrition wheel with all spokes extending into the optimum range (middle white band). The center of the band represents the composition figures shown in Table 11-2. Thus it is possible to illustrate the composition of an individual leaf sample by properly adjusting the length of each nutrient spoke. A single spoke may terminate in the deficiency range (inner black), the hidden deficiency range (inner light), the optimum range (center white), the approaching excess range (outer light), or the excess range (outer black). The point of termination of each spoke in the wheel is deter-

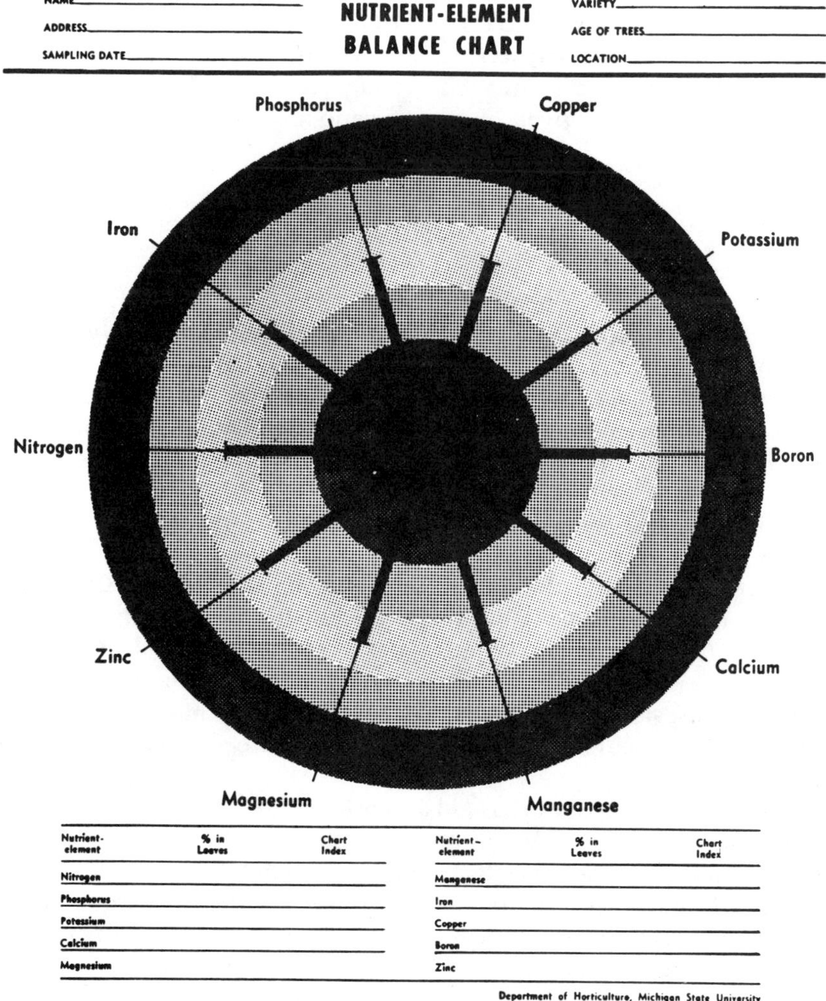

Nutrient-element	% in Leaves	Chart Index	Nutrient-element	% in Leaves	Chart Index
Nitrogen			Manganese		
Phosphorus			Iron		
Potassium			Copper		
Calcium			Boron		
Magnesium			Zinc		

FIGURE 11-2 Plywood chart showing the nutrient elements and a nutrition wheel of optimum balance in place. (From A. L. Kenworthy. Wheels of nutrition — A method of demonstrating nutrient balance. *Proc. Am. Soc. Hort. Sci.* 54:47–52, 1949.)

mined by calculating the percentage variation from the optimum values given in Table 11-2.

Beaufils (1973) has recommended the use of a "diagnosis and recommendation integrated system" (DRIS) in incorporating tissue analysis with other yield parameters. He emphasizes the importance of nutrient balance and suggests that maximum yield can never be obtained unless the proper balance or ratio of nutrients are maintained in plant tissue. This system has certain advantages in that nutrient ratios in plant tissue are rather constant throughout much of the growing season. This avoids the problem of absolute concentration of a plant nutrient changing with stage of growth. The calculations of indexes for each of the elements for the DRIS system is laborious unless a computer is used. Letzsch and Sumner (1983) have discussed the calculations of the indexes and offered for a nominal cost copies of a program to do the computation. Norms (i.e., indexes for maximum yields) have been developed for corn, soybeans, wheat, sugarcane, potatoes, and sorghum; references to these are given by Sumner and associates (1983).

TESTING GREEN PLANTS

Green-tissue tests play a considerably different diagnostic role than do the dry-tissue tests already discussed. They indicate, in a roughly quantitative way, whether or not a plant is getting sufficient nutrients to satisfy its needs **at the moment the test is made.** The tests are for soluble nutrients in the plant, and they are possible because plants take in more nutrients, when supplies are ample, than they currently assimilate. This allows an accumulation in the plant. This tendency to store nutrients is spoken of as "luxury consumption." It is illustrated by Fig. 11-1.

When the test for a certain nutrient, nitrate for instance, is strongly positive, the conclusion is that the supply of nitrate in the soil is, at that time, ample for the needs of the particular plant or plants tested. On the other hand, when the test is blank, we must conclude that the plant is assimilating the nitrate (building it into plant tissue) as fast as it takes it in and that all stored nitrate is used. The conclusion then is that either the supply of nitrate in the soil is just sufficient for the needs of the plant or that it is inadequate. A soil test or the appearance of the plant may indicate which of the two conditions exists.

Tests on green tissue are rapid and relatively easy to perform. They can be used to verify deficiency symptoms or to reveal "hidden hunger" before symptoms appear. The tests may be made in the field, so they are very useful to those working with farmers in an advisory capacity. Nitrate tests may be used to determine the need for nitrogen sidedressings or topdressings.

Methods of Testing Green Tissue

Two methods of green-tissue testing have been widely used. The Purdue method (Thorton, Conner, and Fraser, 1934) is suitable and is recommended for those who prefer to use it. In the opinion of some investigators, it has the advantage of having been designed for use with color charts provided with the testing outfit.

The Spurway Simplex soil-testing outfit may be used for testing green tissue. The reagents and glassware provided for testing soil serve equally well for the testing of tissue. The tests are simple, but there is a possibility of error in interpretation. Perhaps a word of warning is advisable regarding one common error.

Plants are generally found to be lower, relatively, in one element than in all others. That element, of course, is the one lowest in the soil in available form and is generally spoken of as the "first limiting factor" in crop production. Say, for instance, that on some particular soil the first limiting factor is nitrogen and the plants are showing symptoms of nitrogen starvation. The tissue test for nitrate nitrogen will be blank, but the tests for phosphorus and potassium are likely to be even higher than in normal rapidly growing plants. The tests do not mean, though, that the soil contains sufficient phosphorus and potassium for a normal crop, but only that it contains enough for a crop stunted by a shortage of nitrogen. As soon as nitrogen is applied under such conditions, growth is stimulated, and often another element, perhaps phosphorus, becomes the next limiting factor.

Manganese, an element often deficient in alkaline soils, may be determined in green leaves, but the method requires laboratory facilities and so is not considered a quick test that may be performed with the Simplex outfit. By a process of elimination, and because symptoms of manganese deficiency are rather specific and easily distinguished, it is usually possible to avoid the necessity of testing for manganese.

Directions for Making Green-Tissue Tests with Spurway Simplex Soil Testing Kit

Although Spurway soil-testing kits may not be available to you, the reagents and glassware can be found in most school chemistry laboratories or are easily purchased from chemical supply houses. Directions for making the tests may therefore be in order. Furthermore, it seems like an appropriate way of explaining the general principles of the tests. Reagents are referred to by number. Directions for preparing them may be found at the end of this chapter. Incidentally, the numbers are the same as those used in the soil-test bulletin (see Spurway and Lawton, 1949).

Tests should be made on fresh tissue. By keeping plants moist and in a refrigerator, they may be stored for a day or two. Dry plants cannot be used. Leaf blades contain too much pigment, so it is better to use thinly sliced sections of stems or leaf petioles. The mobility of the element within the plant should be considered in deciding what portions of the plant to use.

Nitrogen

Place thinly cut sections of plant tissue in the depression of a spot plate. Apply several drops of reagent 2. With large pieces of plant tissue, like those obtained from the stalk of a corn plant, simply drop the reagent on a freshly cut surface. If nitrate is present in the tissue, the reagent will turn blue. If the blue color is faint and slow to form, the test is low and the plant is about to run out of nitrate. If the blue color develops quickly and is very dark, the test is very high and the nitrate supply of the plant, as of that date, is sufficient. One should not confuse a brown color with a blue color. Brown indicates very low or blank nitrate.

Phosphorus

Place 1 cc of thinly sliced plant tissue in a small glass vial. Add 5 cc of distilled water and shake for 1 min. Add 5 drops of reagent 3, shake, and add a speck of stannous chloride, perhaps about the size of a pinhead. Shake and wait about 2 min., not longer than 5 min., for the color to develop. It is well to add a little more stannous chloride to see whether the color deepens. Too much stannous chloride causes the color to be green instead of blue. A deep blue color means a high test and is an indication that the plant was obtaining sufficient phosphorus at the time the sample was taken. A very light blue indicates that the phosphorus supply is not sufficient.

Potassium

Place 1 cc of thinly sliced plant tissue is a small glass vial. Add 5 cc of cold (below 20°C) distilled water. Shake for 1 min. Add 3 drops of reagent 5, shake, and add 2 cc of reagent 6. The reagent 6 should be cold, unless the distilled water is cold enough so the mixture of distilled water and alcohol (reagent 6) will be below 20°C. Shake and estimate the density of the precipitate by holding a heavy black line on white paper back of the vial. If the line is not visible through the solution, the test may be considered high and the indication is that the plant was obtaining sufficient potassium at the time the sample was taken. If the line is sharply distinct, the test is blank. This, of course, indicates that the plant was starved for potassium at the time the sample was taken. If the black line is easily visible but blurred in outline, the test is low. A faintly visible line through the solution may be considered an indication of a medium supply of potassium.

Portion of Plant to be Used in Tests

In deciding what portion of a plant to use for green-tissue tests, it should be remembered that all three elements, nitrogen, phosphorus, and potassium, are readily translocated from old to new tissue. For that reason, it is sometimes desirable to test both old and new tissue, although usually it is sufficient to test the old tissue. If the test there is high, it is safe to assume that a high test will also be obtained on the new tissue. If the old tissue tests low, however, there is a chance that new tissue may test high. Such results indicate that the plant is just on the verge of becoming deficient in that nutrient element.

The parts of the plants most suitable for green-tissue testing for a few crops are as follows:

Beets	Leaf petioles	Soybeans	Leaf petioles
Corn	Leaf sheath[a]	Potatoes	Leaf petioles or stem
Grains	Stem	Tomatoes	Lower leaf petioles
Alfalfa	Stem		
Beans	Leaf petioles	Geranium	Leaf petioles

[a] Use the stalk in the case of very young plants.

Importance of Comparative Tests

> ### *Soil Management Principle*
>
> Since "quick test" are somewhat relative, it is always advisable to make test on deficient and normal plants at the same time.

Results obtained from green-tissue tests are most useful when they are expressed on a comparative basis. Plants vary somewhat in the quantity of nutrient they may contain at the time when they show indications of being starved. For this reason, it is always advisable to make tests on deficient and normal plants at the same time. In other words, if a test is to be made for phosphorus on a plant where phosphorus deficiency is suspected, test at the same time a plant known to contain plenty of phosphorus. This is usually possible by taking plants from other fields or from treated plots. After many tests have been made, such a comparison becomes less important.

Directions for Making Green Tissue Test Reagents

Nitrate Reagent

No. 2-Diphenylamine solution.

Dissolve 0.03 g diphenylamine in 25 cc of pure, nitrate free sulfuric acid. This solution is strongly corrosive and must not contact hands or clothing. A pink, brown, or blue color indicates that it is unfit for use. Use plastic screw-top dropper bottles.

Phosphorus Reagents

No. 3-Molybdate solution.

Dissolve 5 g of ammonium molybdate, free from arsenic and phosphorus, in 50 cc of distilled water, warming gently to hasten solution. If the solution is turbid, filter. Pour this solution slowly, with stirring, into 50 cc of pure nitric acid, then add 100 cc more of distilled water. Again, use plastic screw-top dropper bottles.

No. 4-Stannous chloride.

Use the pure, dry salt as it comes from the chemical supply house.

Potassium Reagents

No. 5-Sodium cobalti-nitrite solution.

Prepare solution A as follows: Dissolve 25 g of cobaltous nitrate in 50 cc of distilled water in a 500-cc Erlenmayer flask and add 12.5 cc of pure acetic acid. Now prepare solution B as follows. Dissolve 120 g of sodium nitrite in 180 cc of distilled water, making a volume of about 220 cc. Add 210 cc of solution B to all of solution A (under a laboratory hood or out-of-doors). The poisonous, reddish brown fumes of nitric oxide are then removed from the solution by means of an aspirating bottle and a suction filter pump. The gas must be removed completely from the solution by drawing air through it for 3 or 4 hours. Filter the reagent before use, and keep in a cool place. Use the plastic screw-top dropper bottles.

No. 6-Ethyl alcohol, 95 percent.

Use pure, without adulteration.

Note: All reagents should be prepared and stored in resistant laboratory glassware. Chemical should be special grade suitable for microtesting.

You are again warned that reagent No. 2 is strongly corrosive. It will burn hands and destroy clothing. In case of contact, wash at once with plenty of water.

REFERENCES

Baker, A. S., and R. L. Cook (1956). Need of boron fertilization for alfalfa in Michigan and methods of determining this need. *Agron, J.* 48:564–568.

Bear, F. E. (1950). Cation and anion relationships in plants and their bearing on crop quality. *Agron J.* 42:176–178.

Bear, F. E., and A. L. Prince (1945). Cation-equivalent constancy in alfalfa. *Agron J.* 37:217–222.

Beattie, J. W., and C. W. Ellenwood (1950). A survey of the nutrient status of Ohio apple trees. *Proc. Am. Soc. Hort. Sci.* 55:47–50.

Beaufils, E. R. (1973). *Diagnosis and Recommendation Integrated System (DRIS).* Soil Science Bulletin 1, University of Natal, Pietermaritzburg, South Africa.

Bowen, J. E. (1978). Plant tissue analysis: Costly errors to avoid. *Crops and Soils.* 31:6–11.

Dowd, O. J. (1949). Observations on boron deficiency in apples in southwestern Michigan. *Proc. Am. Soc. Hort. Sci.* 53:23–25.

Duncan, C. W. (1955). Effects of fertilizer practices on plant composition. Effect of fertilizer practices on nutritive value of feed for four successive generations of dairy cows II. Nutrition of Plants, Animals, Man Symposium, College of Agriculture, Michigan State University.

Emmert, F. H. (1959). Chemical analysis of tissue as a means of determining nutrient requirements of deciduous fruit plants. *Proc. Am. Soc. Hort. Sci.* 73:521–547.

Godfrey, W., S. Aggrey, and M. J. Garber (1979). Bark analyses as a guide to cassava nutrition in Sierra Leone. *Commun. Soil Sci. Plant Anal* 10:1079–1097.

Jones, J. B., Jr. (1970). Soil and plant analysis of soil extracts and plant tissue ash by plasma emission spectroscopy. *Commun. Soil Sci. Plant Anal.* 1:263–272.

Jones, J. B., Jr. (1977). Elemental analysis of soil extracts and plant tissue ash by plasma emission spectroscopy. *Comm. Soil Sci. Plant Anal.* 8(4):349–365.

Kenworthy, A. L. (1949). Wheels of nutrition — A method of demonstrating nutrient balance. *Proc. Am. Soc. Hort. Sci* 54:47–52.

Kenworthy, A. L. (1953). *Nutritional Condition of Michigan Orchards: A Survey of Soil Analyses and Leaf Composition.* Michigan State University Agricultural Experiment Technical Bulletin 237.

Kwong, S. S., and D. Boynton (1959). Time of sampling, leaf age, and leaf fraction as factors influencing the concentrations of nutrient elements in strawberry leaves. *Proc. Am. Soc. Hort Sci.* 73:168–173.

Lawton, K., and R. L. Cook (1954). Potassium in plant nutrition. *Advances Agron.* 6:253–303.

Letzsch, W. S., and M. E. Sumner (1983). Computer program for calculating DRIS indices. *Commun. Soil Plant Anal.* 14:811–815.

Lucas, R. E., G. D. Searseth, and D. H. Sieling (1942). *Soil Fertility Level as it Influences Plant Nutrient Composition and Consumption.*" Purdue University Agricultural Experiment Station Bulletin 468.

Macy, P. (1936). The quantitative mineral nutrient requirements of plants. *Plant Physiology* 11:749–764.

Spurway, C. H., and K. Lawton (1949). *Soil Testing, a Practical System of Soil Fertility Diagnosis.* Michigan State University Agricultural Experiment Station Technical Bulletin 132(4th rev.).

Sumner, M. E., R. B. Reneau, Jr., E. E. Schulte, and J. O. Arogun (1983). Foliar diagnostic norms for sorghum. *Commun Soil Plant Anal.* 14:817–825.

Thomas, W. (1945). Present status of diagnosis of mineral requirements of plants by means of leaf analysis. *Soil Sci.* 59:353–374.

Thorton, S. F., S. D. Conner, and R. R. Fraser (1934). *The Use of Rapid Chemical Tests on Soils and Plants as Aids in Determining Fertilizer Needs.* Purdue University Agricultural Experiment Station Circular 204.

Tyner, E. H. (1946). The relation of corn yields to leaf nitrogen, phosphorus, and potassium content. *Proc. Soil Sci. Soc. Am.* 11:317–323.

Plant Symptoms of Nutrient Deficiency

The most recent method of measuring the productive power of a soil is that of using the appearance of the plant as an indication of the supply of nutrients in the soil. Underfed plants often grow slowly and develop abnormally in much the same manner as do starved animals. Nutrient deficiencies may result in off-color leaves, abnormally shaped leaves or stems, and sometimes in actual disintegration of various parts of the plants, including the roots. Care must be taken, of course, to avoid confusion between nutrient deficiencies and irregularities brought about by other causes.

Nutrient shortage in the soil may be intensified by such abnormal weather conditions as drought, excessive moisture, or unseasonably low temperatures. In other words, the nutrients may be present in the soil in quantities sufficient for normal growth when conditions are ideal but not sufficient during periods of adverse weather conditions.

Fortunately, symptoms of nutrient deficiency tend to be similar in different plants. If this were not so, it would be rather difficult to become familiar with the symptoms as they occur in a variety of crops. Plants within a family are similar in their nutrient needs. Members of the pea family, for instance, are very sensitive to a deficiency of potassium, whereas those of the goosefoot and mustard families have a marked need for boron. Furthermore, certain deficiencies occur only, or at least more often, under certain soil conditions. For instance, symptoms of boron deficiency are not likely to be found in plants growing on acid soils. The same thing is true with manganese deficiency. Also, plants deficient in manganese are more likely to be found on outwash sandy soils than on soils containing a large percentage of clay, especially if they are of glacial till origin.

Usually soils are lower, relatively, in one element than in all others. The influence of that first limiting factor then dominates the metabolism of the plant, and it shows the characteristic deficiency symptoms. The investigator should keep in mind, of course, that once the first limiting factor is eliminated by addition of the corrective fertilizer, a second limiting factor may come into play.

Since most cases of nutrient deficiency sufficiently severe to cause an abnormal appearance can be traced to a shortage of a single element, it seems logical to consider the symptoms by individual elements, always remembering, of course, the possibility that the abnormal appearance may be entirely due to disease or insect injury. Cook and Millar (1955), and Follet, Murphy, and Donahue (1981) have discussed the relationship between tissue and soil testing and diagnosis from deficiency symptoms.

Soil Management Principle

Careful consideration of the mobility of each nutrient in a plant will aid in distinguishing plant deficiency symptoms.

NITROGEN DEFICIENCY

Nitrogen exists in the soil largely as a constituent of organic matter. Through decomposition of the organic matter by soil organisms, the nitrogen is changed into forms available to plants, ammonium (NH_4) and nitrate (NO_3).

Since the decomposition is caused by living organisms, nitrogen availability in soil is influenced by temperature, moisture, and aeration. Thus, the supply is extremely variable.

This explains why plants sometimes are found to be starving for nitrogen on soil high in organic matter. This occurs generally during cold, wet seasons, sometimes during dry periods, sometimes on the most fertile soils.

On farms where nitrogen starvation is common, even during seasons that favor the activity of soil microorganisms, attention should be directed first to the organic content of the soil. If organic matter is found depleted, it should be replenished by additions of stable manures and/or by the production of green manures, preferably leguminous in nature.

Nitrogen starvation may occur on soils throughout the range of acidity and alkalinity. Strongly acid soils are more likely, however, to be low in easily decomposable organic matter, because legumes do not thrive on such soils and the nitrifying bacteria may not function as efficiently as they do on soils well supplied with lime.

Soil Management Principle

A deficiency of nitrogen causes stunted growth and a loss of chlorophyl.

Nitrogen particularly affects the vegetative growth of a plant. When deficiency occurs, the leaves first become light green and gradually yellow. The oldest leaves on the plant, those nearest the ground on an upright plant, are first affected. After the leaves become yellow, they die. Even after the oldest leaves are dead from lack of nitrogen, the new leaves may be green. This shows that when the nitrogen supply in the soil is low, the new growing tissue has priority on what is taken into the plant. Eventually, even the new leaves lose their chlorophyl, growth slows down, and in extreme cases the plant dies.

Corn

The "firing" of corn is probably the most widely recognized symptom of nitrogen starvation. For years farmers considered this condition to be the result of excessive drought, and it is true that dry weather accentuates the deficiency. When soil becomes dry, conditions are less favorable for nitrification, and the nitrate present is transferred with greater difficulty from the soil or the soil solution to the plant. Also, during such a period, the rate of root penetration decreases, and there is very little opportunity for the formation of the nutrient-absorbing root hairs.

When drought conditions become sufficiently severe, corn will, of course, dry and die regardless of the quantity of nitrogen available. The leaves do not turn yellow in the pattern so commonly considered as dry-weather firing, however, unless there is a deficiency of nitrogen accompanying the drought. As the supply of available soil nitrogen becomes exhausted or reaches a level too low for the demands of the corn plant, the entire plant becomes light green in color and the oldest leaf, the one nearest the ground, starts to turn yellow at the tip. The yellowing then proceeds along the leaf toward the stalk, moving faster in the midrib area and more slowly along the edges. This makes it possible to distinguish nitrogen deficiency from potassium deficiency, as in the latter the yellowing progresses faster along the edges of the leaf whereas the midrib remains green. The difference in the pattern of the yellowing is illustrated in Fig. 12-1. As the supply of available nitrogen in the soil continues to decrease, the second and then the third leaves are affected as the yellowing moves up the stalk and the sheaths of the lower leaves turn red. By the time the third leaf shows distinct signs of nitrogen deficiency, the first leaf has become entirely yellow and almost dead.

Oats and Wheat

The symptoms of nitrogen deficiency in oats and wheat are exactly the same as in corn. At first, the plants develop a light green color, then the lower leaves turn yellow, the yellowing on each leaf starting at the tip. It is important to remember that the yellowing starts at the tip, as that point will be referred to again in differentiating between nitrogen and manganese deficiency.

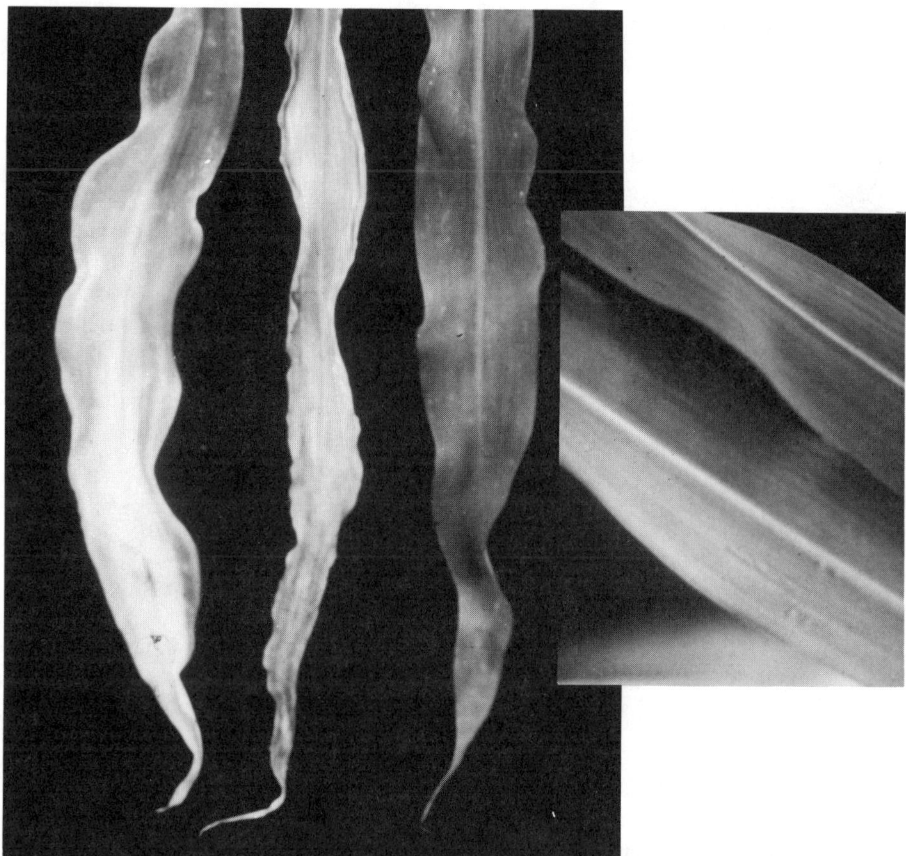

FIGURE 12-1 Nitrogen-, potassium-, and phosphorus-deficient corn leaves. Consider first the three in a group. The right leaf was deficient in nitrogen. The yellowing started at the tip and proceeded down the midrib, whereas in the potassium-deficient leaf in the center, the yellowing started at the tip and moved down the edges of the leaf, leaving the midrib area green. The left leaf was normal. The inset shows a section of normal and phosphorus-deficient leaves. In the field, a reddish purple edge is seen on an otherwise dark-green leaf.

Soybeans and White Beans

Even though the soybean is a leguminous plant, it is sometimes found to be suffering from nitrogen starvation. This is often true on poorly drained, heavy soils where the bacteria are not able to function normally because of insufficient air. This condition is probably due as much to the inactivity of the nitrifying bacteria as to the failure of the legume bacteria to perform their task of changing atmospheric nitrogen to combined nitrogen. As soybeans become deficient in nitrogen, the older leaves gradually change in color to a light green and finally to yellow. The change in color is uniform over the entire leaf. In Fig. 12-2, a nitrogen-deficient leaf is compared with one from a normal plant. The nitrogen-deficient leaf came from a poorly drained, clay loam soil.

FIGURE 12-2 Nitrogen deficiency is common in soybeans. The yellowing of the left, nitrogen-deficient leaf is uniform over the entire leaf. The veins do not remain green as they do where there is a lack of manganese in the soil.

The symptoms on white beans are almost identical—a gradual yellowing of the lower leaves and eventually of the entire plant if the deficiency becomes sufficiently serious. It is very common with beans when the weather is cold and wet during early growth. On many fields, beans become deficient in nitrogen at about the time they start to set pods. This is especially true on what might be considered the poor bean soils. The green-tissue test for nitrate works very well on beans and soybeans.

Sugar Beets

Sugar beets do their best on soils well supplied with organic matter, partly because they require large quantities of nitrogen. From midsummer until the end of the growing season, symptoms of nitrogen deficiency in sugar beets are very common. As the nitrate supply in the soil becomes depleted, the leaves become light green and eventually yellow as the deficiency becomes more serious. The loss of chlorophyl and the resultant yellowing occurs uniformly over the entire leaf. This is an important point, since in the yellowing caused by manganese deficiency the veins and the leaf tissue close to the veins remain green, which gives the leaf a mottled appearance. Another characteristic of nitrogen starvation is the direction in which the leaves grow out from the crown. Instead of standing erect, as they do in a normal plant, they grow out in a horizontal position to give the beet the appearance of having been stepped on.

The green-tissue test is very useful in checking the supply of nitrogen in a sugar-beet soil. When the crop is growing rapidly, a soil test for nitrate is always low, because the plants take it up as fast as it is formed in the soil. Still, there might be enough to satisfy the requirements of the plants. In other words, under those conditions the soil test might not be a true measure of the power of the soil to produce nitrates. A test of the green tissue, however, really shows whether or not the plant, at that particular time, are getting sufficient nitrogen.

Tomatoes

As tomatoes become deficient in nitrogen, the chlorophyl gradually disappears from the leaves. This action is more rapid in the older leaves but may soon affect the entire plant. The new leaves, however, are always darker in color than are the older leaves, because the nitrogen is translocated from the old to the new leaves when the supply in the soil is inadequate for the needs of the plant.

As the chlorophyl fades and the leaves become gradually lighter green to yellow, the veins become purple in color (see Fig. 12-7). Where the deficiency becomes very severe, the older leaves become entirely yellow and the purple color becomes very pronounced even to the smallest veins. The network of the veins after the purple color develops is visible from both sides of the leaf but is more pronounced on the lower side. The purple color extends also to the petiole of the leaf and is somewhat noticeable on the main stem of the plant.

We should be careful to avoid confusion between the purple color of petioles and veins caused by nitrogen deficiency and the purple of the underside of the leaves caused by phosphorus deficiency. In the latter case, the entire undersurface, including the veins, becomes purple.

Cucumbers

When cucumbers are deficient in nitrogen, the oldest leaves lose their chlorophyl first. The yellowing then progresses from leaf to leaf toward the growing tip. It is important to remember that the yellowing is uniform over the entire area of each leaf. The veins do not remain green. The oldest leaves die first, and death of the leaves then follows in the order in which they turned yellow.

Cineraria

Potted plants require large quantities of nitrogen. When the element becomes deficient, the lower leaves gradually turn yellow over their entire area. The yellowing is not confined to leaf edges as is true where the starvation is for potassium. The contrast is shown in Fig. 12-11. The yellowing gradually works up the plant until finally the entire plant has become very light green to yellow.

Coleus

The coleus plant, grown for its beautiful foliage, may be varied in color by regulating the supply of nitrogen. Plants were grown in an experiment to determine the effect of various organic materials on growth and appearance. Four pots were filled with Miami silt loam soil (Typic Hapludalfs) in which had been incorporated a large quantity of chopped straw. The plants made less growth than did those grown on soil containing manure, and their color was a much brigher red. Green-tissue tests showed the bright red plants to be devoid of nitrate nitrogen. Apparently the bacteria had used so much nitrogen in decomposing the straw that little was left available to the coleus.

At the termination of the experiment, two of the four nitrogen-starved plants were supplied with soluble nitrogen in the form of ammonium nitrate. Their color soon became a very deep red. The contrast between the two colors was very striking. Tissue tests at that time showed the **deep** red plants to be very high in nitrate nitrogen. The **bright** red plants were entirely lacking in nitrate.

Begonia

The red begonia is a striking example of a plant that varies in foliage color according to soil nitrogen levels. When soil nitrogen is high, leaves and stems are dark green. Under deficient nitrogen levels, leaf edges and petioles turn red. The color is so striking that the entire appearance of the plant is changed. Some customers prefer the nitrogen-starved plants to those of normal color. It is necessary, however, to feed the plants sufficient nitrogen to obtain satisfactory size and shape before the period of starvation.

White begonias do not develop anthocyanin pigment, so they simply turn light green to yellow when nitrogen-starved. As is typical for nitrogen deficiency, the older leaves are first to lose their chlorophyl.

Grasses

Like their close relatives, corn, oats, and wheat, the common grasses require large quantities of nitrogen. Most persons have had occasion to witness the quick change in color that results from the application of soluble nitrogen fertilizers on lawn and pasture grasses. A close examination of nitrogen-deficient bluegrass, timothy, Sudan grass, or smooth bromegrass shows the symptoms to be exactly the same as those described for oats and wheat, a general light green color with yellowing starting at the tips of the lowest leaves. Bromegrass responds very quickly to applications of soluble nitrogen, which is why it does so well when grown in a mixture with alfalfa. Timothy is very responsive to soluble nitrogen fertilizer.

Fruit

It has been common knowledge for years that fruit tress must be well supplied with nitrogen. Where there is a deficiency, apples make a small set of fruit and leaves mature while yet small. There is a gradual loss of chlorophyl, noticeable first on the oldest leaves on the current year's growth. An increase in development of anthocyanin pigment may be associated with nitrogen deficiency. Twig and spur elongation stops early, and the twigs are stiff and woody.

Peaches are especially sensitive to a deficiency of nitrogen. As with apple trees, there is a gradual loss of chlorophyl, first apparent on the oldest leaves of the current year's growth. Red spots appear on the leaves. Nitrogen-deficient and normal leaves are shown in Fig. 12-3. During the early part of the season, symptoms may be corrected by an application of soluble nitrogen. They may be prevented by applying adequate amounts of nitrogen before growth starts in the spring.

NITROGEN SURPLUSES

After all this discussion about nitrogen deficiency, it seems unwise to leave the subject without mentioning the fact that nitrogen excesses may be harmful and that for many crops it would not be desirable to maintain a continuously high level of available soil nitrogen (nitrate) throughout the season. This is especially true of fruit. If nitrate leaves are too high, too much new

FIGURE 12-3 Peaches are very sensitive to nitrogen deficiency. Normal leaves are here compared with those which are nitrogen deficient that have a light-green color and the red spots on the leaves. (Photo courtesy Wesley P. Judkins, Ohio Agricultural Experiment Station.)

vegetative growth is likely to occur with too little flower-bud differentiation. The accumulation and storage of carbohydrates may be inhibited. If this condition continues until late in the season, much winter killing may result.

A continuously high nitrate level in the soil tends to delay flowering and fruiting. For this reason, it is desirable in the production of certain flowering plants to maintain a high level during the period of rapid vegetative growth, then withhold or remove soluble nitrogen to induce flowering.

A very late application of soluble nitrogen for such crops as strawberries, tomatoes, potatoes, or sweet corn may stimulate vegetative growth and delay fruiting or tuber formation to the extent of actual injury to the crop. In other words, nitrogen deficiency at certain times is actually desirable.

PHOSPHORUS DEFICIENCY

In the raw materials from which soils are formed, phosphorus exists as a component of the mineral **apatite.** As a result of the weathering processes, which have changed soil materials into soil, much of the apatite has been broken down. The released phosphorus has combined with other elements to form secondary minerals and complex chemical compounds. Since the quantity of apatite in the soil-forming rocks is generally low, it follows that the resulting soils are low in phosphorus.

During the years of soil formation, a portion of the released phosphorus was taken up by the plants and returned to the soil as a part of the organic matter. This means that at the present time soils high in organic matter contain more phosphorus readily available to plants than do soils low in organic matter. It is true, of course, that the organic matter must be in a state of decomposition for the phosphorus to be available to growing plants.

With the release of phosphorus in the soil, either from the primary and secondary minerals, from the decomposition of organic matter, or from other sources, some is absorbed by the clay particles. In fact, this absorption is so great that very little fertilizer phosphorus is lost from a soil by leaching unless rate of application was very high, much higher than the need indicated by soil tests. In other words, phosphorus ions are quite immobile in soils relatively low in available phosphorus. This slow mobility means that plant roots must move to the phosphorus, as the movement of the phosphorus ions toward the roots is too slow for satisfactory growth in soils needing fertilizer phosphorus. In fact, phosphorus is probably removed from the clay particle largely by direct contact between soil particle and plant root. This slow mobility explains why corn soon shows signs of deficiency when soils are wet and cold, as in early spring.

Another source of phosphorus in the soil is that often termed the **difficultly available phosphorus.** This is in the form of iron and aluminum phosphate compounds, so slowly available that profitable crops cannot be produced if other sources are not available. It is desirable to raise the pH of mineral soils above 6.5 to discourage the formation of these difficultly available phosphorus compounds.

Easily available phosphorus exists in the soil in the water-soluble form and as calcium and magnesium phosphates. The quantity existing in water-soluble form at any one time is indeed very small, from none in less fertile soils to perhaps 40 lb/acre in very fertile soils. A considerable quantity may exist as calcium and magnesium phosphates, provided the pH of the soil is above 6.8. Below this pH these phosphates are unstable.

Soluble phosphorus combines readily with various compounds in the soil. In commercial fertilizer the phosphorus is usually in the form of calcium or ammonium phosphate. If the pH of the soil is above 6.8, calcium phosphate remains stable and the phosphate ion is thus held in union with the calcium, a form in which it is readily available to plants. It is true, however, that if the pH is too high (probably above 7.5), phosphorus intake by plants may be slow. In other words, very high alkalinity may be just as harmful as strong acidity. If the pH of the soil is below 6.8, calcium phosphates break down and the phosphorus combines with hydrated iron oxides to form iron phosphates that are not easily available. Combinations with aluminum may also take place. Under such conditions, phosphorus starvation may exist even though phosphate fertilizers have recently been applied.

Phosphorus starvation is usually the most pronounced during the early growth of plants, especially if growing conditions during that time are not ideal. Corn, for instance, during a cold, wet spell early in the season, may show marked signs of phosphorus deficiency, even on soil that tests rather high in available phosphorus. This is because at that time, owing to the slow growth, the roots are not extending to the phosphorus and, because of its immobile condition in the soil, the phosphorus is not moving to the roots. Furthermore, during periods when growth is almost completely at a standstill, new root hairs are not being formed. It is through the new root hairs that nutrients are taken into the roots. When weather conditions improve, growth increases and the phosphorus deficiency symptoms often disappear. Even though the specific symptoms disappear, however, the plants may have been stunted and final yields depressed.

Phosphorus is essential for cell division. When a shortage exists, new growth is depressed and plants become stunted. Since the plant attempts to furnish phosphorus for new growth, the element is transferred from old to new tissue. For this reason, it is desirable to make tissue tests for phosphorus on both the old and newer growth. For instance, the new center leaves of sugar beets may be high in phosphorus when the older outside leaves are low. This shows that phosphorus has been transferred from the older leaves to the central growing tissue. At such a time, the plant may not yet have suffered from phosphorus starvation, but the time is rapidly approaching when the deficiency will slow up growth. The outer leaves may become so depleted in phosphorus that death results. Whenever sugar beets are found to have a lot of dead leaves around the outer edges, it is a sure sign of some nutrient deficiency, unless, of course, some disease is responsible.

Small grains may fail to tiller when phosphorus is lacking, and an excess tends to induce sucker formation in corn. Phosphorus is said to stimulate root production and to increase resistance to disease, which is probably due to a stimulation of cell division and a more vigorous metabolism. It has been shown that black root of sugar beets causes much less damage on plots heavily fertilized with phosphorus than on unfertilized plots.

Phosphorus is essential in seed formation. With all conditions normal, a plant stores up phosphorus to be moved toward the fruiting region as the seeds being to develop. If the supply of phosphorus during the vegetative stage of growth has been only sufficient to keep the plants alive, there will be no reserve supply for this extra demand at fruiting time. Such a condition results in yellowing and dropping of older leaves while seeds are forming.

Corn

In young corn plants starved for phosphorus, nitrogen is usually high and the plant has an unusually dark green color. Phosphorus-starved plants also accumulate sugars. When the sugar concentration becomes abnormally high, anthocyanin pigment accumulates, which turns the leaves to a reddish purple. This usually happens during the early growth of the plant and is the most pronounced during periods of adverse weather when growth processes are slow. The reddening starts at the tip of the leaf and proceeds along the edges toward the stalk. A badly deficient plant is shown in Fig. 12-4. A close-up view of an individual leaf is shown in Fig. 12-1.

As the plants become older and root development increases, the reddening usually disappears and no definite symptom of phosphorus deficiency remains except that the plant may have been stunted in growth. As seed time

FIGURE 12-4 A phosphorus-deficient corn plant.

approaches, such plants turn yellow, provided the starvation for phosphorus has remained serious. The yellowing starts on older leaves in much the same manner as does nitrogen deficiency. Confusion between the two may be avoided by use of tissue tests. In some cases, symptoms of phosphorus deficiency appear in young plants on soil rather well supplied with phosphorus, usually during periods when growth is restricted because of cold, wet weather. If such a condition is suspected, it should be verified by **soil** tests for phosphorus.

Grains

The small grains are very responsive to phosphorus fertilizers, especially during early growth, but deficiencies cannot be readily detected by any definite leaf symptoms. When phosphorus is lacking in the soil, the plants start slowly and do not tiller well. Some reddening similar to that of corn is sometimes evident. In cases of extreme deficiency and when the season is cold and wet during early growth, there is a general yellowing of the leaves, which may be confused with nitrogen deficiency. It is easy to avoid such confusion, however, by making green-tissue tests for phosphorus and nitrate on the stalks of the young plants.

Alfalfa

Alfalfa is very responsive to phosphate fertilizer, especially on fine-textured soils. When the available supply in the soil is inadequate, the plants start slowly and remain small. At blossom stage the lower leaves turn yellow and drop off. Lack of leaves results in poor quality hay.

If phosphorus deficiency is suspected because of slow growth, and symptoms of potassium deficiency are absent, a test for phosphorus is usually sufficient to confirm the suspicion.

Sugar Beets

A deficiency of phosphorus delays the emergence of sugar-beet seedlings and makes them more subject to disease. Once the plants have emerged, they grow slowly and become stunted. The stunted plants (see Fig. 12.5) have small, dark green leaves that, in extreme cases, are fringed with red. If the red color is present, the symptom is very reliable. If the red fringe is not in evidence, it is well to make a tissue test for phosphorus. A low-phosphorus test, of course, confirms the suspicion. Such leaves always test high in nitrate. The test should be made on the petioles of the older leaves.

Small, dark green leaves on phosphorus-deficient plants stand more erect than on normal plants. This is in contrast to the horizontal position of the light green leaves on nitrogen-deficient plants.

After about the middle of the growing season, if phosphorus deficiency continues to be serious, the beet leaves gradually lose their dark green color

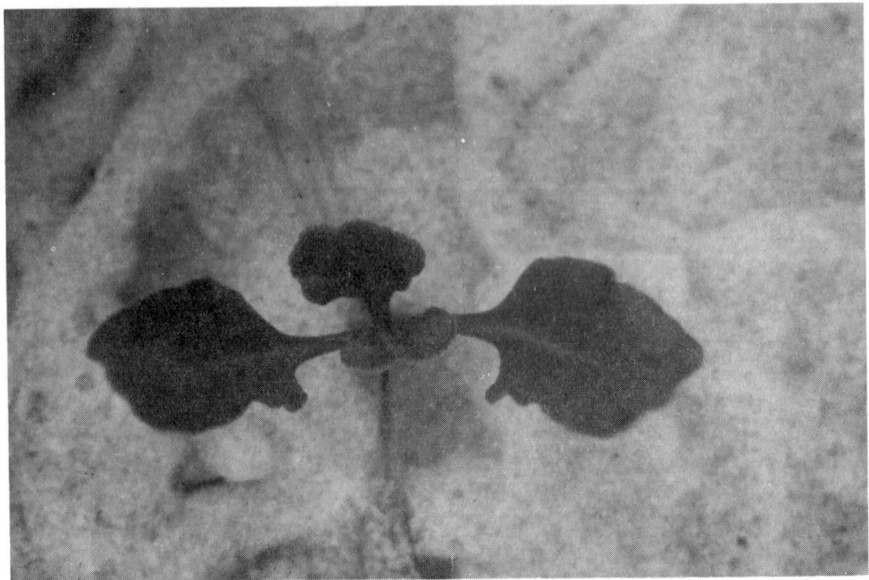

FIGURE 12-5 Sugar beets deficient in phosphorus grow slowly and are very dark green in color during early growth. The leaves may be fringed with red.

and finally become very light green or yellow. The older leaves die, and the condition very closely resembles nitrogen deficiency. Such plants, however, have a high content of nitrate, so the tissue test is very useful in avoiding a mistake. During the latter part of the growing season, it is always desirable to use the nitrate test to distinguish between these two deficiencies.

Beans

In extreme phosphorus starvation, bean leaves turn yellow and die. This occurs first on the older leaves, and there is no definite pattern, as illustrated in Fig. 12-6. It apparently is the result of rather complete transfer of the phosphorus from the older leaves to the growing tissue. Since the symptom is less common than is the yellowing of bean leaves from deficiencies of nitrogen, potassium, and manganese, care must be taken to avoid confusion. Perhaps the safest precaution is the green-tissue test and the soil reaction test. If the yellowing is really due to a lack of phosphorus, the affected leaf petioles will test low in phosphorus and high in nitrate and potassium. There is a possibility of eliminating manganese deficiency as a contributing factor by determining soil reaction. If the pH is below 6.8, there is little possibility of a shortage of manganese. Furthermore, yellowing due to a shortage of manganese does not start on the lower leaves but appears throughout the plant, and it is uniform over the entire area of each leaf.

Tomatoes

The growth of tomatoes is seriously affected by a deficiency of phosphorus, especially if the shortage occurs during early growth. Young seedlings become very dark green, and the lower sides of the leaves, including the veins, turn purple. This need not be confused with nitrogen-deficiency symptoms because in the latter only the veins and petioles become purple, with the leaf tissue between the veins showing very light green or yellow. A comparison is shown in Fig. 12-7. The dark green color is a result of the intake of an excess of nitrogen while plant growth is held back by the phosphorus shortage.

After root systems become well established, tomatoes, like corn, sometimes are able to obtain sufficient phosphorus to eliminate the characteristic purple discoloration. For that reason, the symptom is not so common on older plants. In the case of nitrogen deficiency, the purple color of the veins usually remains until maturity unless nitrogen fertilizer is applied.

On strongly alkaline soils or on soils high in calcium, as occurs in some greenhouse soils watered with hard water, phosphorus deficiency symptoms are common. Apparently, the high alkalinity lessens the intake of phosphorus by the plant.

The green-tissue test for phosphorus works well on tomatoes. Blank or low tests mean that phosphate fertilizer should be applied at once. Tests remain low even after the purple color disappears. Apparently, the purple shows up only when the deficiency is severe.

FIGURE 12-6 Bean leaves turn yellow when phosphorus deficiency becomes serious. The green veins would not indicate manganese deficiency because of the lack of uniformity in the yellowing of the other leaflets. The left leaf was from a normal plant.

FIGURE 12-7 Tomato leaflets. At the right is the end leaflet from a lower leaf of a seriously nitrogen-starved plant. The prominent purple veins will show on an almost yellow leaf. The center leaflet is from a similar position on a phosphorus-starved plant with purple color on the entire under surface of the leaflet. The left leaflet is from a normal plant.

POTASSIUM DEFICIENCY

Potassium occurs in the soil largely in the mineral form. During the processes of weathering, the minerals undergo changes that result in clay formation and the liberation of potassium in carbonate form. As potassium carbonate is easily soluble, the potassium ion is free to unite with secondary compounds to form insoluble silicates, or it may be removed by growing plants, or it may become attached to the clay colloids. In this latter role, the ion has given up its positive charge and is called an **exchangeable ion.** As weathering is constantly going on, some potassium is always in solution and is in danger of being lost in drainage water.

The potassium required by plants must come largely from that which is absorbed by the clay. Such ions, held against loss by leaching, are readily removed from the soil particle on contact by the plant root. The potassium thus removed from the clay is replaced by that which is release during the weathering of minerals. As this replacement is slow, there may develop, during a vigorous cropping program, a serious depletion of the exchangeable potassium. Eventually, the time comes when the crop cannot obtain enough for maximum growth.

Potassium is released from decomposing organic matter. In soils well-supplied with organic material, a considerable portion of the potassium used by plants comes from this source. Cover crops take up potassium that might otherwise be lost by leaching and return it to the soil through decomposition.

A small quantity of soluble potassium exists in soil. In clay soils, the affinity between clay and potassium ions is so high that a water extract of the soil may be almost devoid of potassium. However, enough may be available from the organic matter and the clay to grow a crop. On the other hand, a sandy soil may contain a considerable quantity of soluble potassium but with so little in the exchangeable form that crops may soon show symptoms of starvation.

Symptoms of potassium deficiency are common on sandy soils because of the scarcity of potassium-bearing minerals and the low content of clay. They are also common on organic soils, formed largely from organic materials low in potassium. In high lime soils, the intake of potassium is apparently inhibited by the high concentration of calcium. On such soils, symptoms of potassium deficiency are more common than they are on neutral and slightly acid soils.

The role of potassium in plants is not known, but experiments have shown that it is essential. That most of the potassium remains soluble in the plant is shown by the fact that it may be removed from dried plant tissue by leaching with water. There is some evidence that potassium sometimes moves from maturing plants back into the soil. Young tissues contain more potassium than do older tissues. This shows the element to be essential for growth. As the supply from the soil becomes inadequate, potassium moves out of the older leaves into the growing tissue. It apparently moves out of the edges of the leaves first, as yellowing starts at the tips and edges. By the time the yellowing has reached the central area and base of the leaf, the edges have usually turned brown and may have shattered. As the deficiency becomes more severe, the potassium moves out of the next older leaves and so on until even the newest leaves sometimes show signs of potassium starvation. Under certain conditions, green veins are prominent on potassium-starved leaves. In such cases, the leaf yellowing is not uniform; which makes it possible to avoid confusion with symptoms of manganese, calcium, or iron deficiencies.

A word of caution seems desirable at this point. Certain virus diseases may produce similar symptoms. In some crops, the ravages of insects, such as leafhoppers, may produce an appearance similar to that resulting from a deficiency of potassium. This is true of alfalfa, and the condition has been reported for apples. It is possible to confirm or disprove the suspicion of potassium deficiency by the green-tissue test.

Corn

Symptoms of potassium deficiency may occur in corn at any time during the growing season. The condition is most common, however, after the corn has made several weeks' growth. The pattern of yellowing is different from that of nitrogen deficiency, as shown by Fig. 12-1. The yellowing starts at the tip of the leaf and proceeds along the edges toward the stalk, instead of along the midrib. The edges soon turn brown and become dry, which has led to the term "leaf scorch." The rustling caused by the wind soon frays the edges and gives them the ragged appearance so common when the deficiency becomes serious.

As is the case with nitrogen deficiency, potassium deficiency is apparent first on the lowest leaves. When the oldest leaves are entirely yellow, the newest leaves may be normal.

When the pattern of yellowing is indefinite, conclusions may be verified by the use of tissue tests. If the yellowing is really due to potassium deficiency, high nitrate and phosphorus tests will be obtained, and the potassium test will, of course, be low, usually almost blank.

Small Grains

With small grains, if symptoms alone are relied on, there is danger of confusion between potassium and nitrogen deficiency. There is a distinct tendency for the yellowing to proceed faster along the edges of the leaves than in the middle, exactly as potassium deficiency develops in corn. With the grains, however, the leaves are so narrow that this characteristic is likely to be unnoticed.

When potassium deficiency is suspected in the grains, it should be confirmed or disproved by tissue tests. With the grains, the tests are very clear and dependable.

Alfalfa and Clover

Leguminous crops as a whole require rather high levels of available potassium. As the supply in the soil becomes exhausted, the older leaflets start to turn yellow at the tips and along the edges. At the same time, white dots appear in the yellowed region or along the boundary between the yellow and green areas, as shown in Fig. 12-8. The leaf in Fig. 12-8 was grown in the greenhouse on Plainfield sand (Typic Udipramments) deficient in available potassium. In those pots fertilized with potash, the alfalfa leaflets did not turn yellow and did not contain the white dots.

As already suggested, injury to alfalfa by leafhoppers must not be confused with potassium deficiency. When leafhopper injury is serious, as it often is during the second growth, the leaflets turn yellow at the tips and along the edges in the manner so characteristic of potassium deficiency, but the white dots are not present. Another very good way to differentiate between leafhopper injury and potassium deficiency is to make a green-tissue test for potassium. If the injury is actually due to leafhoppers, the potassium test will be even higher than normal. This is because, as the leafhoppers restrict the growth of the plant, potassium tends to accumulate as it does when growth is restricted by a lack of some other plant nutrient.

Beans, Soybeans, and Cowpeas

The symptoms of nutrient deficiency for beans, soybeans, and cowpeas are very similar. Symptoms of potassium deficiency usually appear while the

FIGURE 12-8 Alfalfa leaf grown in the greenhouse on Plainfield sand. The yellow edges on the leaflets and the white dots arranged in this pattern are sure symptoms of potassium deficiency.

plants are still in the early stages of growth. The symptoms first appear on the older leaves and work gradually toward the newer growth. The newest leaves are usually normal in appearance. As is invariably true with potassium deficiency, the yellowing appears first at the tip and edges of the leaflet and gradually spreads toward the center and base. As time goes on, the tissue along the edges become necrotic (brown and dry), and eventually the entire leaf dies. Apparently, as the leaflet edges become yellow, growth becomes slower in the yellow areas and continues at a more normal rate in the remaining green portions. The result is a decidedly crinkled appearance.

During cold, wet periods in the early part of the growing season and during unusually dry periods in midsummer, potassium-deficiency symptoms are very common in white field beans and soybeans. On one experimental field, white beans followed soybeans. The beans on all plots that had not received potash fertilizer were decidedly yellow. In fact, only the very newest leaves were normal in color. The plots that received only phosphate fertilizer contained plants even yellower than did the unfertilized plots.

In several instances, white beans and soybeans have been profitably sidedressed with potash fertilizer after symptoms appeared. It is better, of course, to avoid the occurrence of such a condition by applying the proper fertilizer at the time of planting.

Cowpeas growing on a soil deficient in available potassium develop deficiency symptoms not unlike those described for soybeans and white beans.

Green-tissue tests may be made to good advantage on beans, soybeans, and cowpeas. When potassium deficiency becomes so serious that symptoms develop, the petioles of deficient leaves always test low. Usually the test is blank. At the same time, the new green leaves may give a rather high test, because the potassium from the old leaves is being translocated to the new leaves.

Potatoes and Tomatoes

Large quantities of potassium are required by potatoes and tomatoes. When grown on organic and sandy soils where the supply of available potassium is rather limited, they must be liberally treated with potash fertilizer. When potassium is deficient during the early stages of growth, it first shows up as a yellowing of the tips and edges of the oldest leaflets. The deficiency shows up earlier in the stage of growth when soil nitrogen and phosphorus levels are high, and when nitrogen levels are high the yellowing is commonly accompanied by the formation of white dots in a manner already described for alfalfa and clover.

As the season progresses, potassium-deficiency symptoms on tomatoes do not disappear as do those of phosphorus deficiency, but the change somewhat in appearance and become more pronounced as the plants produce fruit. The symptoms of deficiency on the newer leaves of the older plants are much as described for the old leaves of the young plants, but the old leaves of the old plants (those old enough to bear fruit) present a considerably different appearance. The yellowing occurs as a mottling, with the area between the main veins much more yellow than the area directly adjacent to the veins. The mottling starts near the leaflet edges and progresses inwardly until it includes almost the entire area, as illustrated in Fig. 12-9.

On potassium-deficient plants, the newest leaves are of a much darker green color than they are on those severely deficient in nitrogen. In fact, some of the older leaves remain dark green so that the general appearance is a contrasting dark green and yellow as compared with a very light green and yellow on nitrogen-deficient plants.

Sugar Beets

A considerable quantity of potassium is contained in sugar beets. Therefore, it is not surprising that a deficiency is very soon reflected in the appearance of the plant. The oldest leaves are affected first. They start turning yellow first at the tip and along the edges. Gradually, the yellow areas work toward the center of the leaf and the edge turns brown. Sometimes the chlorotic area becomes yellowish gray rather than a distinct yellow as is characteristic of bean leaves yellowed by potassium starvation.

Tissue tests for potassium should be made on the leaf petioles. It has sometimes been observed that plants showing slight to medium symptoms of potassium starvation test low to medium in potassium. In other words, plants with such symptoms do not always test blank. This probably means that for maximum growth, the petioles of sugar-beet leaves should contain potassium sufficient for a high test.

FIGURE 12-9 A mature tomato leaf deficient in potassium.

Cucumbers

The characteristic symptom of nitrogen deficiency in cucumbers is a gradual fading of the chlorophyl over the entire leaf. Contrast such an appearance with that of the leaf shown in Fig. 12-10, where the yellowing was distinctly around the leaf edges. In cucumber plants suffering from potassium starvation, the portion of the leaf blade nearest the petiole remains green long after the tip and sides of the blade have turned yellow and died.

Green-tissue tests work very well on cucumbers. Testing may indicate the need for potassium fertilizer in time to avoid the appearance of symptoms.

Cabbage and Celery Cabbage

Both cabbage and celery cabbage are quick to show symptoms of potassium starvation. The oldest outside leaves are affected first. The newest center leaves may be perfectly normal on plants very severely starved for potassium.

FIGURE 12-10 A cucumber leaf deficient in potassium. The yellowing is distinctly around the edges of the leaf. The appearance presents a striking contrast to that of a leaf deficient in nitrogen where the yellowing is uniform over the entire leaf. In both cases the oldest leaves are the first ones affected.

Cineraria

Florists are finding that it is essential to maintain nutrients at the correct levels if plants carrying green healthy foliage are to be produced. This is especially true of cineraria. A plant having the most beautiful blossoms will be classed as second-rate if the leaves are off-color. When the level of available soil potassium is low, the older leaves of the cineraria turn yellow around the edges. Soon the edges become brown, and portions drop away, giving the leaf a ragged appearance. In Fig. 12-11, a plant that had received very little potash but plenty of nitrogen is shown in comparison with a normal plant and one that received plenty of potash but very little nitrogen. The picture shows that a deficiency of either nitrogen or potassium nearly always causes delay in flowering. Likewise, of course, an excess of either element will delay flowering. Deficiencies in cineraria are readily detected by the green-tissue test.

Stocks

Stocks grown in soil low in available potassium tested very low in potassium, and very early in their stage of growth showed symptoms of deficiency. The lower leaves first turned yellow at the tips. The yellowing progressed rapidly toward the base of the leaf and worked from leaf to leaf up the plant. The very newest leaves, however, were always normal in color.

Tests showed the yellowed outer leaves to be devoid of soluble potassium, whereas the very newest green leaves contained enough to give a positive test. Potassium, of course, was being moved from the old to the new leaves. It may be assumed, perhaps, that in such a plant the same potassium is used several times.

Geranium

Geraniums seem to be very sensitive to a lack of available soil potassium. When the element is extremely deficient, the leaves turn yellow around the edges. Tissue tests work very well on geraniums, and when potassium deficiency is suspected, it is well to perform a test on the petioles of the leaves to confirm the symptoms.

Paddy Rice

More people depend on rice as a main source of food than on any other grain. Since it is a grass, it is not surprising that nitrogen, furnished in the ammonia form, is an important nutrient. However, other nutrients are also essential. The "suffocation disease" ("Akiochi disease" in Japan) described by Ming-huei Wu Sheng in her Master's thesis at Michigan State University (1966), accentuates the need for potassium. The symptoms are shown in Fig. 12-12. The extremely reduced condition around the roots affects the physiology of the plant and hinders the uptake of potassium.

FIGURE 12-11 Ceneraria plants, potassium-deficient on the left, normal in the center, and nitrogen-deficient on the right. It is very important that the foriage of flowering plants be of a normal color. Nutrient levels must be properly maintained if healthy plants are to be grown. These plants were grown by John Gartner, former graduate student at Michigan State University.

FIGURE 12-12 Rice yellowing with brown spots indicates potassium deficiency, accentuated by the "suffocation disease" described by Professor Ming-huei Wu in Taiwan. In Japan, the condition is called "Akiochi disease." The disease occurs when the growing medium is reduced to the point of hydrogen sulfide production around the rice roots. Dr. Kosakuro Ono, National Kokuriku Agricultural Experiment Station, Takada, Japan, described the disorder as "Akagare" or stifle disease in 1957. See Japanese Potassium Symposium, International Potash Institute, Berne, Switzerland.

MANGANESE DEFICIENCY

Manganese occurs in soils in several forms — as exchangeable manganese, as a cation attached to the clay, as a part of organic matter, and as a constituent of certain minerals, chief of which are pyrolusite (MnO_2), rhodonite ($MnSiO_3$), and rhodochrosite ($MnCO_3$).

To be available to plants, manganese must exist as exchangeable manganese, as a part of organic matter, or as inorganic, readily reducible manganese. Manganese is readily oxidized to the manganic form. Under certain conditions, manganic manganese may be reduced to the manganous form. Briefly, it is affected by relative acidity or alkalinity, content of lime and phosphate, aeration, temperature, clay content, moisture, and the presence of certain reducing and oxidizing compounds. On alkaline soils high in lime, the equilibrium swings toward the manganic side, and plants are unable to obtain sufficient manganese for normal growth. Likewise, any manganese applied in a soluble form, such an manganese sulfate, is quickly oxidized to the manganic form.

Acid conditions favor the formation of manganous manganese. Since this form of manganese is readily available to plants, we do not expect to encounter manganese deficiency on acid soils. It is believed that pH 6.5 to 6.8 marks the borderline for manganese deficiency on mineral soils in Michigan.

Manganese is rather easily leached from acid soil; thus the total levels sometimes become rather low. When such soils are limed above pH 6.5, it is more likely that plants grown on them may suffer from lack of manganese than would be true on soils with an original pH above 6.5.

Since manganese deficiency usually occurs on alkaline soils, it seems logical that the reaction of the soil should be first determined when it is necessary to make tests to confirm some suspected deficiency.

The role of manganese in the plant is not clearly understood. McHargue (1923) has suggested that manganese performs a function in photosynthesis. Others contend that it increases the activity of oxidizing enzymes in plants. Some consider it a regulator of the intake and state of oxidation of certain other elements in the plant, especially iron. There is considerable evidence, for instance, that manganese must be present in the plant to keep iron in an oxidized state so it will not become toxic in the plant. In other words, manganese deficiency may actually be iron toxicity. There is need for more research along this line.

Symptoms of manganese deficiency first appear in relatively young, but not always the youngest, growth. When the symptoms start early in very young plants, practically all leaves may be affected, but when the symptoms first appear after a plant has reached a considerable size, the older leaves may remain green while the upper portion of the plant becomes chlorotic. This shows that the element is not readily translocated from old to new tissue, so a constant supply is necessary for normal growth.

On most plants, manganese-deficient leaves are distinctly mottled. The leaf tissue between the veins becomes increasingly lighter green until it is definitely yellow, but the veins remain green. The color pattern is uniform over each individual leaf. By the green veins, it is rather easy to differentiate between chlorosis due to manganese deficiency and that caused by a lack of nitrogen. With some plants, the leaf tissue directly adjacent to the veins may also remain green.

Care must always be exercised to avoid confusion between the symptoms of certain virus diseases and those of manganese deficiency. Also, there are some plants with normally variegated foliage. It is difficult to diagnose manganese deficiency correctly by the appearance of such a plant. In such cases it may be necessary to resort to the use of a laboratory test for manganese in fresh plant tissue.

Small Grains and Grasses

Many members of the grass family have been shown to be sensitive to a lack of manganese. Perhaps oats are the most sensitive of the grains. When oats are grown in a medium deficient in manganese, they develop the "disease" called **gray speck.** This is not a pathologic disease, but is simply a physiologic breakdown of the leaf tissue. It was first reported in this country by Sherman and Harmer in 1941.

Gray speck starts as a gray oval-shaped spot on the edge of a new leaf. It usually appears when the plant has reached the stage where it has three or four leaves. The gray speck appears some distance back from the tip and gradually enlarges until it spreads across the entire leaf. As the spot enlarges, the gray color gradually gives way to yellow. The yellow color finally extends over a considerable length of the leaf. At about the time the affected area has reached the full width of the leaf, the leaf droops, leaving the yellowed portion sticking up where it helps to give the field, or perhaps a spot in the field, a yellow appearance. During all this time the tip of the leaf remains green. This is an important point to remember, because it positively proves that the symptom is not that of nitrogen or potassium deficiency.

An examination of Fig. 12-13 shows that oats are sensitive to a deficiency of manganese in the soil. The same is true for wheat and barley, but the symptoms are somewhat different from those in oats. Instead of the appearance of the characteristic gray speck, the upper leaves simply start turning yellow in streaks parallel to the length of the leaf. Some of the streaks become wider in places, so that the leaf has a spotted appearance. The leaf tip does not remain as distinctly green as do the tips of oat leaves. The yellowing does not, however, start at the tip as it does on plants deficient in nitrogen or potassium. The leaves of barley and wheat plants droop in much the same manner as do those of oats.

FIGURE 12-13 Oats grown on Granby sandy loam, an alkaline sandy soil rather high in organic matter. The left pot received manganese sulfate at the rate of 100 pounds per acre. Note the light color of the oats which did not receive manganese. Also note the manner in which the leaves droop where manganese is deficient.

On alkaline soils, especially if they are sandy, sudan grass may be very responsive to applications of manganese. The symptoms of deficiency resemble closely those in wheat and barley, a yellow striping and spotting of the upper leaves with a general light-green color over the entire plant.

Beans

When grown on soils insufficiently supplied with available manganese, beans, including white field beans, soybeans, and garden beans, are similar in appearance. Shortages of manganese for these crops occur on soils with a pH above 6.5 and are more serious on sandy than on clay soils. The leaf shown in Fig. 12-14 is typical of those deficient in manganese. The yellowing is uniform over the entire leaf and the veins remain green.

Manganese deficiency shows up very early, often on the first pair of leaves, in white field beans and soybeans grown on the very sandy alkaline soils. On some of the better alkaline soils, the symptoms appear later in the stage of growth, and here it is the new growth that is affected. This is in contrast to nitrogen deficiency, which shows up first on the lower, older leaves.

In fields where symptoms have appeared early in the season, sidedressings of manganese sulfate at the rate of 100 lb/acre have caused very profitable increases in yield. Applied beside the seed at planting time, smaller applications are satisfactory. Spraying deficient plants with as little as 5 lb/acre in 200 gal of water has furnished manganese sufficient for normal growth.

Garden beans are very sensitive to a lack of manganese. Many garden soils are alkaline because they have been irrigated with hard water. In such soils, some effort should be made to supply beans with soluble manganese.

Sugar Beets

Sugar beets are commonly grown in Michigan on soils deficient in available manganese. The deficiency may show up in the plant at any time during the growing season and is first noticed as a mottling of the new growth. The green color gradually fades from the leaf tissue between the veins. The veins themselves and the area closely adjacent to the veins remain green for a considerable period after the rest of the leaf is yellow. The leaf shown in Fig. 12-15 is typical of beet leaves badly deficient in manganese. Sometimes the veins are much less distinct. In fact, it is occasionally difficult to distinguish the mottling, and there is a possibility of confusion between manganese and nitrogen deficiency. The leaf petiole should then be green-tissue tested for nitrate nitrogen. If the test is positive, it is safe to assume the yellowing is due to something other than nitrogen. Whenever manganese deficiency is suspected, it is always wise to determine the reaction of the soil.

FIGURE 12-14 Bean leaves yellowed from deficiency of manganese and potassium compared with a normal leaf. The left leaf was normal. The center leaf was deficient in potassium. It yellowed first at the tips and along the edges of the leaflets. Notice also the crinkled appearance, brought about by the continued growth of the interior portions of the leaflets, after growth had ceased along the edges. The right leaf was deficient in manganese. The yellowing was uniform and the veins remained green.

FIGURE 12-15 Sugar-beet leaves deficient (1) in manganese, (2) in potassium, and (3) in nitrogen. The number 4 leaf was normal. Compare the uniform yellowing of the manganese-deficient leaf, with prominent green veins, with the strikingly different pattern of yellowing on the leaf deficient in potassium. Nitrogen deficiency differs from manganese deficiency in that the veins do not remain green when the leaf is nitrogen deficient.

Other Crops

Manganese deficiency has been observed in several other crops and in various shrubs and weeds. With all plants the pattern is the same, light green to yellow areas between dark veins. The trouble is always associated with neutral, or only very slightly acid, to alkaline soils.

Alfalfa and the clovers are much less sensitive to manganese deficiency than are oats, beans, and sugar beets. Good crops of alfalfa have been observed on fields where oats failed from a lack of manganese during the previous season.

Manganese deficiency has been observed on tomatoes, potatoes, strawberries, and peaches, but more experimental work needs to be done before recommendations regarding these crops may be complete.

BORON DEFICIENCY

Boron occurs mostly as a constituent of the mineral **tourmaline** and in organic matter. The total quantity of boron in a soil seems of little importance. Even soil tests for boron are not dependable as a measure of the need for boron applications as fertilizer. The important point, as far as plants are concerned, is its availability. Observations and experiments have shown that boron deficiency in humid regions occurs only on soils high in lime, either on those naturally alkaline or where too much lime has been applied. It is possible to induce boron starvation by applying lime in excessive amounts to an acid soil, or to prevent the deficiency from appearing by adding sulfur or sulfuric acid to a high lime soil.

In previous publications, these authors and others have discussed the subject of boron deficiency and toxicity. Boron is usually not needed on soils that need lime; quoting from Gerber (1980), "Other environmental and plant factors have a greater effect on the occurence of boron toxicity than the actual concentration of boron in the soil."

The exact role of boron in plants is somewhat obscure. It seems to function chiefly as a regulator of the intake of other ions. Various investigators have reported certain ratios between quantities of boron taken in by plants and the intake of such elements as calcium and potassium. It is known, of course, that calcium affects the intake of certain other elements, so if boron affected calcium intake it would indirectly affect the intake of all ions affected by calcium.

A deficiency of boron causes a breakdown of the young growing tissue. Growth slows down, and in many plants the cells actually disintegrate, causing the tissue to crack, blacken, and become abnormally shaped. With alfalfa, internodes are shortened and the buds fail to open. Although boron is essential for plant growth, only small quantities are needed, and the amount varies greatly for different plants. An excess is extremely toxic and may affect different plants in different ways. With some plants, soybeans for instance, the growing tissue may be overstimulated so the plant makes a tall, spindling growth. Characteristic brown spots develop on the edges of the leaves, and the older leaves are shed early. Corn injured by an excess of boron is stunted, and the leaves turn gray along the edges.[1] Barley, so injured, develops brown spots about one-eighth of an inch long and about one-half as wide well scattered over the leaf. The spots are actually dead tissue, and they persist to the time of maturity. Peas turn brown at the tips and around the edges of the leaflets, and alfalfa leaves turn white across the ends when borax is applied in excess. Under certain conditions, 40 lb of borax per acre will be toxic to alfalfa, whereas a much smaller quantity will injure beans, barley, peas, and corn. Crandall, Chamberlin, and Garth (1981) found that pears have a "relatively high B requirement" and "were able to tolerate very high levels of B in the soil and in the blossom tissues." We believe that most plants that are sensitive to a lack of boron can tolerate relatively high levels in the soil. This would be an important fact to remember in selecting crops to be grown in areas where irrigation is with water high in boron.

Sugar Beets

Boron deficiency of sugar beets has for many years been termed **heart rot.** It was so-called long before it was realized that a lack of boron caused the "disease." Investigators in Germany and France spent many years trying to isolate an organism responsible for the death of the heart tissue. Of course they were not successful, and in 1931 Brandenburg of Germany showed that the "disease" could be prevented by adding soluble boron to the soil. Heart rot was observed in Michigan by Kotila in 1934.

It is easy to distinguish heart rot in the field. There are several rather specific and different symptoms of boron deficiency, all of which may or may not occur in the same plant. Some plants may have both top and root symptoms whereas others may have only root or only top symptoms.

The most noticeable symptom, of course, is the dead heart, illustrated in Fig. 12-16. This is first noticed during midsummer, after the beet has attained considerable size. The leaves may all die. On severely affected fields, the crop may appear almost entirely defoliated by the end of the summer. Some beets actually die, but most of them make new growth during the early fall months, until new leaves from around the edge of the crown almost cover the dead heart.

Perhaps the first symptoms of heart rot are cross-checked petioles and misshapen leaves, illustrated in Fig. 12-17. The leaves seem to grow unevenly on the two sides, which causes the petioles and midribs to twist and the leaf to develop only on one side. There is also a tendency, as shown in Fig. 12-18, for the development of a large number of small leaves, few of which ever reach normal size.

Boron-deficient beets also give one the impression of having been stepped on. The leaves grow out in a horizontal rather than in the normal

---

[1]We are indebted to H. J. Guist for specimens of corn injured by an excess of borax.

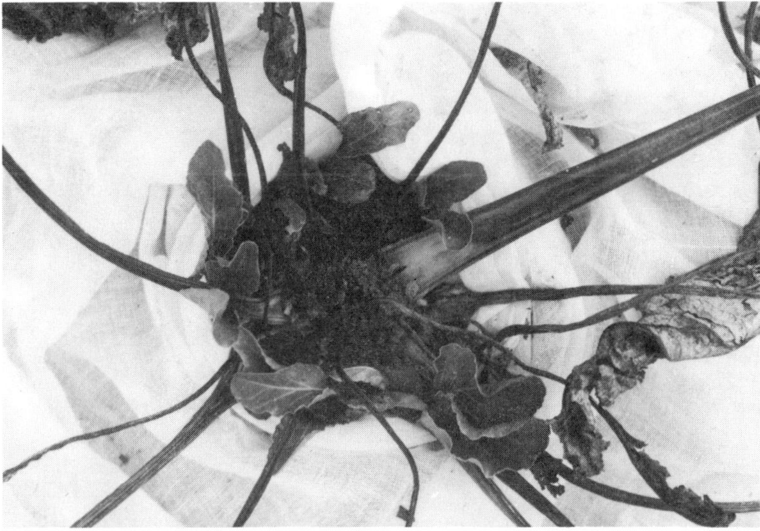

FIGURE 12-16 Boron-deficient sugar beet with a typical dead heart. Only one old leaf is still alive. Notice the new leaves coming out from the edge of the crown. Such new leaves may attain normal size, or they may become deformed and remain small, depending on how seriously the beet is affected.

FIGURE 12-17 Cross-checked petioles and misshapen leaves, curved to one side, are early symptoms of heart rot of sugar beets.

FIGURE 12-18 Sugar beets with heart rot sometimes produce numerous small leaves, few of which reach normal size. Most of the leaves on this beet were misshapen.

vertical position. Their color is dark green until they start to disintegrate; then they turn yellow, brown, and black.

The roots of sugar beets affected by heart rot vary greatly in the **extent** to which they break down. In fact, some plants show by their leaves that a deficiency of boron exists while their roots appear perfectly normal. Usually, however, the root tissue turns black and disintegrates to various degrees. In some cases, the whole crown breaks down, whereas in others the disintegration is scattered throughout the beet. When the disintegrate areas break through the surface, they appear as external cankers. A good idea of the appearance of sugar-beet roots deficient in boron may be obtained from Fig. 12-19. As little as 10 lb of borax per acre may prevent the deficiency.

Table Beets

Table beets are also very sensitive to a deficiency of boron. The symptoms are not unlike those of sugar beets, although the dead heart is not so common. The cross-checked petioles and twisted leaves may be observed on the plant shown in Fig. 12-20.

The roots of table beets deficient in boron commonly develop what have been termed **internal black spots.** These spots occur anywhere throughout the fleshy portion of the root, and if they happen to occur very close to the surface they break through the epidermis to form external cankers. The blackened internal area contain cork tissue, developed apparently by the plant to set off the disintegrated cells from the healthy cells. When the beets

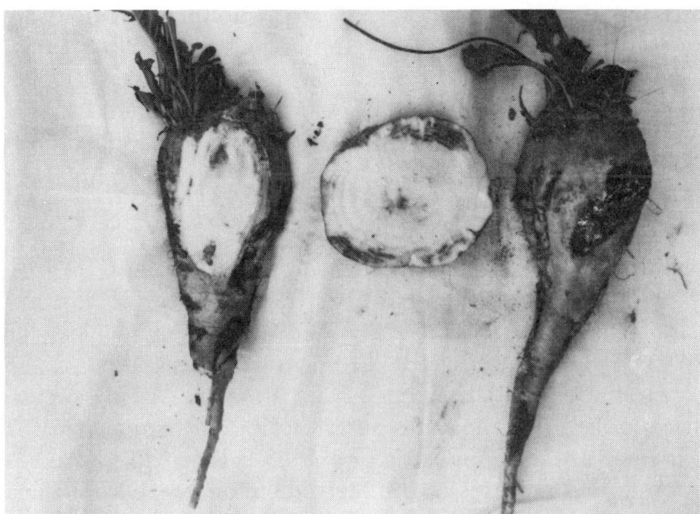

FIGURE 12-19 Boron-deficient sugar-beet roots. Such roots vary greatly in the degree to which they break down. In some cases, the internal tissue turns black but remains firm, whereas in others it disintegrates considerably. Sometimes the crown disintegrates to the extent that the plant dies. Where the disintegration is near the surface, it breaks through the epidermis of the root and forms an external canker, as is shown on the beet at the right.

FIGURE 12-20 Table beet deficient in boron. Note the cross-checked petioles and twisted leaves so common on sugar beets deficient in boron.

are cooked, the black spots are even more pronounced than before cooking, and when chewed the black tissue actually feels like cork.

Boron deficiency of table beets, as with sugar beets, occurs on alkaline soils. Where the fertilizer for beets on such soil is applied broadcast before planting, it is recommended that the quantity of borax applied be 40 lb/acre.

Spinach

The spinach plant is closely related to the sugar beet and, in its manner of growth, closely resembles the sugar beet. The symptoms of boron deficiency are also similar in the two plants.

In an experiment conducted on boron-deficient, Thomas sandy loam soil (Histic Humaquepts), the spinach from plots that did not receive borax in the fertilizer looked like those already described for sugar beets. Individual leaves were small and developed on one side, and the roots were blackened in much the same manner as is characteristic of sugar beets with heart rot.

Mangels

A close relative of the sugar beet, the mangel is also very sensitive to a deficiency of boron. The symptoms of deficiency are the same except that the mangel does not develop the cross-checked petioles as does the beet. Normal and boron-deficient mangel roots are shown in Fig. 12-21.

Head Lettuce

In the experiment on Thomas sandy loam referred to, lettuce on the plots that did not receive borax failed to head. The new center leaves became black

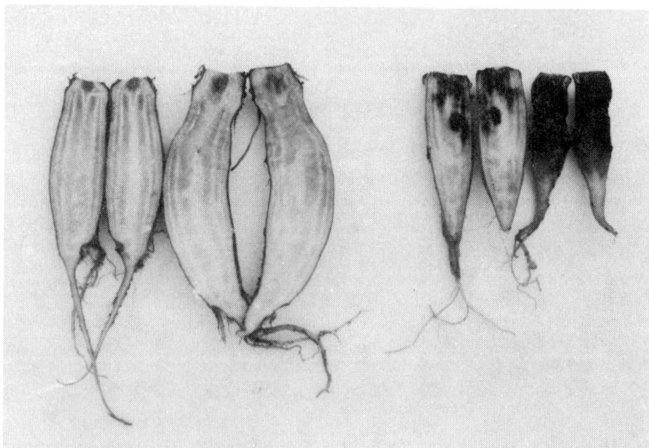

FIGURE 12-21 Mangel roots, normal on the left and deficient in boron on the right. The tissue breakdown is not unlike that observed in sugar beets and red table beets.

and ill-shaped while yet small. On the plots where 10 lbs or more of borax was applied, the heads were normal in appearance and of good quality.

Cauliflower

Members of the mustard family, of which cabbage and cauliflower are common examples, need plenty of boron. When it is lacking, the insides of the stalks disintegrate, and the heads turn brown. The stalks of cabbage also break down at the heart when boron is deficient.

Celery

Crack stem of celery, first called "marl disease" because it was associated with the growth of celery on Histosols (organic) underlaid by marl, is caused by a lack of boron. Recommendations for the use of boron to prevent crack stem on the Histosols of Michigan were published by Harmer (1935).

Other Crops

Boron deficiency has been observed on several other crops. Rutabagas and turnips develop water-soaked areas where the cells break down. The disorder has been called **water core.**

Chicory leaves turn red and tend to grow with twisted midribs when boron is lacking. Apparently, this tendency for petioles and midribs to twist is common with many plants when boron is deficient.

Alfalfa has been shown by investigators in many states to be very sensitive to boron deficiency. The deficiency is indicated by a yellowing and bronzing of the leaves and by a restriction of the terminal growth, which produces very short internodes and gives an umbrella-like appearance to the upper part of the plant. Affected plants fail to flower. Leafhoppers cause yellowing and bronzing of alfalfa leaves but do not cause the internodes to shorten.

MAGNESIUM DEFICIENCY

Magnesium occurs in the soil as primary and secondary minerals and as exchangeable ions. During the process of soil formation, the magnesium, along with other basic cations (calcium, potassium, and sodium), is released from the minerals. The magnesium ions then unite with other ions to form secondary insoluble compounds, or they may be used by plants or become attached to the exchange sites. As soils become excessively leached, the total basic cation concentration becomes reduced and the concentration of hydrogen ions is correspondingly increased. Such soils are said to be acid. The addition of large quantities of calcium lime, such as calcium carbonate or marl, brings about an unbalanced condition with calcium high in proportion to magnesium and potassium.

Excessive applications of potassium fertilizer have been known to cause a deficiency of magnesium, as has also the application of sodium, either as the nitrate or the chloride. Walsh and O'Donohoe (1945) found that applications of 500 lb sulfate of potash per acre significantly reduced the percentage of magnesium in potatoes, sugar beets, and tobacco, and caused serious magnesium deficiency symptoms. Hale, Watson, and Hull (1946) showed by pictures that sodium chloride induced magnesium deficiency symptoms in sugar beets. Prince, Zimmerman, and Bear (1947) observed that magnesium constituted less than 6 percent of the exchange cations in the soil and that the ideal amount of magnesium in the soil was an amount equal to 10 percent of the exchange capacity. Thus, we may expect to encounter magnesium deficiency on very acid soils, on soils limed with calcium lime, or on soils heavily treated with potassium- or sodium-bearing fertilizers.

Magnesium is the only mineral element contained in chlorophyl, the pigment that makes plants green. It is not surprising then that magnesium deficiency should result in chlorosis.

Truog and associates (1947) have emphasized the role that magnesium plays as a carrier of phosphorus in plants. They obtained a greater increase in the phosphorus content of peas by adding magnesium sulfate than by increasing the concentration of phosphorus in the culture medium. In the field, they obtained marked increases in the phosphorus content of dry peas by adding dolomitic limestone. In their conclusions they state that, "Failure in many experiments to produce crops of higher phosphorus content through phosphate fertilization has undoubtedly been due to a lack of available magnesium."

Symptoms of magnesium deficiency first appear on the older leaves. This shows that the element, like nitrogen, phosphorus, and potassium, is readily translocated from old to new tissue. Yellowing appears in patches between the veins and and around the leaf edges. The edge yellowing is not so prominent in magnesium deficiency as in potassium deficiency, especially from the point of view of demarcation between yellow and green. Leaf edges usually roll slightly.

As the magnesium-deficient leaf becomes older, the yellow areas scattered over the leaf blade become necrotic until finally the tissue may disintegrate and leave holes through the leaf. Prominent green veins may accompany the mottling, in much the same pattern as has been described for potassium deficiency. There need be no confusion with manganese deficiency because in the latter the yellowing is uniform over the entire leaf or leaflet.

Celery

Magnesium deficiency in celery begins as a mottling on the tips of the older leaves. Yellowing then progresses around the leaf margins and inward between the veins. Veins stand out clearly, and necrotic spots soon appear between them. The appearance of such leaves is illustrated by Fig. 12-22.

FIGURE 12-22 Celery leaflets: magnesium-deficient (left) and normal.

Some varieties are more susceptible than others to magnesium deficiency. Utah 10-B is very susceptible, whereas Utah 15 is quite resistant to the deficiency. Emerson Pascal and Utah 52-70 varieties are intermediate in this respect. The deficiency may be controlled by an application, as a spray, of 10 lb of magnesium sulfate per acre every 10 days during the growing season.

Sugar Beets

Magnesium-deficiency symptoms in sugar beets are quite similar to those of potassium deficiency. The chief difference lies in the fact that chlorotic areas appearing within the central portion of the leaf soon become necrotic, as shown in Fig. 12-23. As the leaf is moved by the wind, the necrotic tissue breaks out, with holes resulting.

FIGURE 12-23 Sugar-beet leaves. Numbering from left to right: (1) normal, (2) magnesium deficient (note the necrotic patches throughout the leaf and slightly yellow leaf margin), (3) iron deficient (note green veins on a uniformly yellow background), (4) manganese deficient.

CALCIUM AND IRON DEFICIENCIES

The symptoms of deficiency of calcium and iron are almost identical, so these two elements are discussed together. Calcium occurs in the soil in the same kind of combination as does magnesium. The chemistry of the two elements is similar. They should be in the soil, in available form, in a rather definite ratio, variable to some extent for different crops, but generally in the neighborhood of 1 part of magnesium to 10 parts of calcium. The elements differ widely, however, in their functions within the plant and in the way their absence affects the plants.

Calcium becomes a part of cell walls and plays an important role in regulating the pH of the cellular tissue. Accordingly, the first symptoms of deficiency occur on the youngest leaves. Even before the leaves unfold, they are chlorotic, uniformly so, with prominent green veins. The spinach plant shown in Fig. 12-24 illustrates in a characteristic fashion the appearance of a calcium-starved plant. The calcium deficiency was induced by an excessive application of magnesium sulfate.

Iron occurs in soil quite largely as the oxide or hydrated oxide. When highly oxidized, the iron is not available to plants. Iron, like calcium, is not

FIGURE 12-24 Calcium-deficient spinach plant, the deficiency having been induced by an excessive application of magnesium sulfate.

translocated from old to new tissues. The visible symptoms of deficiency are identical with those for calcium deficiency. Iron deficiency in sugar beets is illustrated by one of the leaves shown in Fig. 12-23. A pH test on the soil makes it easy to differentiate between calcium and iron deficiencies. In a humid region, calcium deficiency occurs only on acid soils, unless an excessive quantity of some other basic cation has been applied. Iron deficiency, on the other hand, occurs only on alkaline soils. It is the most common on shrubs, garden crops, and greenhouse plants watered with hard water. Growers of gardenias, greenhouse roses, and other acid-loving plants find it necessary to spray their plants with ferrous sulfate to avoid iron deficiency, commonly called **lime chlorosis.**

In arid and semiarid regions, iron deficiency is common on soils high in soluble salts (saline soils and saline-sodic soils).

OTHER DEFICIENCIES

Zinc

Deficiency in zinc was first commonly reported on citrus where the soils have a pH above 6.0. Yellow streaks in narrow leaves are the common symptoms. In Florida, the disorder is called "frenching," in California — "mottle-leaf." In the past 20 years, zinc deficiency has been reported in many states on a variety of crops.

Zinc deficiency in corn has been reported widely in Florida and is common on the Histosols in Michigan and on mineral soils with pH above 7.2. Characteristic symptoms are yellow streaks between the veins of the older leaves and deposits of iron in the nodes (see Fig. 12-25).

Onions are also sensitive to a lack of zinc, particularly during the first year or two after a Histosol is first broken. After a few years of weathering, enough zinc is liberated to supply the needs of the crop.

Response of white pea beans to zinc is common on high pH soil in Michigan. The plants first become light green. Then the area between the leaf veins become yellow. Some of the leaves are dwarfed and crumpled and take on an appearance resembling sun scald. High levels of soil phosphorus tend to accentuate zinc deficiency. Varieties of beans vary with respect to response to zinc fertilization.

Zinc may readily be applied as a foliar spray. Citrus growers find this to be the most convenient and satisfactory method.

Copper

Many crops respond favorably to the application of cooper as a spray or in fertilizer. A lack of copper results in lemon yellow rather than brownish yellow onions and pale yellow instead of orange yellow carrots. Copper deficient spinach and lettuce leaves are fired around the edges in a manner similar to those starved for potassium. Copper-starved carrots, lettuce, pars-

FIGURE 12-25 Zinc-deficient corn. Note yellow streaks in older leaves and iron deposits in the nodes. Grown in Michigan on organic soil.

nips, and beets are low in sugar content, a characteristic that unfavorably affects flavor.

Copper-deficient grains (wheat, oats, and barley) are yellowish green, and the leaves fail to unroll as they emerge. Very similar symptoms occur in corn.

Citrus has long been known to be sensitive to lack of copper, especially after heavy fertilization with nitrogen. In fact, the first symptoms are not unlike those of an excess of nitrogen. Leaves are larger than normal, and shoots are long and soft. As deficiency becomes more serious, smaller than normal leaves develop and are quickly shed from twigs that are in the process of slowly dying. The early symptoms have been called **ammonia disease** and the later **die-back.**

Copper-starved citrus trees develop gum pockets between the wood and the bark and in the rind of the fruits. Some of the pockets break to allow excretion of the gum. The excretion is sometimes unnoticed because it is readily dissolved by rainwater.

Wide use of copper as a fungicide has no doubt lessened the incidence of copper deficiency. Persons growing copper-reponsive crops should study the copper status of their soils. Information based on soil type is available from local Cooperative Extension personnel.

Molybdenum

Molybdenum is the most recent element found to be essential to the growth of microorganisms and higher plants. It is especially interesting since very small quantities are used by plants. Applications are measured in ounces, rather than pounds, and a single application to the soil may last for years. Tomatoes, cauliflower, and oats have responded to molybdenum applications when grown in experimental cultures. Perhaps the most spectacular use of the element as a nutrient in the field has been on pasture grasses in Australia. The deficiency in the grasses resembles nitrogen deficiency to the extent that some investigators feel that the molybdenum, used in such small quantities, is feeding the symbiotic nitrogen-fixing organisms, which in turn are stimulated to fix more nitrogen for use by the clovers and grasses.

Clover in one location in Michigan responded favorably to molybdenum application. The symptom resembled that commonly known as nitrogen starvation on broad-leaf nonleguminous plants. New Jersey scientists have shown that lime increases the availability of molybdenum to alfalfa. It is one of the few nutrients made more available by lime.

Whiptail of cauliflower is the most common disorder known to be surely due to molybdenum starvation. The leaves elongate abnormally and are often twisted, thus the term "whiptail." In cases not sufficiently acute to cause whiptail, the heads fail to fill or may be only partly filled, and thus yields are greatly reduced. Other crops known to be sensitive to molybdenum deficiency include sweetclover, citrus, cantaloupe, lettuce, broccoli, cabbage, table beets, and kale.

Sulfur

Sulfur deficiency resembles nitrogen deficiency. A light green color of all leaves first develops. As deficiency becomes more severe, the lower leaves turn yellow, indicating translocation from old to new tissue. Leaf tips may become more intensely yellow.

Most of the sulfur in soil is in the organic matter, as, of course, is nitrogen. Soils high in organic matter should contain plenty of sulfur. The green-tissue test for nitrate nitrogen is the diagnostic tool to use. If the test for nitrate is high, the chances are good that the yellowing is due to a lack of sulfur.

REFERENCES

Cook, R. L., and K. Lawton (1944). A rapid laboratory method for the detection of manganese in fresh plant tissue. *Soil Sci. Soc. Am. Proc.* 8:327–328.

Cook, R. L., and C. E. Millar (1939). Some soil factors affecting boron availability. *Soil Sci. Soc. Am. Proc.* 4:207–301.

Cook, R. L., and C. E. Millar (1940). The effect of borax on the yield, appearance, and mineral composition of spinach and sugar beets. *Soil Sci. Soc. Am. Proc.* 5:227–234.

Cook, R. L., and C. E. Millar (1955). *Plant Nutrient Deficiencies Diagnosed by Plant Symptoms, Tissue Tests and Soil Tests.* Michigan Agricultural Experiment Station Special Bulletin 355 (1st rev.).

Crandall, P. C., J. D. Chamberlain, and J. K. L. Garth (1981). Toxicity symptoms and tissue levels associated with excess boron in pear trees. *Commun. Soil Sci. Plant Anal.* 12:1047–1057.

Davis, J. F., and W. W. McCall (1953). Occurrence of magnesium deficiency in celery on the organic soils of Michigan. Mich. State Univ. Agr. Exp. Sta. Quart. Bull. 35:324–329.

Follett, R. H., L. S. Murphy, and R. L. Donahue (1981). *Fertilizers and Soil Amendments.* Prentice–Hall, Englewood Cliffs, N.J.

Fujimoto, C. K., and G. D. Sherman (1945). The effect of drying, heating, and wetting on the level of exchangeable manganese in Hawaiian soils. *Soil Sci. Soc. Am. Proc.* 10:107–112.

Gerber, J. M. (1980). The relationship between boron soil test results and boron uptake by bean plants. *Commun. Soil Sci. Plant Anal.* 11:525–532.

Hale, J. B., Watson, M. A., and R. Hull (1946). Some causes of chlorosis and necrosis of sugar-beet foliage. *Ann. Appl. Biol* 33:13–28.

Harmer, P. M. (1941). *The Muck Soils of Michigan, Their Management and Uses.* Michigan State University Agricultural Experiment Station Special Bulletin 314.

Kotila, J. E., and G. H. Coons (1935). Boron deficiency disease of beets. *Facts About Sugar* 30:373–376.

McHargue, J. S. (1923). Effect of different concentrations of manganese sulfate on the growth of plants in acid and neutral soils and the necessity of manganese as a plant nutrient. *J. Agr. Res.* 24:781–794.

Parks, R. Q., and B. T. Shaw (1944). Possible mechanisms of boron fixation in soil: 1 chemical. *Soil Sci. Soc. Am. Proc.* 6:219–223.

Prince, A. L., M. Zimmerman, and F. E. Bear (1947). The magnesium-supplying power of 20 New Jersey soils. *Soil Sci.* 63:69–78.

Purvis, E. R., and Ruprecht, R. W. (1937). *Cracked Stem of Celery.* Florida Agricultural Experiment Station Bulletin 307.

Robertson, L. S., D. R. Christensen, and D. D. Warncke (1976). *Essential Secondary Elements: Calcium.* Bulletin E-996. Cooperative Extension Service, Michigan State University.

Robertson, L. S., and R. E. Lucas (1976). *Essential Micronutrients; Manganese.* Bulletin E-0131. Cooperative Extension Service, Michigan State University.

Robertson, L. S., and R. E. Lucas (1976). *Essential Micronutrients: Zinc.* Bulletin E1012. Cooperative Extension Service. Michigan State University.

Robertson, L. S., M. L. Vitosh, and D. D. Warncke (1976). *Essential Secondary Elements: Sulfur.* Bulletin E-997. Cooperative Extension Service, Michigan State University.

Robertson, L. S., D. D. Warncke, and B. D. Knezek (1981). *Iron: An Essential Plant Nutrient.* Bulletin E-1520. Cooperative Extension Service. Michigan State University.

Sherman, G. D., and P. M. Harmer (1941). Manganese deficiency of oats on alkaline organic soils. *J. Am. Soc. Agron.* 33:1080–1092.

Sherman, G. D., and P. M. Harmer (1942). The manganous-manganic equilibrium of soils. *Soil Sci. Soc. Am. Proc.* 7:398–405.

Truog, E., R. J. Goates, G. C. Gerloff, and K. C. Berger (1947). Magnesium-phosphorus relationships in plant nutrition. *Soil Sci.* 63:19–25.

Vitosh, M. L., D. D. Warncke, B. D. Knezek, and R. E. Lucas (1981). *Secondary and Micronutrients for Vegetables and Field Crops.* Bulletin E-486. Cooperative Extension Service, Michigan State University.

Walsh, T., and T. F. O'Donohoe (1945). Magnesium deficiency in some crop plants in relation to the level of potassium nutrition. *J. Agr. Sci.* 35:254–263.

Soil Acidity and Liming

Rainwater absorbs carbon dioxide (CO_2) from the air and becomes carbonic acid (H_2CO_3) before it reaches the soil. Within the soil it absorbs more CO_2, a by-product of organic matter decomposition, and certain organic acids. Thus soil water has such an acidifying effect that the entire soil profile may become acid, even though the parent material was strongly basic. Krug and Frink (1983) have calculated that 1 m of rainwater in equilibrium with atmospheric CO_2 will require 400 to 500 kg of $CaCO_3$ per hectare to neutralize. Thus, it is the nature of soils in humid regions to become acidic. The "Bluegrass" soils of Kentucky provide a good illustration of the results of such soil formation processes.

THE NATURE OF SOIL ACIDITY

Solution Acidity

The acidity of a solution is measured by the quantity of "active" hydrogen ion (H^+) in the solution and is usually expressed as pH — pH being the negative logarithm of the H^+ activity in solution. In many discussions, the concentration of the H^+ is used interchangeably with the activity of H^+, but the chemist will correctly point out that the true activity is usually somewhat less than concentration because the other ions in solution affect the H^+ activity. Because water dissociates very slightly, the product of the H^+ and the OH^- activity must remain constant (1.008×10^{-14} at 25°C). Therefore, at neutrality, both the H^+ and OH^- activity are equal to 1.004×10^{-7} and by definition pH = 7.0 and pOH = 7.0. This means that a solution at pH 7.0 contains 0.0000001 g H^+/l and an equivalent amount of OH^- (0.0000017 g OH, since 1 g of H^+ is equivalent to 17 g OH^-). As pH becomes lower, the H^+ concentration increases and OH^- decreases until at pH O the H^+ concentration is 1 g/l and the OH^- concentration is 1×10^{-14}—an amount that is almost neglible.

Forces Causing Acid Soils

In the beginning, most soils contained some basic minerals although some formed from acidic rocks. Some even contained free calcium carbonate. But during the soil-forming processes H^+ ions (acid) from the rainwater in equilibrium with the atmosphere dissolves minerals and the resulting bases compete with H^+ for sites on the exchange complex of soils. As basic cations are removed from carboxyl groups of soil humus, they are replaced by H^+ ions, leaving undissociated organic acids in the soil. The basic cations may then be removed from the soil by leaching or by crop removal. As basic cations become depleted from the soil, more H^+ are left in the soil solution, the soil pH becomes lower, and we say that we have an "acid soil." Indeed, some might define soil acidity as a lack of bases.

Soil Management Principle

The driving force of nature is to produce acid soils!

In most temperate-zone soils, the competition is between H^+ and basic cations—for example, Ca, Mg, and K—for exchange sites on the organic matter. But what happens as the soil becomes more acidic? Other minerals, including layer silicate clays, are dissolved by the acid and release acidic cations. Generally, these would include both iron and aluminum, but we will dwell on aluminum here. The aluminum released is a trivalent cation when fully hydrated with water. Since it is trivalent, it competes very strongly for exchange sites both on the organic matter and on clay mineral surfaces and rather easily displaces H^+ and other more desirable cations such as Ca^{2+}. Thus we find that our acid soils become high in exchangeable aluminum and aluminum bound to organic matter. Under extreme weathering conditions (e.g., in tropical soils), nearly all bases have been lost and the soil is saturated with exchangeable aluminum. Such a soil is very poor for crop growth.

The question may be asked, "Why are some ions such as aluminum acidic and others such as calcium basic?" Acidic ions will react with water (hydrolyze) to release hydrogen ions. For example, if aluminum is displaced from soil organic matter into solution, it will react with water and the water associated with aluminum ions is easier to dissociate than pure water, as is shown in the following equation:

$$Al^{3+}(6H_2O) \longleftrightarrow Al(OH)_3 + 3H^+$$

This tendency to hydrolyze is so strong that if aluminum chloride is mixed with water the pH will be less than 5.0. Calcium, on the other hand, does not have this strong tendency to hydrolyze, and calcium ions in water give a neutral pH. Of course, the reaction can go the other way and hydrogen ions will react with aluminum hydroxide to remove the hydrogen from solution by forming soluble aluminum with six waters of hydration.

$$Al(OH)_3 + 3H^+ \longleftrightarrow Al^{3+}(6H_2O)$$

Many acidifying effects in soils are produced by humans. Nitrogen fertilizers containing ammonia or substances that revert to ammonia cause a loss in bases. Heavy applications of nitrogen as NH_4NO_3 have been shown to reduced the pH of the topsoil and increase that of soil below a depth of 10 in., showing that bases had moved down in the profile. As we continue to apply larger and larger amounts of ammonia nitrogen, this effect is serious unless soil pH is managed through lime applications. For instance, continuous corn production may call for as much as 200 lb of nitrogen per acre per year. At the end of 5 years, such a program would increase soil acidity enough to neutralize a ton of 90-percent limestone.

Leaching removes bases and thus acidifies fallow soil. It must be remembered that the cations that are most competitive for cation exchange sites are the least likely to leach. Soils therefore retain aluminum at the expense of calcium, magnesium, and potassium because aluminum is more strongly attracted to exchange sites. And cropping intensifies the process. Alfalfa hay, for instance, contains enough basic cations per ton to be equivalent to 67 lb of calcium carbonate. The corresponding figure for red clover is 54 lbs, and for 25 bu of wheat including the straw it is 97 lbs. Soils low in clay and organic matter may drop appreciably in pH during a single crop rotation. Such soils should be tested frequently.

Bases are lacking in humid-region soils, roughly where rainfall is in excess of 15 to 25 in. (375–625 mm), depending on average yearly temperatures, — in other words, where leaching removes bases as they are released from organic matter and minerals and/or are displaced from the exchange complex. In semiarid and arid regions, the basic cations are not removed because rainfall is insufficient to cause leaching. As cations, and also anions, accumulate sufficiently to impair productivity, soils are said to be **saline.** As degree of alkalinity becomes very high and/or exchangeable sodium increases to toxic levels, the term **alkali** is used. Saline and alkali soils are discussed at greater length in Chapter 16.

Organic acids dissociate to furnish H-ions (H^+). Most organic materials contain enough bases, however, to be equivalent to the hydrogen, once decomposition is complete. This depends, of course, on the nature of the organic matter. Many sedges and grasses that give rise to peat soils are so low in bases that the resulting soil is very strongly acid.

In a sense, soil acidity is an abstract term. When we determine degree of acidity, we are indirectly determining several soil characteristics. Because this is true, it follows that there are several ways of determination. The most common method has already been mentioned, that of measuring hydrogen-ion activity, commonly done with a pH-meter using a glass electrode. The quantity of active or ionized hydrogen in a soil at any particular time is relatively small. In fact, 2,000,000 lb of soil (acre plow depth) containing 20 percent moisture at pH 5.0 would contain 1.816 g H^+. At pH 7.0, the same soil would contain 0.018 g H^+, a difference of 1.798 g.

This quantity of H^+ is equivalent to 89.8 g of $CaCO_3$. In other words, 89.8 g of pure limestone will raise the pH of the soil solution from 5.0 to 7.0. Everyone acquainted with liming practices knows, of course, that this cannot be demonstrated. As soon as the 89.8 g of limestone neutralizes the 1.798 g of ionized hydrogen, an equal amount of hydrogen moves from the colloidal complex into the soil solution and calcium takes its place. This would continue without measurable change in soil pH until the reserve hydrogen was reduced to the level where the exchange slows down. This tendency to resist change in pH is known as **buffering.** Soils high in colloidal material (clay and organic matter) and exchangeable aluminum (including organically bound) are highly buffered.

Soil Management Principle

Soils high in clay, organic matter, and exchangeable aluminum are highly buffered.

Indicator dyes have long been used for determination of soil acidity. The oldest of these is litmus paper. All involve the principle of changing color at different pH values. Indicators are not sensitive to differences closer than about one-half a pH unit, but many fieldworkers still use them to advantage.

Determine Exchangeable Bases, Aluminum, and Percentage Base Saturation

If soil acidity may be defined as replacement of bases with hydrogen or acidic cations, it follows that tests for those bases should give an indication of acidity. In other words, the relationship between bases (base saturation) attached to the clay and organic colloids and the concentration of active hydrogen is inverse. This is because the active hydrogen (H^+) in the soil solution reflects the amount of hydrogen and aluminum bound by organic exchange sites and clay. Since the total exchange capacity is the sum of all exchangeable bases and acidic cations (i.e., hydrogen and aluminum), one can estimate acidity by measuring the percentage base saturation (total milliequivalents of bases divided by the cation exchange capacity).

Exchangeable aluminum is usually determined by extraction with 1 N KCl. Experience has taught us that this procedure does not extract all the aluminum that contributes to soil acidity. In particular, aluminum bound by organic matter may not be extracted by this method. An extraction with 1 N NH_4OAc buffered at pH 4.8 will remove both the exchangeable aluminum and organically bound aluminum.

Chemical tests have been designed to estimate the total acidity in soils that would need to be neutralized with lime to increase the pH to a desired level. The soil-testing laboratory at Michigan State University uses the SMP buffer method of determining lime requirement. It is especially adapted to soils of the North Central region of the United States, where soils contain some exchangeable aluminum unless they are primarily sands. For other methods of determining lime requirement, refer to *Recommended Chemical Soil Test Precedures for the North Central Region,* Bulletin 499 (rev.), North Dakota State University, Fargo.

In many tropical soils, the quantity of aluminum to be neutralized is rather high and the cost of limestone is very high. For these soils, the lime requirement is related to neutralization of exchangeable aluminum present.

THE EFFECT OF SOIL ACIDITY

Agronomists rather generally accept the idea that hydrogen, per se, is not particularly toxic to plants. An excess of hydrogen ions in a system decreases the number of basic cations that may be present and, of course, lowers soil pH. Thus it may be said that the indirect effects of soil acidity really determine the pH preference of a plant. These indirect effects are rather numerous.

Acid Soils May Contain Toxic Aluminum and Manganese

When soil pH drops below 5.5 aluminum ions, released from clay minerals, hydrolyze to become exchangeable cations with the results of the hydrolysis adding hydrogen ions that further increase soil acidity. As pH continues to drop, aluminum concentration becomes sufficient to be toxic to many plants. Iron has similar reactions but is less toxic. As shown by the chart in Figure 13-1, manganese is highly available from 6.5 to 5.0 and may be toxic to some plants if the concentration in the soil is high.

Acid Soils are Low in Plant Nutrients

Plants obtain their hydrogen from water, whereas calcium, potassium, and magnesium are needed by all plants and must be absorbed as cations from the soil solution. In other words, acid soils are infertile because the nutrients essential to plant growth are missing or are in very low supply. The effects of such deficiencies on the growth of plants are rather completely discussed in Chapter 12, you should remember that many basic nutrients are deficient in acid soils and that correction of the acidity by application of calcium carbonate (limestone) leaves the soil still deficient in all other basic nutrients. In fact, the deficiency of some of the others may be intensified by adding only calcium. Soluble fertilizers may be used to correct the unbalanced condition brought about by the limestone application.

Effect of Soil Reaction on Nutrient Availability

Agricultural soils vary in pH range from 4.0 to 10.0. All common agricultural plants exhibit some degree of preference within that range. Availability of nutrients is one of the reasons. Figure 13-1 illustrates graphically the effect that degree of acidity or alkalinity has on this availability.

Calcium, Magnesium, Potassium

As shown by the width of the bands in Fig. 13-1, the availability of calcium, magnesium, and potassium to plants drops rapidly as soil pH falls below 7.0. As has been pointed out, this is simply because these nutrients are absent. Either the soil was developed largely from acidic rocks or, where the parent rocks were basic, these minerals were dissolved and removed by leaching. Furthermore, cation exchange removed most of the basic ions in the exchange complex, leaving aluminum and hydrogen in their places.

Nitrogen

The effect of pH on availability of nitrogen is indirect. The microorganisms that change ammonia to nitrite, then to nitrate, do not function in strongly acid soils. Small amounts of nitrate in such soils are probably the result of small localized zones where pH is higher as a result of the disintegration of basic minerals.

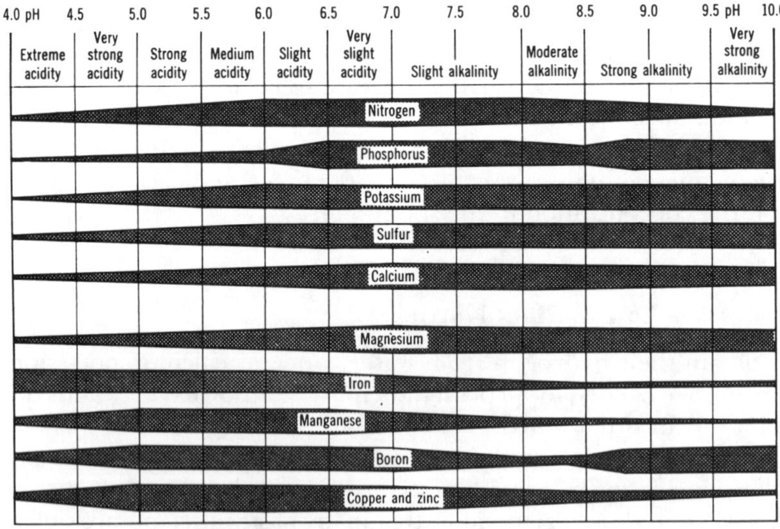

FIGURE 13-1 Agricultural soils range in pH from 4.0 to 10.0. The width of the bands shows relatively the effect that soil reaction has on the availability of 11 different nutrients. Note that extreme acidity results in low availability of all nutrients except iron. (From *Michigan Extension Bulletin* 314, revised 1955. See also *Michigan Extension Bulletin* E-471, 1981.)

Bacteria and protozoans prefer the pH range 6.0 to 8.0; actinomyces, in particular, are even more specific, preferring 7.0 to 7.5. Only the fungi seem to prefer acid conditions as low as 4.0 to 5.0; this may be due to lack of competition from other microorganisms in that pH range.

The two great groups of nitrogen-fixing organisms fail to function in strongly acid soils. The nonsymbiotic, aerobic nitrogen-fixing azotobacters do not function effectively in soils more acid that 6.0, although they are found in soils having somewhat lower pH. The anaerobic, free-fixing nitrogen bacteria clostridia are able to function at lower pH levels but are not very effective in well-aerated agricultural soils. The symbiotic nitrogen-fixing organisms (genus *Rhizobium*) are often spoken of as the most important single group of soil organisms. These organisms are found in soils where specific legumes have been grown. To make certain of their presence around the roots of the host plant, the seeds are inoculated with the organisms at the time of planting. The bacteria function in nodules on the legume roots. It is believed that rhizobia are not so specific as far as actual pH is concerned, but they are strong feeders on calcium, an element usually lacking in acid soils. Thus they are commonly classed as acid-intolerant organisms.

Perhaps the preferences of the nitrogen-fixing organisms and the nitrification organisms are largely responsible for the general conclusion, as shown in Fig. 13-1, that pH 6.0 to 8.0 is the most desirable range for nitrogen availability.

Phosphorus

Plants prefer the acidic phosphate ion $H_2PO_4^-$. Arizona workers have shown that plants starve for phosphorus in Arizona soils (see Fig. 13-2) because the pH is so high that most of the soluble phosphorus is ionized as HPO_4^{2-} rather than $H_2PO_4^-$. They found that above pH 7.6 plants starve for phosphorus even in solutions high in soluble phosphorus. They could not use the mono-hydrogen phosphate.

Monocalcium phosphate, $Ca(H_2PO_4)_2$, and the corresponding magnesium phosphate are most stable at pH 6.5 to 7.0. As pH rises owing to calcium or magnesium, the monohydrogen, then the basic phosphate (PO_4^{3-}) are formed, the latter above pH 8.5 as shown in Fig. 13-2. Availability to plants decreases accordingly.

You may wonder, if plants are unable to use HPO_4^{2-}, how are they able to grow at all within the 8.0 to 8.5, or even higher, range. The answer is simple. Plants are able to adjust soil pH around their roots because they excrete CO_2 that combines with water to form H_2CO_3. If the soil mass has a pH below 8.5, plants are generally able to excrete enough CO_2 to keep the pH down where the roots obtain sufficient $H_2PO_4^-$ for normal growth but above pH 8.5, alkalinity may be just too great. Plants vary, of course, in the amount of CO_2 excreted. Those that excrete the most are able to obtain sufficient phosphorus to survive. They are said to be "alkali" tolerant. No doubt other

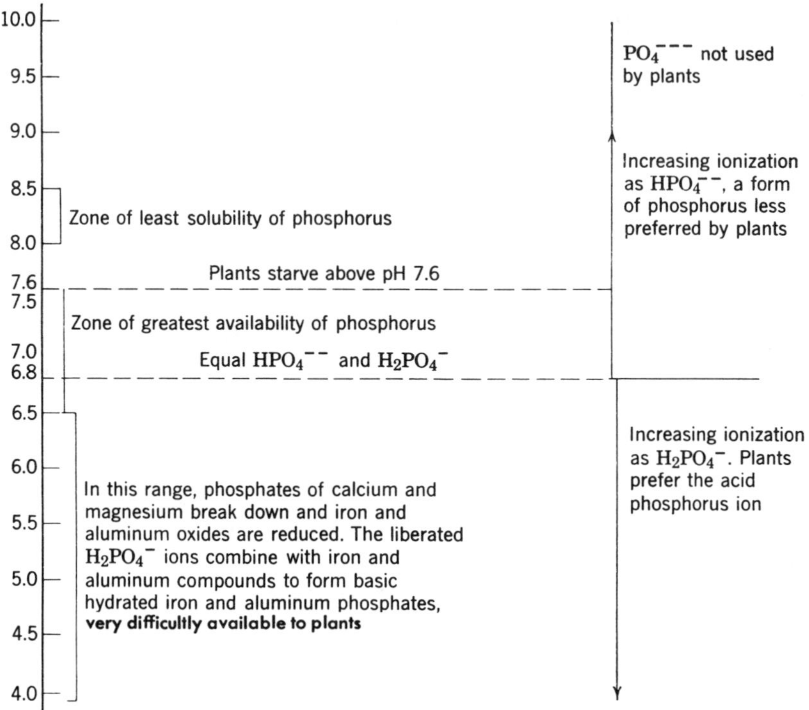

FIGURE 13-2 The effect of soil reaction on the solubility of phosphorus compounds in soil and the ability of plants to take in different phosphate ions. (From W. T. George and J. F. Breazeale. *Phosphate Solubility Studies on Some Unproductive Calcareous Soils,* Arizona Agricultural Experiment Station Technical Bulletin 35, 1931, adn W. T. George. The relation of potential alkalinity to the available phosphate in calcareous soils. *Soil Sci.* 39:443–452, 1935.)

nutrients such as manganese and iron are also made available by the lowered alkalinity around the roots.

We have discussed only the pH range above 6.8. Why not solve the phosphorus problem by allowing the pH to fall below 6.8, perhaps to 5.0, or even 4.0, where essentially all phosphorus would ionize as $H_2PO_4^-$? In water solution or in organic soils this might be a good answer, but in mineral soils other complexities enter the picture. As the pH falls (see Fig. 13-2), combinations occur between the $H_2PO_4^-$ and soluble compounds of iron and aluminum to form basic hydrated phosphates so slowly available that huge quantities are necessary to furnish subsistence to growing plants.

In the first volume of the *Journal of Agricultural Science* (1905), Director A. D. Hall of Rothamsted listed "Shenington" soil among those on which rutabagas responded favorably to phosphate fertilizers even though it was very high in total phosphoric acid. He stated "The explanation is to be found in the fact that the soil is derived from the Marlstone (Lower Lias) and is largely made up of hydrated ferric oxide, containing as much as 28.16 percent Fe_2O_3 in the air-dried soil, whereas ordinary soils only yield 2–4 percent to hydrochloric acid. This tends to show that an excess of ferric oxide, like calcium carbonate, keeps the soil phosphoric acid in an undissolved state and the soil water low in phosphoric acid, so starving the crop as regards this particular constituent." This phenomenon, termed **phosphorus fixation,** is largely responsible for the vast amount of research on methods of applying fertilizer. Many studies have been directed toward an attempt to place phosphorus in the soil so plant roots might pick it up but in a form or location so that the chemical reaction between phosphorus and iron and aluminum compounds would be minimized.

As shown by Fig. 13-2, attention to the entire picture seems to indicate that a soil pH between 6.5 and 7.0 is best for phosphorus availability in mineral soils. On low-fixing soils, that is, those low in iron and/or aluminum, it might be safe to allow the level to drop as low as 6.0, but there should never be an attempt to raise pH above 7.0. In organic soils, iron and aluminum oxides are not present to cause excessive fixation, so phosphorus availability may be higher at pH levels as low as 5.5.

The question is less clear for highly weathered soils. Here liming to pH's above 5.5 does not appear to improve phosphorus availability. Sims and Ellis (1983), on highly weathered soils, found that phosphorus uptake was equal to or greater when the soil was limed to precipitate exchangeable aluminum than when limed to that recommended by SMP.

Jenny, Valmis, and Martin (1950) presented some results obtained by growing romaine lettuce in pot tests. They found in this biological test that about one-half as many California soils having a pH between 6.4 and 7.9 were deficient in phosphorus as were the more acid or more alkaline soils. All were mineral soils.

Iron, Manganese, Boron, Copper, and Zinc

These nutrients, commonly called **trace elements,** are shown by Fig. 13-1 to be the most readily available in the pH range between 5.0 and 6.5. Availability of iron continues high until the pH falls below the so-called agricultural plant range. Note also that iron starvation may occur at pH 6.0, and the chance of its occurring increases rapidly above that pH. In fact, an excess of phosphorus fertilizer applied to soil above pH 7.0 induces iron starvation in a number of plants. This occurs frequently with lawn grasses, where irrigation with hard or sodic water raises the pH. This is an interesting situation because it is the reverse of what happens at low pH, where an excess of soluble iron is responsible for the fixation of applied phosphates and the inducement of phosphorus starvation.

Manganese also becomes less available in soil having a pH higher than 6.5, and the deficiency may become serious above pH 7.0. Both manganese and iron are unavailable because of high degree of oxidation. Manganic manganese and ferric iron are not readily used by plants. In high pH soils, both manganese and iron may be needed as fertilizers. They can be most efficiently applied as leaf sprays because they are quickly oxidized to unavailable forms when placed in alkaline soil.

Boron, copper, and zinc, are also likely to be deficient in alkaline soils, above pH 7.0. In sodic soils, however, boron compounds again increase in solubility, because sodium borates are more soluble than calcium and magnesium borates. In fact, boron toxicity is quite common in desert areas where sodic soils are common. Copper and zinc become increasingly insoluble as the pH levels rise to 10.

Manganese, boron, copper, and zinc are sometimes deficient in very strongly acid soils for two reasons, extreme leaching of soluble forms and the formation of certain acid-insoluble compounds. This has been observed particularly with manganese in citrus production in Florida and with copper in the production of several crops grown in strongly acid organic soils in Michigan.

Soluble Substances

Sufficient aluminum to cause injury to plants may be in the soil solution where the soil pH is below 5.0. Spurway realized this when he included an aluminum test with his "Simplex" soil test kit in 1932.

Manganese has been widely shown to be an essential element and, as pointed out in Chapter 12, many instances of widespread deficiency are known. However, an excess of manganese is toxic, and a few such cases have been reported. The best-known example is in the production of potatoes on certain strongly acid soils of Wisconsin.

LIME AS A CONTROL FOR ACIDITY

Lime is the cheapest and most satisfactory material for the correction of soil acidity. Agriculturally speaking, the term **lime** refers to all compounds of calcium and/or magnesium capable of neutralizing acidity. Only the oxides, hydrates, silicates, and carbonates of these two essential nutrients fall in this classification.

Sulfates and chlorides cannot be used as liming materials because on hydrolysis they form strong acids (H_2SO_4 and HC1). At the other pH extreme, carbonates or hydrates of sodium and potassium are harmful to soils. They break down organic matter and soil minerals and destroy soil structure. The "alkali" or sodic soils of the arid regions are, in fact, the result of accumulation of sodium and potassium. Soil pH goes as high as 10, and few plants survive.

Action of Lime in the Soil

The addition of bases to an acid soil brings about a reaction that in effect is a reversal of the force that brought about the acidity.

Liming, however, is not truly reversing the reaction that occurred when the soil became acid, which is one of the reasons why liming is sometimes less effective than the farmer expects. As soils become acid through natural processes, all bases are gradually reduced in concentration on the exchange complex and aluminum or hydrogen ions replace them. A large amount of calcium is then mixed into the soil, and, by the law of mass action, calcium replaces mostly hydrogen (and/or aluminum in very acid soils) because it is the most plentiful ion, but perhaps it also replaces some magnesium and potassium. The result is a soil with corrected pH but with an unbalanced mineral nutrient condition.

Soil Management Principle

Magnesium deficiencies are best corrected by long-range planning of a liming program.

Soil testing makes it possible to evaluate the mineral balance and make corrections through the application of commercial fertilizers. The use of dolomitic limestone instead of entirely calcic material may prevent inducement of a magnesium deficiency. Magnesium may also, of course, be included in commercial fertilizers.

Advantages of Lime

Lime eliminates the ill effects of soil acidity. Calcium and magnesium are supplied as nutrients, and phosphorus and molybdenum availability are increased. An abundance of calcium and lowered acidity favor soil organism activity that hastens general soil-formation processes. Improved conditions for higher plants are the direct results.

Lime makes it possible to grow soil-building legumes. This is almost a must for the livestock farmer, and it greatly simplifies the soil conservation plans for even the cash-crop farmer who wishes to grow legumes for green manure and protective cover.

Lime improves soil structure for temperate-zone soils. But highly weathered soils that are saturated with exchangeable aluminum have excellent structure. In this case, if the aluminum is replaced with calcium, the structure will not be improved. The action is direct as long as there is free lime in the soil because the excess of calcium in the soil solution causes flocculation of the soil colloids. Indirectly, however, the effect is even greater. Soil organism activity is really responsible for soil aggregation, and lime, as mentioned previously, greatly stimulates organism activity, hastening organic matter decomposition and thus bringing about all the good effects accruing therefrom. The farmer must be careful, of course, to frequently replenish the organic matter being decomposed. In other words, the farmer must feed his soil organisms with crop residues, cover crops, and green manures or animal manures.

Lime applied to an extremely acid soil raises the pH enough to make aluminum, iron, and manganese less soluble and so prevent their being toxic to plants. There is always a possibility of toxicity from compounds of these elements in soil having a pH below 5.5.

In mineral soils, possibly excepting Oxisols, the efficiency of fertilizers containing soluble forms of phosphorus is increased by liming up to a pH of 6.8. This is the result of lessened fixation into difficultly soluble compounds of iron and aluminum. It is in effect a result of the reduced solubility of iron and aluminum compounds. Liming keeps them in forms with less affinity for $H_2PO_4^-$ ions.

FORMS OF LIMING MATERIALS

Chemically, lime is CaO (calcium oxide). The farmer thinks of lime as any material that will correct soil acidity. The agronomist thinks of calcium and magnesium compounds.

Limestone

The leading liming material is limestone, a sedimentary rock. According to Whittaker, Anderson, and Reitemeier (1959), ground limestone made up 90 percent of all liming materials used in the United States in 1955. Limestone may be mainly calcium carbonate ($CaCO_3$), or it may be a mixture of calcium and magnesium carbonates. Such mixtures are commonly called **dolomite** or dolomitic limestone.

Calcium carbonate in the sedimentary form is so widely distributed that it is used as a standard for rating the neutralizing value of all other liming materials. Calcite, a pure $CaCO_3$ mineral, is used for determining the normality of acids. Since the compound has a molecular weight of 100, calculation of the neutralizing value of other forms of lime, based on $CaCO_3$ as 100 percent, is simple.

Neutralizing Value and Fineness

The value and effectiveness of limestone as a soil amendment depend on neutralizing value and degree of fineness. Most limestone deposits range in neutralizing value from 75 to 95 percent. Those higher in purity than 95 percent usually contain magnesium and are therefore dolomitic limestones. Those having less than 75 percent neutralizing value are usually not quarried for agricultural lime.

Large limestone particles raise soil pH slowly and have less effect on crop yields than do small particles. This is to be expected, since small particles have more surface for chemical action, plants cannot use calcium until it is liberated from the limestone particles, and the value of lesser acidity is not realized until the H^+ ion concentration is actually reduced.

Barber (1958) summarized the results from 18 limestone-fineness experiments in Alabama, England, Illinois, Iowa, Kentucky, Maryland, Michigan, Missouri, Ohio, Ontario, and Rhode Island. The results showed that in 16 of the experiments, yields were greater where the finer materials were used. As an average for the 18 experiments, 60-mesh limestone resulted in yields more than 50 percent higher than those resulting from limestone with only 10 percent fine enough to pass the 60-mesh sieve.

Stated in another way, it was necessary to apply 7.5 tons of the coarsest material (10 percent through 60 mesh) to bring about yields equal to those obtained where 3 tons of the 100 percent 60-mesh material was used.

The importance of fineness of grinding as far as the value of limestone is concerned is recognized by the State Agricultural Stabilization Committees in establishing practice specifications for cost-sharing in the purchase of liming materials. They also, of course, specify minimum neutralizing values. As an illustration, the specifications for Michigan are as follows:

> Neutralizing value shall be 80 percent or higher; 85 percent through a U.S. standard 8-mesh sieve; 25 percent through a U.S. standard 100-mesh sieve; and moisture content not to exceed 8 percent

You are urged to obtain similar specifications from your Cooperative Extension advisor, the ASC in your own state, or, in other countries, from your department of agriculture.

Regulations Regarding Liming Materials

Most of the U.S. states where agricultural lime is used have a law governing its sale. The laws vary but have some features in common. One typical law defines "liming material," then specifies that each package, lot, bag, or container be labeled with the following information.

1. Net weight of the contents, except for materials sold only by the cubic yard.
2. The name, brand, or trademark.
3. Name and principal address of the manufacturer or person responsible for placing the commodity on the market.
4. The minimum neutralizing value in terms of calcium carbonate, or, for materials sold by the yard, in terms of calcium carbonate equivalent per cubic yard.
5. The degree of fineness expressed as
   (a) Minimum percentage passing through 8-mesh sieve.
   (b) Minimum percentage passing through 60-mesh sieve.
   (c) Minimum percentage passing through 100-mesh sieve, except that in the case of marl, beet-sugar-factory refuse lime, paper-mill refuse lime, carbide-plant refuse lime, water-softener refuse lime, wood ashes, and other forms of by-products or refuse limes, there need be no guarantee relative to fineness.

State laws make no regulations regarding minimum materials that should be used in agriculture; they are simply written to provide enforcement regarding guarantees of the product.

Considerable confusion exists regarding the terms used for commercial grades of limestone. Perhaps the Ohio nomenclature, promoted by the Department of Agronomy of Ohio State University, is most widely used by the agricultural lime trade. Specifications are as follows.

Grades	Passing 8-mesh (percent)	Passing 20-mesh (percent)	Passing 60-mesh (percent)	Passing 100-mesh (percent)
Screenings	80	—	—	10–19
Coarse meal	80	—	—	20–29
Fine meal	80	—	—	30–39
Ground limestone	95	—	—	40–59
Pulverized limestone	—	95	—	60–79
Superfine limestone	—	—	95	80–100

Other Liming Materials

The purchase of other liming materials should be based on a comparison with limestone, both as to neutralizing value and cost uniformly spread on the land. A comparison of the neutralizing values of oxides, hydroxides, and carbonates of calcium and magnesium with calcium carbonate is shown in Table 13-1. They should be purchased accordingly. We should remember, of course, that oxides and hydroxides are very fine. They would compare agronomically with very finely ground limestone. Furthermore, they are more soluble than limestone, so their rates of activity in the soil are much greater. Oxides are very disagreeable to handle, and they hydrate rapidly when exposed to air. They are seldom used in agriculture except as they occur in certain by-product materials.

TABLE 13-1 Relative Neutralizing Value of Different Forms of Lime

Forms of lime	Molecular weight	Neutralizing value (percent)	Amount necessary to be equal to 1 ton $CaCO_3$ (lb)
$CaCO_3$	100	100	2000
$MgCO_3$	84	119	1680
$Ca(OH)_2$	74	135	1480
$Mg(OH)_2$	58	172	1160
CaO	56	178	1120
MgO	40	250	800

Source: Michigan State University Extension Bulletin 314, 1955.

Hydrated lime is sometimes recommended where limestone must be shipped long distances or where a rapid, short-lasting liming effect is desired. The best illustration of this situation is on the farm where lime is needed for alfalfa but where potatoes are also grown in the rotation. A relatively small application of hydrate, or very finely ground limestone, applied just before alfalfa is planted will make that crop possible but will be used or leached out before potatoes come in the rotation. On such soil, alfalfa, once established, is usually left down for several years. A common rotation is alfalfa for 3 or 4 years, rye green manure, and then potatoes. The rye is planted in late summer of the last alfalfa year.

Marl

In areas where soil developed from glacial materials, marl is common. It is precipitated calcium carbonate with small amounts of magnesium carbonate, which, over the centuries, have leached from surrounding uplands. At the time of leaching, the bases were in ionic or bicarbonate forms. Marl is usually overlaid by a deposit of organic soil or is at the bottom of shallow lakes. Impurities of organic soil, silt, and clay are always present. Since organic matter shrinks when drying, marl should always be purchased on the basis of pounds of $CaCO_3$ equivalent per cubic yard and not on the basis of the neutralizing value of the dry product. In fact, all liming materials purchased by the cubic yard should be specified in this manner.

The importance of purchasing marl in this manner was strikingly illustrated by the test results from 311 consecutive samples sent to the Michigan laboratory for testing. They were tested in two ways, neutralizing value on a dry basis and pounds $CaCO_3$ per cubic yard. One-half of the samples that tested below 60 percent in neutralizing value graded good or high in pounds $CaCO_3$ per cubic yard. At the other extreme, 4 percent of the samples testing over 90 percent in neutralizing value were actually of low grade, containing less than 800 lbs $CaCO_3$ per cubic yard.

Marl is a very good liming material, but is less uniform than dry, finely divided materials. ASCs are likely to approve cost-sharing for marl on the basis of amount of $CaCO_3$ per cubic yard. In Michigan, the minimum is 800 lb.

Slag

Blast-furnace slag is a by-product of the steel industry. The water-quenched or granulated form is recognized by many experiment stations as being quite suitable as a liming material. It consists of silicates of calcium and magnesium, is generally marketed as a dry, finely divided product, and should be purchased and applied on the basis of neutralizing value. The Michigan ASC requires it to have 80 percent neutralizing value if it is to qualify for cost-sharing.

More recent methods in steel production have resulted in a different slag product called **BOF slag** (basic oxygen furnace slag) that has a much higher calcium carbonate equivalent than does the old blast-furnace slag. It contains more than 90 percent neutralizing value as compared with 70 percent for the old material, making it equal to the better grades of limestone.

Carbide, Sugar-factory, Water-softener, or Paper-mill Refuse

These materials are all perfectly satisfactory for correcting soil acidity, as long as they contain sufficient lime per cubic yard to justify the original cost and the expense of hauling and spreading. Again, a clue to minimum quality may be obtained from the Michigan ASC's willingness to allow cost-sharing if the materials contain 800 or more pounds $CaCO_3$ equivalent per cubic yard. They usually test well over this minimum. As is the case with marl, the purchaser should be sure to obtain uniform distribution. Most by-product liming materials are high in moisture, and some lack uniformity in particle size.

Chalk, Oyster Shells, and Such

Any form of calcium oxide, hydroxide, or carbonate, natural or by-product, may be used as long as the material does not contain toxic substances. Chalk is soft calcium carbonate rock, widely used in England but found in only a few localities of the United States. Like limestone, it must be ground before using.

Oyster shells and other seashells, mostly composed of calcium carbonate, are satisfactory for liming if ground sufficiently fine. They are harder than chalk and should be handled more like limestone.

AMOUNTS OF LIME TO APPLY

There is no correct pH level for all crops. Whittaker, Anderson, and Reitemeier (1959) have published the pH preference range for 51 crops grown widely in the United States. The list is shown in Fig. 13-3. Note that only four crops have a range extending as high as 7.5 and three of that group have a wider range below than above the neutral (7.0) point.

Note further that 41 crops do well across the full 6.0 to 6.5 range, whereas only 27 have their preferred range running as high as 7.0. In fact, more plants prefer the full 5.5 to 6.0 range (the number is 32) than the range between 6.5 and 7.0. In other words, plants generally prefer slight to medium acidity.

Field applications of limestone seldom bring pH levels up to those expected from the amount of limestone applied, because much of the limestone is coarser than the lime used in the lime requirement test, or where only a pH test was made, the buffer capacity of the soil was underestimated. Furthermore, as pH rises, the remaining coarse particles of limestone dissolve more

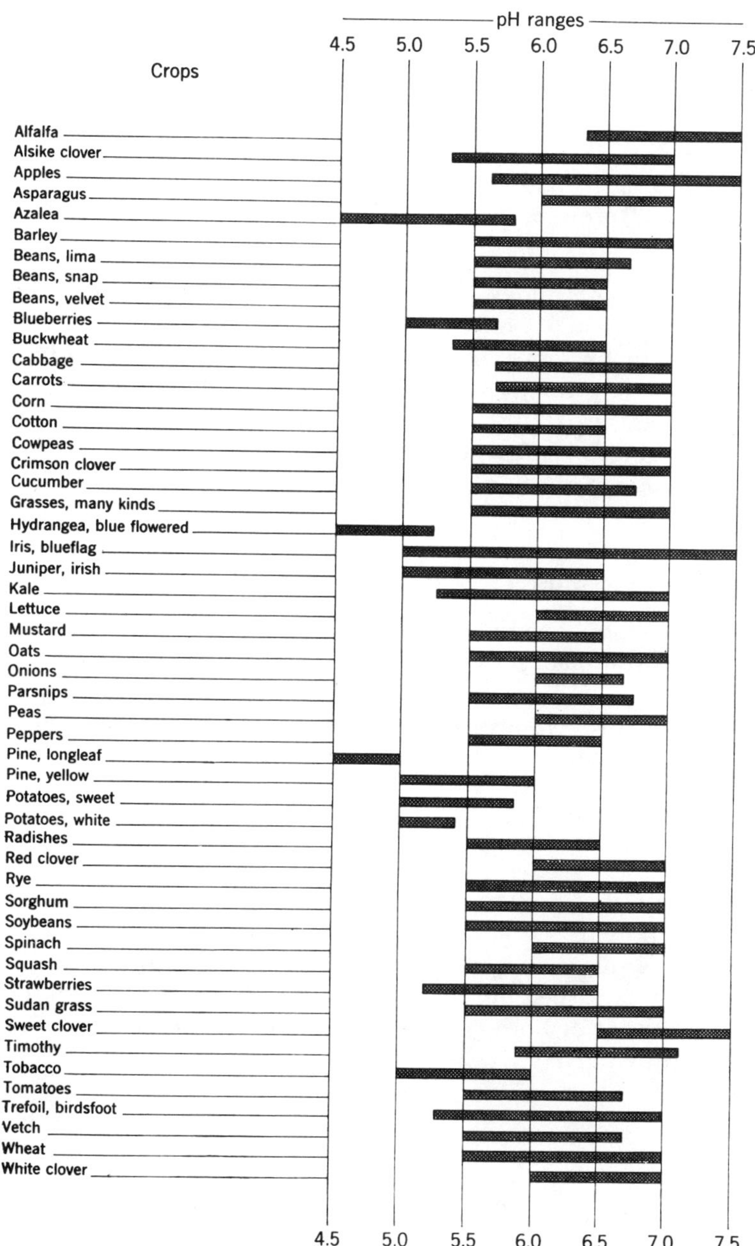

FIGURE 13-3 Suitable pH ranges for various crops and ornamental plants. (From C. W. Whittaker, M. S. Anderson, and R. F. Reitemeier. *Liming Soils, an Aid to Better Farming.* USDA Farmers Bulletin 2124, 1959, by permission.)

and more slowly until finally there is very slow solution effect. Thus it becomes very difficult to raise pH levels above 6.5 with limestone. This is especially true with dolomite, which is harder and less soluble than is calcic limestone.

Organic Soils Present a Special Problem

Crops seem to have a lower pH preference range when grown on organic soil than when grown on mineral soil, because the availability of calcium from organic colloids is high and little iron and aluminum are available to fix phosphorus. Furthermore, many of the crops more commonly grown on organic soil are low in preference range. Lime is seldom recommended for these soils unless the pH drops below 5.0. Where lime is needed, rates must be high because the soils are very highly buffered. Chapter 22 gives a more complete discussion of soil reaction and liming of organic soils.

Factors That Determine Lime Requirement

A pH determination is simply a measure of the activity of H^+ ions in the soil solution. This so-called active acidity, as noted earlier in this chapter, is negligible as far as liming is concerned. It is necessary to measure the reserve acidity — the hydrogen associated with the clay, organic colloids, and aluminum — to be able to calculate lime requirement.

Buffer Capacity

Clay and organic matter determine buffer capacity and have a direct bearing on lime requirement. Organic colloids exert more buffering than do clay colloids. As a result, well-decomposed organic soils are the most highly buffered of all soils. Washed quartz sand without clay or organic matter is not buffered. As either or both clay and organic matter increase, buffering increases and with a given pH, lime requirement increases.

Nature of the Clay

The 2:1 clay minerals that include montmorillonite (a smectite), hydrous mica (illite), and vermiculite hold calcium against release to plants much more strongly than do the 1:1 clays such as the kaolinites. The 2:1 clays must be 70 to 90 percent saturated with calcium to be sure of sufficiently easy release to plants, whereas with the 1:1 clays, 40 to 50 percent saturation is sufficient. Furthermore, the exchange capacity of 2:1 clays is much higher than that of 1:1 clays. In short, the soils high in content of 2:1 clays require much more lime at a given pH than do those high in 1:1 clays. The soil series classification with taxonomic families and subgroups furnishes a clue to clay type.

Degree of Saturation with Basic Cations

Percentage saturation with basic cations has a marked effect on the amount of lime necessary to raise soil pH. As percentage base saturation increases, calcium release to plants is easier, so the need for lime application becomes less acute.

Subsoil Lime Content

In soils developed from limestone or high lime glacial till, the parent materials are high in lime. Free calcium carbonate exists at shallow depths in many such soils. If soil structure is good, plant roots penetrate rapidly to the lime-rich subsoil. Alfalfa may thrive on such soil even though the surface is quite acid. High calcium starter fertilizers are especially valuable on such soil.

Crops to be Grown

Figure 13-3 shows how plants vary in pH preference. It is well to consider this when deciding on a cropping system. A farmer growing alfalfa, corn, oats, and wheat should plan on liming to pH 6.8. After a lime requirement test, if the farmer follows the local county agricultural agent's recommendation for lime, the resulting pH of the soil will probably be about 6.5, quite satisfactory according to the chart.

Assume the crops to be grown are alfalfa, corn, tobacco, and oats. Tobacco does better if the pH is held below 6.0. Corn and oats will grow satisfactorily at 6.0, but it is not so with alfalfa. It is well then to apply a relatively small amount of very fine limestone or hydrated lime when the oats are planted. The rise in pH will be quick and should be just enough to ensure a satisfactory alfalfa crop. When tobacco comes again in the rotation, the pH will again be down to the preferred level.

Form and Grade of Lime

Lime recommendations are usually based on the use of agricultural meal or medium-grade materials. Finer materials may be applied at lower rates and coarse materials at higher rates.

By-product liming materials and marl should be applied at rates equal to the tested limestone requirement of the soil divided by their pounds of calcium carbonate equivalent per cubic yard.

Hydrated lime should be applied at three-fourths the rate recommended for limestone if the goal is complete acidity correction. Lesser rates may be used in special cropping systems where acid-loving crops are included.

Determining Lime Requirement

In the past, several methods of arriving at lime requirements have been used. Many of them related lime requirement to soil pH, soil texture, and organic matter content. An example of this is shown in Table 13-2 with data from Oklahoma. These factors do not consider the important contribution of aluminum to soil acidity and therefore give erroneous estimates for soils high in exchangeable aluminum and nonexchangeable acidic aluminum. For that reason, we feel that lime requirement should be based on a chemical test.

Soil pH may be used to determine if a soil needs liming. If it does, then a chemical test should be run to determine the proper amount of lime.

Chemical methods for determining lime requirement are described by McLean (1980). Tisdale and Nelson (1956) have suggested a clever way of determining lime requirement in the laboratory.

Perhaps the most accurate of all lime-requirement tests is the actual liming of small samples, say 500 g each, with the same lime to be applied in the field. If coarse limestone is to be used, it is necessary to wait several months for the small samples to react with the lime. Fine limestone, hydrated lime, or finely divided by-product limes react with soil rather rapidly. A few weeks is long enough for equilibrium to be established. By applying variable amounts of lime to the 500-g samples, and then determining pH after equilibrium is established, we may arrive at a fairly accurate recommendation for the field.

Danger of Overliming

As shown by Fig. 13-3, few plants have pH preference ranges extending above 7.0. We have noted further that phosphorus, manganese, iron, boron, copper, and zinc are less available to plants at pH levels above than below 7.0. There seems to be plenty of evidence then that excessive lime applications should be avoided.

As has already been mentioned, limestone is so difficultly soluble in soil with pH near 7.0 that there is actually little danger of injury from overliming with limestone, especially with the coarser grades usually applied on the farm.

Marl, hydrated lime, and certain by-product limes are finer and more

TABLE 13-2 Limestone Requirement for Soils of Varying Texture and Degree of Acidity

| | | Tons per acre | | |
Soil reaction	pH range	On sand soils	On sandy loams, loams, and silt loams	On clay loams and clay
Very slightly acid	6.6–6.7	None	None	None
Slightly acid	6.1–6.5	None	1	1 to 2
Moderately acid	5.5–6.0	1	1 to 2	2 to 3
Strongly acid	5.0–5.4	1 to 2	2 to 3	3 to 4
Very strongly acid	Less than 5.0	2	3 to 4	4 to 5

Source: Oklahoma Extension Circular 625.

soluble than is limestone. Furthermore, some of these materials are cheaper and therefore are sometimes applied in larger quantities. Also, they sometimes are not distributed uniformly, which may result in overliming of small areas. It is advisable to regulate rates and methods of application to avoid excesses. Economy also favors such a precaution.

METHODS OF APPLYING LIME

In the early days, most of the limestone used in the United States was applied with a two-wheel spreader similar to a grain drill. Such spreaders are still used on experimental plots and where small amounts of lime are applied frequently, as in a rotation that includes both lime- and acid-loving plants. In such a cropping system, it is often advisable to apply as little as 500 to 1000 lb of pulverized limestone or hydrated lime per acre. Fertilizer drills may also be used.

At present, most of the liming material used in the United States is applied with a truck spreader like that shown in Fig. 13-4. Farmers purchase the lime for a price quoted to include spreading at a designated rate per acre. The hoppers on truck spreaders have a conveyor in the bottom that carries the lime to the back where it falls on spinners or a transverse conveyor. The spinners throw the lime out in a fan-shaped pattern several feet to each side of the truck. By using two spinners running in opposite directions, it is possible to get fairly even distribution. The greatest error results from particle size variation. The large particles are thrown farther than the fine material, and

FIGURE 13-4 A truck spreader with transverse conveyor for even distribution of liming material. A canvas apron may be added to prevent blowing.

dust-size particles are easily blown away by the wind. Overlapping may correct for some of this variation due to particle size.

Truck spreaders equipped with a transverse conveyor do a more uniform job of spreading. Hoods are often used on such spreaders to prevent blowing. They can do a very nice job if the driver is reasonably careful with adjustments and driving.

Marl and Wet By-product Limes

Some truck spreaders will handle wet marl; others will not. Large amounts of marl and by-product limes have been spread with manure spreaders. A convenient method is to cover each load of manure with a few inches of lime. A little experience enables the farmer to adjust to the desired rate of application. Only carbonate forms of lime should be applied with manure. Hydrated lime or oxides react with the ammonium compounds in the manure to liberate ammonia so significant losses of nitrogen may result.

Broadcast versus Row Applications

As the previous discussion indicates, most of the lime applied in this country is broadcast and worked into the soil. This is generally the most desirable method to use since all soil particles should come in contact with lime particles, as much as is possible. This is advisable because lime does not move far in a horizontal direction. Good mixing with the soil is very important. The mixing is brought about by tillage operations during preparation of the soil for crops. A farmer practicing minimum tillage should be extra careful to obtain uniform lime distribution. To some extent, lime does move down and up in the soil with movements of soil moisture.

Row applications of lime at the time of seeding legumes have been tried experimentally, and with good results. They are difficult to manage, however, because the farmer should also apply fertilizer. To apply both materials, it is necessary to go over the land twice or mix the fertilizer and lime. The latter alternative is permissible if limestone is being used. Cement mixers do a satisfactory job of mixing. The application of small amounts, 300 to 500 lb, of finely ground limestone at the time of seeding sweetclover has been quite successful in Kansas. But in general it is better to broadcast larger amounts at less frequent intervals; particularily, alfalfa has done much better where the latter method was used.

Generally, the broadcasting of enough lime for the entire soil mass is recommended over row applications. Specialty farming, however, where it is desirable to keep the soil quite acid, may be simplified by using lime in the row for a single lime-loving crop in the rotation. In general farming, row applications would be more costly than less frequent and larger broadcast applications, largely because of labor costs.

Depth of Application

In soils well supplied with lime in the subsoil, surface applications should be mixed to plow depth. Where the amount to be applied is 4 tons or more per acre, it is wise to make two applications, one before, one after plowing. A year may intervene, if the crop to be grown that year is tolerant of acidity. Deeper than normal plowing may be advisable when two applications of lime are being made. Thus the reaction of a good deep layer is corrected.

TIME OF APPLYING LIME

There is no best time to apply lime, unless it is "right away!" If lime is needed, it should be applied as soon as the farm program permits. Extremely acid soils are nonproductive, except for a few acid-loving crops, and their improvement without lime is difficult or at least impractical.

Perhaps one rule should be followed. Apply lime several months in advance of the seeding of lime-loving crops such as alfalfa and clover. In the Northern states, this usually means the previous season, or early spring for a summer seeding.

Farmers who plan to lime an entire farm should set up a definite liming schedule. If the soil is not too acid for corn, oats, and wheat in the rotation, a good plan is to apply lime 1 year or possibly 2 before the alfalfa is to be seeded. If a split application is advisable, one may be made before the land is plowed for corn, a second before plowing for the following grain crop.

Avoid Soil Compaction

Trucks loaded with lime are so heavy that serious damage to the structure of clayey soils may result if lime is applied when soils are wet. Such soils should be limed during the summer when moisture is usually at a minimum. In the winter wheat areas, this can be during the time when wheat soil is being fitted. Alfalfa may then be seeded in the spring.

Winter applications on frozen ground are permissible if the soil is covered with a dense cover of some kind, that is, meadow, stubble and straw from grain, or green manure. Very steep slopes should be avoided at that time of year. Perhaps winter applications on frozen ground would be best for clayey, poorly drained soils.

As far as soil moisture is concerned, light sandy soils may be limed at any convenient time.

See your county agricultural agent, U.S. Soil Conservation Service representative, or Department of Agriculture official for timely suggestions appropriate to your climatic conditions and farming plan.

REFERENCES

Barber, S. A. (1958). The Effect of Fineness of Grinding on the Quality of Agricultural Limestone. A review. Mimeograph publication from Purdue University, West Lafayette, Indiana.

Christenson, D. R. (1982). Lime, lime materials, and other soil ammendments. In Kilmer, V. J. (ed). *Handbook of Soil and Climate in Agriculture*. CRC Press, Boca Raton, Fl.

Christenson, D. R., D. D. Warncke, and R. Leep (1981). *Lime for Michigan Soils*. Bulletin E-471. Cooperative Extension Service, Michigan State University.

Ellis, B. G. (1966). BOF slag, new way to lime. *Crops and Soils* 18:7–8.

Ellis, R. Jr. (1954). *Liming Kansas Soils*. Kansas Agricultural Experiment Station Circular 313.

Jenny, H., J. Valmis, and W. E. Martin (1950). Greenhouse assay of fertility of California soils. *Hilgardia* 20:1–8.

Krug, E. C., and C. R. Frink (1983). Acid rain on acid soil: A new perspective. *Science*. 221:520–525.

McGeorge, W. T. (1935). The relation of potential alkalinity to the available phosphate in calcareous soils. *Soil Sci.* 39:443–452.

McGeorge, W. T., and J. F. Breazeale (1931). *Phosphate Solubility Studies on Some Unproductive Calcareous Soils*. Arizona Agricultural Experiment Station Technical Bulletin 35.

McLean E. O. (1980) *Recommended pH and Lime Requirement Tests*. Bulletin 499, North Dakota State University, Fargo.

Reeve, N. G., and M. E. Summer (1970). Lime requirement of Natal Oxisols based on exchangeable aluminum. *Soil Sci. Soc. Am. Proc.* 34:595–598.

Shoemaker, H. E., E. O. McLean, and P. F. Pratt (1961). Buffer methods of determining lime requirement of soils with appreciable amounts of extractable aluminum. *Soil Sci. Soc. Am. Proc.* 25:274–277.

Sims, J. T., and B. G. Ellis (1983). Adsorption and availability of phosphorus following the application of limestone to an acid, aluminous soil. *Soil Sci. Soc. Am. J.* 47:888–893.

Spurway, C. H., and K. Lawton (1949). *Soil Testing, A Practical System of Soil Fertility Diagnosis*. Michigan State University Agricultural Experiment Station Technical Bulletin 132 (4th rev.).

Tisdale, S. L., and W. L. Nelson (1956) *Soil Fertility and Fertilizer*. Macmillan, New York.

Volk, G. M. (1956). Efficiency of various nitrogen sources for pasture grasses in large lysimeters of Lakeland fine sand. *Soil Sci. Soc. Am. Proc.* 20:41–45.

Whittaker, C. W., M. S. Anderson, and R. F. Reitemeier (1959). *Liming Soils, an Aid to Better Farming* USDA Farmers Bulletin 2124.

Plant Nutrients and Fertilizers

Mineral plant nutrients in a soil to which fertilizers have not been applied are made available to plants as a result of the disintegration and dissolution of the soil's parent material. Granites, for instance, disintegrate into minerals, one of which is the feldspar orthoclase, an important potassium-bearing mineral. Orthoclase is acted on by carbonic acid (rainwater and carbon dioxide) to form clay and potassium carbonate.

Likewise, other minerals dissolve to liberate potassium and other nutrients. Sand soils may be largely quartz (SiO_2), whereas fine-textured soils have a high percentage of basic minerals. Thus, the mineral composition, particularly with respect to the nature of the clay minerals, determines to a considerable extent the possible nutrient-supplying power of the soil. Because of the role of CO_2 in forming carbonic acid (H_2CO_3), nutrient release from minerals is greatly affected by rate of soil organic matter decomposition, total CO_2 release being a function of organic matter content and rate of soil organism activity. As was mentioned in Chapter 3, the latter is largely a function of drainage and the nature of the cultural practices.

In an active, productive soil, plants obtain some of their mineral nutrients and most of their nitrogen from the decomposition products of organic matter. In many soils, however, the total amount of nutrients available during a season, from both mineral and organic sources, is insufficient to produce the large crops now necessary for economical production. It becomes desirable then to supply **additional** nutrients as fertilizers, commonly called **commercial fertilizers.**

NUTRIENT REQUIREMENTS OF PLANTS

In the early days, only 10 elements were known to be essential for plant growth. They were carbon, hydrogen, oxygen, phosphorus, potassium, nitrogen, sulfur, calcium, iron, and magnesium. During modern times, others—boron, manganese, copper, zinc, molybdenum, chloride, and vanadium—have been added to the list. Sodium is known to increase the growth of certain plants but perhaps only in the absence of sufficient potassium. Cobalt is needed by nitrogen-fixing organisms. Chlorine is abundant in plants but the requirement is very small. Other nutrients may be required by animals (e.g., cobalt and chromium) but are nonessential to plants.

Nutrient Balance

Plants have only limited power of selection as far as nutrient uptake is concerned. Composition, particularly of seeds, tends to be constant as long as the nutrient status of the soil is in balance. If the balance is at a low level, yields will be low, but the plants will be normal in appearance and in chemical composition. If the nutrient status of the soil is balanced at a high level of intensity, yields will be high, if it is assumed that growth factors other than nutrients are favorable, and the plants will still be normal.

Constancy of Ions

Within a given climatic area, plants within a species tend to be constant in total cation composition (on an equivalent basis), in total anion composition, and in the relationship between cations and anions. These tendencies have a marked bearing on the effect that nutrient imbalance may have on plant composition and crop yields.

Assuming the ions are added together on an equivalent basis, we find that

1. $Ca + Mg + K + Na + Mn + NH_4 = $ a near constant. Likewise,
2. $NO_3 + SO_4 + Cl + H_2PO_4 + HCO_3 = $ a near constant and

3. $$\frac{Ca + Mg + K + Na + Mn + NH_4}{NO_3 + SO_4 + Cl + H_2PO_4 + HCO_3} = \text{an absolute constant}$$

Soil Management Principle

The ratio of sum of the cations/sum of the anions in a plant is constant if both are expressed as milliequivalents.

It follows, then, that an overabundance of one cation in the root medium, and consequent increased uptake of that cation, tends to depress the intake of one or more of the remaining cations. Plant analyses indicate there is some range in the figure indicated in equation 1 or 2 as a "near" constant, but that the range is not large. This means that an excess of one mineral nutrient may soon result in a deficiency of another mineral nutrient. Such relationships are often observed. Heavy use of potassium fertilizer in Maine, for instance, intensified the need for magnesium in potato fertilizers. The same thing happened on Long Island and has occured to a lesser extent in Michigan. A similar relationship between calcium and magnesium has long been recognized.

More recent studies with radioactive phosphorus in planting-time fertilizers have shown that ammonium fertilizers increase the uptake of phosphorus from superphosphate. The foregoing equations show how this can be true. The sum of the cations and the sum of the anions can vary within narrow limits. It follows then that if ammonium ions are suddenly supplied to a plant, more anions must be taken up to keep the cation-anion relationship. If $H_2PO_4^-$ ions are close to the root and in high concentration, they will naturally be taken up to fill the anion need for the balance. This indicates that ammonium and phosphorus fertilizers should be applied as a mixture or as an ammonium phosphate salt.

Major Nutrients

Nitrogen, phosphorus, and potassium are the nutrients most likely to become limiting factors in crop production in all humid region soils. Likelihood of their becoming limiting factors would fall in that order in most light-colored soils, especially where erosion is severe, as in the East Central and Southern states. The same would be true for corn (maize), a nitrogen-loving plant, in even wider areas of the Eastern United States. The almost universal need of crops grown in the Eastern United States for these three nutrients is reflected in the fertilizer grades now used throughout the area. Attention is called, however, to the fact that the three major nutrients may all be deficient but still not be required in equal quantities. Intensity of need may or may not have a relationship to order of need. In addition, phosphorus tends to build-up in soils as a result of fertilization, so current soil tests are showing that much less phosphorus fertilizer is now needed. Hence, future grades of fertilizer should reflect this.

Potassium is plentiful in most semiarid and arid soils, the result of low rainfall and no leaching. Nitrogen is the first nutrient to limit yields in semiarid soils and in arid soils under irrigation. Phosphorus is usually needed after the need for nitrogen has been supplied.

Secondary Nutrients

Calcium, magnesium, and sulfur have been called secondary nutrients, not because they are unimportant in plant growth but because they are more plentiful in soils and because they have been carried free of charge in most complete fertilizers. However, modern fertilizer technology has eliminated much of this source of supply. High-grade dry mixtures, liquid fertilizers, and the presently more popular nitrogen carriers contain much smaller quantities of calcium and sulfur than did the lower-grade mixtures and carriers used earlier.

Trace Nutrients

The trace elements, boron, manganese, copper, zinc, and molybdenum may be deficient in soils to the point where they limit yields and affect the quality of crops. Generally, the pH of the soil can be used as an indication of the ability of plants to obtain adequate supplies of these nutrients.

Molybdenum, for certain crops, is likely to be deficient in medium to strongly acid soils. Firman E. Bear found that part of the value of lime for alfalfa on New Jersey soils is in making the molybdenum more available. Reisenauer (1956) showed that on the Couse soils [Typic Haploxerolls (Pachic)] of northeastern Washington known to be deficient in nitrogen, molybdenum applications as low as 0.8 lbs (363 g)/acre (0.4 ha) increased alfalfa yields, almost to the 5-percent probability level and caused a strongly significant increase in the nitrogen content of the plants. The response to molybdenum did not occur where the alfalfa was not nodulated. This fact and many references cited by Anderson (1956) indicate for sure that molybdenum is necessary for efficient symbiotic nitrogen fixation. Anderson's review, a thorough one of molybdenum as a fertilizer, also cites the work of H. Bortels to the effect that molybdenum also stimulates the activity of such free-fixing organisms as azotobacter, clostridium, and blue-green alga.

Plants do need small amounts of molybdenum to enhance their physiologic activities, but much evidence does point to its chief value being that of increasing nitrogen fixation. We have, of course, long realized that this is also an advantage of liming. In Michigan and many other areas where legumes are grown, liming to a pH of 6.5 is so advantageous in soil management that we can generally do without the application of molybdenum as a fertilizer.

Boron-deficiency symptoms are likely to occur on sensitive crops grown on soils having a pH above 6.9, provided the high pH is the result of high calcium and/or magnesium. Where sodium is the cause of high pH, boron may be available, even to the point of toxicity. This situation is encountered only in arid regions.

Manganese deficiency occurs in high pH soils, 6.3 and above, and occasionally in very strongly acid soils. In the acid soils, the soluble manganese has been leached out, whereas in high pH soils it is highly oxidized (MnO_2), a form slowly available to plants.

Iron, zinc, and copper are all made less soluble by lime application to raise soil pH. Iron deficiency is common in greenhouses and around homes where irrigation with hard water results in high soil pH. At the high pH, the iron is so highly oxidized that plants are unable to assimilate it.

Zinc deficiency commonly occurs on soils where the pH is high. When acid soils are limed, certain crops are almost sure to respond to fertilizer zinc, especially if soil pH goes above 7.2. Also, there may be a relationship with phosphorus levels in the soil. Friesen, Miller, and Juo (1980), working in Nigeria, found that Zinc *activity* in the soil solution declined when soils were limed above pH 5.0 (from 4.3) but was *not* affected by phosphorus applications within the pH range of 4.3 to 7.2. They also obtained growth response to zinc treatments in soil treated with high levels of phosphorus as fertilizer but *not* in soil where phosphorus levels were low.

In those particular soils, the response to phosphorus as fertilizer was very great. The concentration of zinc in the maize tissue was markedly reduced. In other words, phosphorus does affect zinc uptake but perhaps has no affect on zinc activity in the soil solution.

Weather conditions affect zinc availability to sensitive plants. In Michigan, deficiencies are most likely to occur when soils are wet and cold, as in early spring.

Copper deficiency is less dependent on pH but does occur on certain strongly acid organic soils, a result of actual deficiency in the soil. Burned-over peats are seldom deficient in copper because the level is raised by simple concentration of copper in the ashes.

Major Effects of Nutrients on Crops

Nitrogen

Vegetative growth is stimulated by an abundance of nitrogen. When nitrogen is plentiful, plant cells are large and cell walls are thin to give the plants the characteristic called **succulence,** desirable under some circumstances, undesirable under others. High nitrogen levels increase plant protein and result in lowered carbohydrate percentages. This is desirable for feed grains and forages, but not for soft wheat or sugar beets. It is essential, however, that sufficient nitrogen be present to produce the desired vegetative structure and leaf area, or photosynthesis will be limited and maximum yields of roots or grain are not possible.

Sufficient nitrogen results in rapid growth and hastened maturity. Too little nitrogen results in slow growth and a delay in maturity, whereas excess nitrogen keeps growth vegetative, which results also in delayed flowering and maturity. A quantity of nitrogen applied as a fertilizer in proper season to increase yields and hasten maturity might delay maturity and fail to increase yields if applied too late in the season. This frequently happens with wheat and sugar beets. In fact, the effect of excess nitrogen in depressing sucrose percentages in sugar beets is reflected in the low percentages found in beets grown on organic soils.

Phosphorus

That starter fertilizers should be high in phosphorus is a fact shown by the results of experiments. It reflects the role of phosphorus in stimulating the early growth of seedlings. Sugar-beet seedlings actually emerge sooner if well supplied than when deficient in this element. Phosphorus also stimulates the production of fibrous roots, an action that results in better use of soil moisture. Deeper fertilizer placement has merit partly as a result of this stimulation of fibrous root growth. When the fibrous roots are stimulated at greater depths, they are very likely, during drought periods, to be in zones of higher moisture content.

Fungus diseases are less likely to injure young seedlings grown in an abundance of phosphorus. It is thought that this is a result of increased vigor and rate of growth rather than of any depressing effect that the phosphorus may exert on the disease organism. This effect is well illustrated by the tendency of phosphorus fertilizers to prevent the injurious effects of blackroot or damping-off organisms.

Phosphorus hastens maturity, largely as a result of the hastened early growth, as evidenced by the fact that flowering occurs earlier in plants well supplied with this nutrient. Phosphorus-deficient plants are high in carbohydrate, a condition that results in accumulation of the purple anthocyanin pigment. The characteristics of the color patterns are explained in Chapter 12. Phosphorus has a marked affinity for iron. Where excessive quantities are applied as fertilizers, the iron may be so completely tied up as insoluble hydrated iron phosphates that plants become chlorotic. This is common in greenhouse plants and lawn grasses.

Potassium

Because potassium does not enter into permanent organic combinations in plant tissue, its specific physiologic role is obscure. However, certain effects on plants are recognized. Directly or indirectly, potassium affects CO_2 assimilation in plants, and thus it is connected with the synthesis of simple sugars and starch. Experiments have shown that an abundance of potassium is correlated with a high sucrose percentage in sugar beets. All high-carbohydrate-producing crops need plenty of potassium as a nutrient.

High carbohydrate levels are associated with thicker cell walls, which result in stiffness of stems. Thus, abundant potassium has been associated with stiffness of straw and the turgor of plants. This does not always occur, however, because associated high levels of nitrogen may result in increased metabolism and more rapid growth, which tends to keep cell walls thin and the stems weak. The two tendencies may balance each other.

There is some evidence that potassium aids in the transformation of sugar to oil. Potassium fertilizer has been known to cause significant increases in the percentage of oil in soybeans.

Resistance to disease may be greater when plants are grown in a medium high in available potassium. As with phosphorus, the effect may be the result of plant vigor and resulting ability of the plant to withstand the attack of the disease organisms.

The appearance of plants deficient in potassium is described in Chapter 12.

Other Nutrients

Sulfur is an essential plant nutrient, but little is known about its actual role in plants, except that a deficiency results in loss of chlorophyl to cause light green to yellow leaves, similar to those deficient in nitrogen. Sulfur occurs mainly in the proteins of plants but also in other forms. Experiments have shown that application of sulfur as a fertilizer increased sulfur percentage without increasing protein in the plant. Millar (1954) suggested that such increases might be as a constituent of amino acids, compounds very important in animal nutrition. Much sulfur has been applied to soils as an ingredient of fertilizer used to furnish other plant foods. Ammonium sulfate, for instance, contains more sulfur than nitrogen. Present fertilizer technology, however, has changed the sulfur picture. Much less is contained in modern fertilizer, and clean-air requirements have forced industry to avoid the use of high sulfur coal. "Acid rain" is the result of sulfur and nitrogen oxides in the atmosphere. Experiments show the element to be an essential nutrient to which more attention must be paid as supplies in the soil become exhausted. With present farming methods, this may soon happen.

Magnesium is a constituent of chlorophyl. It must then have an effect on photosynthesis. Experiments quickly bear this out as a deficiency that causes yellowing of foliage. It is interesting though to observe that potassium deficiency also causes yellowing in a pattern quite similar in many respects to that caused by magnesium deficiency, and chlorophyl does not contain potassium. It is also interesting that only a small percentage of the total magnesium in plants is in the chlorophyl. In other words, the element plays other significant roles in plant nutrition. It has become a nutrient to be carefully considered in soil management.

Calcium is largely contained in the cell walls of plants. As a result, a constant supply is necessary throughout the growth of the plant. Calcium has been called the balance wheel in plant nutrition, largely because it is plentiful in so many soils and thus it serves to regulate the intake of other less plentiful cations. This is made clear by equation 1 given earlier. Plant nutrition requires the maintenance of a balance between the three cations, calcium, magnesium, and potassium.

Calcium and magnesium levels affect the need of plants for boron, an element particularly essential to the production of meristem tissue. Boron is seldom deficient in acid soils and not in soils where high pH is caused by sodium, as in the saline soils of arid regions. It is deficient, however, for some crops in soils having a pH above 6.9 where calcium and magnesium make up an appreciable fraction of the total exchangeable bases. Experiments have shown that a wide (Ca + Mg)/boron ratio results in boron deficiency. Where that situation exists, boron may be needed for such crops as sugar beets, table beets, mangels, celery, cauliflower, cabbage, and turnips. See Chapter 12 for a discussion of boron-deficiency symptoms.

Manganese and iron affect the new growth of plants. Their role is associated with oxidation within the plant tissues. Manganese is an oxidizing agent that serves to regulate the degree of oxidation of iron. When there is too little manganese, iron is reduced and may become toxic. On the other hand, when manganese is present in excess, the iron may be so completely oxidized that too little of the reduced form is available for metabolic activities and the plant shows what has been recognized as iron deficiency.

Quantities of Nutrients in Crops

Fertilizer management should not be based on the quantities of nutrients removed by the crops. Soil is nature's source of plant nutrients, and through processes of soil formation (weathering) nutrients are continually being made available to plants. To the extent that such nutrients are soluble in the soil solutions, they may be lost through leaching. Furthermore, some nutrients are taken up in excess (luxury consumption) by plants when levels in the soil are high. It is evident, then, that large quantities of nutrients in soluble or readily available form cannot be safely stored in soil. Losses may be avoided by adding for each crop only the amounts of nutrients that may be necessary to supplement those available from other sources.

Farmers need not worry about how much plant food is being removed by crops, but rather how much they must apply to make up the difference between what their crops need and what the soil can furnish. Such information is available from the results of field-plot experiments and soil-test results on the particular soil in question.

Soil Management Principle

In planning rotations, it is useful to understand nutrient removal by crops.

There is value, however, in comparing the relative quantities of nutrients in plants. Such information is useful in fitting the crops to the soil and in arranging crops in their most effective sequence. For instance, a crop such as corn, which removes much nitrogen from the soil, should follow a crop that uses less nitrogen or one that actually adds nitrogen, such as alfalfa or clover.

Another value of considering the nutrient composition of crops is recognized. Sometimes a farmer may wish to calculate the value of crops or crop residues for soil-building purposes. Thus it may be possible to determine whether to sell the crop or return it to the soil for its nutrient content, disregarding, perhaps, its value as a source of organic matter. Many farmers have sold hay and straw for less than the value of the plant food they contain. Others have burned crop residues, thereby losing not only their value as organic matter but also the value of the nitrogen liberated as a gas in the burning process. The nutrient content of the more common farm crops was reported by Foth (1984).

FERTILIZERS DEFINED

Fertilizers are materials applied to supplement the nutrients furnished by the soil. As other agronomic practices are improved and higher-yielding varieties are grown, the demand for nutrients becomes so great that the soil cannot furnish enough to supply crop needs. Then nutrients become limiting factors and fertilizer should be applied. There is no point in applying fertilizer as long as weathering and natural soil-forming processes can furnish all the nutrients needed to grow the size crop desired or possible on that particular soil. But complete soil management includes the practices that set the stage for the higher yields that demand the use of fertilizers.

Grades, Ratios, Mixed Fertilizers

Grade

The farmer or home gardener thinks of fertilizer in terms of three numbers known as the **grade** or **analysis.** A common grade in the Midwest is 6-24-12, written in that manner. The numbers mean 6 percent total nitrogen (N), 24 percent available phosphoric acid (P_2O_5), and 12 percent water-soluble potash (K_2O).

What are the other "percents" in such a fertilizer? This question is often asked by wary persons who do not wish to purchase sand used as "filler." There is no make-weight material in present-day high-grade fertilizers. The other 58 percent in 6-24-12 is largely calcium, sulfur, magnesium, oxygen, and hydrogen. As already suggested, some of these elements are needed as plant foods. The farmer may be getting them free of charge, or perhaps for the cost of hauling them home.

Ratio

Grades of fertilizer that carry the same proportions of N, P_2O_5, and K_2O may be freely substituted, one for another. Thus the determined needs of the soil may be met by adjusting the rates of application to arrive at the same quantities of each nutrient, on an acre basis. For instance, a 5-20-10 grade has a 1 : 4 : 2 ratio. Other common grades in that ratio are 4-16-8 and 6-24-12. To furnish the same quantities of the three nutrients, the three grades should be applied at respective rates in order of 100, 125, and 83.3 lbs. Where transportation costs play an important role, the 6-24-12 would likely be the most economical grade to apply. Other factors, however, sometimes prevent this from being true.

Mixed Fertilizers

The term "fertilizer" in most people's thinking means a mixture of materials, largely mineral, containing the three major plant foods. In the early days, they were built around superphosphate as a base, with sufficient nitrogen and potassium salts to produce the desired grades. Nitrogen was omitted from some mixtures to take care of the situations where farmers might wish to depend on legumes and manure to furnish what their crops needed.

Currently, ammonium phosphates (monoammonium and diammonium) are common sources of fertilizer phosphorus. Applied alone, they may be called **mixed fertilizers** because they contain two nutrients. Polyphosphate materials are common in liquid fertilizers. Much of the ammonium phosphate now produced, however, is mixed with other materials, including potassium salts. Addition of the latter, of course, makes a more complete fertilizer.

Oxides versus Elements

The chemist has for all time expressed the elements as oxides. Even today, geology books express the composition of rocks and minerals as oxides. It is true even when the rock in question is known to be composed of carbonate minerals or hydroxides. That such a practice was adopted by the fertilizer industry now seems absurd, but it really is not strange because chemical and fertilizer companies have always been closely allied.

Muriate of potash is actually an impure KCl salt containing no oxygen, except perhaps as an impurity, but its potassium composition is stated in terms of K_2O, an impossible oxide from KCl.

A little more logic exists in the custom of stating the phosphorus composition of fertilizers in terms of P_2O_5. The most common source of phosphorus is rock phosphate, a massive form of apatite commonly called **phosphorite.** Several forms of apatite are known. A common one is fluorapatite, represented by the formula $(Ca_{10}F_2(PO_4)_6$. In soil and earthy deposits, phosphorite is classed as a secondary mineral. When used directly as a fertilizer, it is sold, as is monocalcium phosphate $(Ca(H_2PO_4)_2)$, on the basis of percentage P_2O_5. This "guaranteed" oxide is actually phosphoric acid anhydride. It can be formed by removing the water from liquid phosphoric acid as follows.

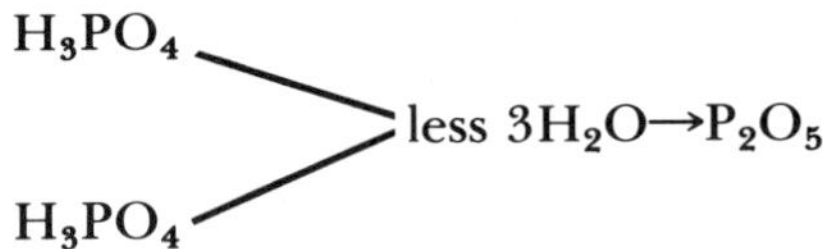

$$H_3PO_4$$
$$\text{less } 3H_2O \rightarrow P_2O_5$$
$$H_3PO_4$$

During the first 70 years of the fertilizer industry in this country (prior to about 1920), nitrogen in fertilizer was stated as percentage of ammonia. This seemed ridiculous because ammonia (NH_3) is a gas and so could not be purchased as such in mixed fertilizers. Gradually, state by state, laws were changed to allow the guarantee of nitrogen as the simple element.

Elemental Guarantees

Sooner or later, all plant nutrients will be stated on the elemental basis, percentage P rather than P_2O_5, and percentage K rather than K_2O. Even now, the secondary and trace elements are being so stated. In Michigan, for instance, this is required by law passed in 1954. Unified, regionwide action has already brought about the expression of soil-test results on the elemental basis, P and K rather than P_2O_5 and K_2O. Much simplicity will result when the percentages of phosphorus and potassium in fertilizers are likewise stated as elements. Then the term "phosphoric acid" can be confined to H_3PO_4 and "potash" can be returned to the iron pot and the business of making soap in the backyard.

Carriers

A fertilizer containing only one nutrient is called a **carrier.** In most instances, such materials are salts or mixtures of salts. Superphosphate, for instance, is a mixture of monocalcium phosphate and calcium sulfate, whereas concentrated superphosphate carries the same form of phosphorus but without calcium sulfate. The latter is a salt of rock phosphate, an alkaline material, and H_3PO_4, an acid.

Some carriers are not salts. Liquid phosphoric acid is now being used as a fertilizer for direct applications and in mixtures. Ammonium hydroxide is a carrier of nitrogen, as is anhydrous ammonia. The latter is truly a gas but is held under sufficient pressure to cause liquification until it is released within the soil where it contacts hydrogen and is absorbed by the soil as NH_4^+.

Solubility versus Availability

Four degrees of solubility are recognized in the fertilizer industry. They can be best illustrated by considering phosphorus carriers. First come those entirely insoluble in water and in ammonium citrate (see A.O.A.C. methods). Some of the rock phosphates on the market in this country fall into this class. They are very resistant to solution, even in strongly acid soils.

Second come the rock phosphates that do contain small amounts of ammonium citrate soluble phosphorus. Three percent of the phosphorus in Florida rock is citrate soluble whereas rock from Africa may contain as much as 8 percent in this form.

Third, there are a number of phosphorus carriers that are quite insoluble in water but are almost fully soluble in ammonium citrate. Such phosphorus is considered to be available to plants even though it is not soluble in water. Calcium metaphosphate, fused tricalcium phosphate, dicalcium and tricalcium phosphates, and certain by-product materials fall in this group. Heavy ammoniation of superphosphate results in the reversion of monocalcium phosphates to the di- and tricalcium forms with resulting sacrifice of water solubility, but the material is still largely citrate soluble.

Citrate soluble, but not water soluble, phosphates should be in pulverant form and should be mixed with soil to bring about solution in the soil acids, largely carbonic acid.

Finally, several phosphorus carriers are largely or completely soluble in water. Superphosphate is an example of a carrier containing most, but not all, of its phosphorus in water-soluble form, whereas the ammonium phosphates are 100 percent soluble. As is explained later, phosphorus from such sources quickly becomes fixed as a constituent of insoluble compounds when fertilizers are broadcast and mixed into soil. Granulating and banding lessens the fixation and results in more uptake by plant roots.

METHODS OF FERTILIZER APPLICATION

Fertilizer placement experiments were conducted annually from 1934 to 1960 by the Michigan Agricultural Experiment Station. Studies with potatoes even preceded that period. The results, reported regularly in the Proceedings of the National Joint Committee on Fertilizer Application, show proper placement increased yields and improved fertilizer efficiency. Many other experiment stations reported similar results.

Young seedlings should be well fed to promote rapid, vigorous growth, but care must be taken to avoid injury to the germinating seeds and young roots. With most crops, this means that fertilizers should not be placed in direct contact with seeds. In the early days, this was accomplished by broadcasting the fertilizer ahead of planting. The advice to "work well into the soil before planting" was common, but not wise.

Band Application Best for Row Crops

Soil Management Principle

Proper placement of fertilizer will improve efficiency.

Certain soil minerals react readily with soluble phosphates applied as fertilizer. The resulting compounds are insoluble and very slowly available to plants. Formulated from superphosphate, the fertilizers of the 1930s were powdered rather than granular and carried a high percentage of their phosphate in water-soluble form. As a result, fixation was high.

Experiments showed that it was necessary to apply several times as much phosphorus as plants actually needed. Fertilizer placement experiments were designed to show where fertilizers might be placed to produce the desired starter effect while developing maximum efficiency. Application in bands below and slightly to the side of the seed proved best for row crops.

The first Michigan experiments conducted cooperatively with the USDA showed that for potatoes it was best to place the fertilizer in two bands, one on either side of the row and on a level with the seed. Practically all of Michigan's potato crop is now fertilized in this manner. Planters have been perfected to do a good job of placing fertilizer in the proper location.

Before further cooperative placement tests were started with sugar beets and beans (1934), these crops were fertilized broadcast or by direct contact with seed. When the latter type of application proved very unsatisfactory for beans, the Michigan Station recommended that the crop be planted without fertilizer.

Experiments between 1934 and 1938, however, showed properly applied fertilizer was very profitable and "proper application" was in a single band 1 in. out and 1 ½ in. below the seed. The same location also proved right for sugar beets. This was fortunate, because the two crops are grown on the same farms and by changing seed plates may be planted by the same drills.

Machinery companies heeded experiment station recommendations by marketing the first modern side-placement beet and bean drills in 1940. World War II delayed volume production, but since 1946 practically all of Michigan's sugar beets and beans have been planted in the recommended manner.

Since 1945, fertilizer placement studies have centered around corn, vegetable crops, legumes, and small grains. During that time, great changes have occurred in experimental methods and in fertilizer technology. Radioactive phosphorus (^{32}P) made it possible to trace the element from the fertilizer into the plant (see Fig. 14-1). Granular and liquid fertilizers have become common, and concentrated materials have replaced the relatively low grades of prewar days.

Furthermore, ammoniation and the development of new carriers have influenced the degree of solubility of the phosphorus in present-day solid materials. All these changes have made it desirable to take a new look at the old subject "Fertilizer Placement."

FIGURE 14-1 A belt-type fertilizer distributor made especially for the experimental application of radioactive phosphorus (^{32}P). Sugar beets are being planted in Clinton County, Michigan. Note the plastic shields that protect the men while they place "hot" fertilizer on the belts.

Corn

Separate band placement had proved so satisfactory for potatoes, beans, and sugar beets that investigators decided to try the method on corn. Accordingly, the USDA Agricultural Engineering Laboratory again responded to a call for assistance, and the machine shown in Fig. 14-2 was constructed. The machine was so designed that the conventional split boot could be used for delivering fertilizer on plots used as controls. From the first, the yields definitely favored placement to the side of and below the seed. In many instances, the results were better where the bands were 4 to 6 in. deep than where the shallower placement was employed.

From the start, it was evident that one band was as good as two. In all the experiments with this USDA machine, continuing through 1952, the two bands were at equal depths. The split-boot applicator was always unsatisfactory. This statement was made in an early issue of the Joint Committee Proceedings and in *Fertilizer Recommendations for Michigan Crops*, Michigan State University Extension Bulletin 159, 1953.

The superiority of deep placement to the side of the seed over the method of plowing fertilizer under is shown in Fig. 14-3.

Later research on that same field showed corn roots were slow to absorb the fertilizer nutrients that had been plowed under. This was due partly to the compaction of the soil between plowing and planting. Later experiments indicate that plowed-under fertilizer may be more efficient when minimum tillage methods are practiced.

FIGURE 14-2 Corn planter for band placement of fertilizer. This machine made it possible to place one or two bands of fertilizer level with the seed or as deep as 6 in. below the seed. It was designed and constructed by Glenn A. Cumings, USDA, and Clarence Hansen, Michigan State University, in 1945 and was used through 1952.

FIGURE 14-3 Fertilizer placement made the difference in this corn experiment in which an 8-8-8 fertilizer was applied at the rate of 1000 lb/acre. *Left:* Fertilizer placed in a band on the bottom of each furrow when plowing the land. *Right:* Fertilizer placed at time of planting in bands 2 in. to the sides of the row and 6 in. below the level of the seed.

A quotation from the 1948 National Joint Committee Bulletin follows: "Fertilizer at rates of application as high as 500 to 600 pounds per acre may be drilled in a continuous band approximately 2 inches to one side of the row in a depth zone about 2 inches below the level of the seed."

The basis for this statement was the work done in Michigan in 1945, 1946, and 1947.

Vegetables

In the early studies on fertilizer placement, vegetable crops were neglected. This is unfortunate because, of all crops, vegetables are the most responsive to fertilizer. And, no doubt, large quantities were applied at the expense of efficiency and crop yields.

Michigan researchers again cooperated with those from the USDA Agricultural Engineering Laboratory to design the machine shown in Fig. 14-4. Experiments with onions and spinach on organic soils showed variations in fertilizer placement methods actually caused yields to vary as much as 103 to 43 percent, respectively.

FIGURE 14-4 Fertilizer placement drill designed especially for experiments with vegetables on organic soil. This drill is equipped to apply liquid and solid fertilizers in bands at various depths below the seed and at various distances to the side of the rows. Note the tank and top-delivery hoppers mounted on the rear of the tractor. The smaller hoppers on the four-row planter are for applying minor elements and insecticides.

Band Seeding for Legumes

Small-seeded legumes (alfalfa, clovers, sweetclover) should be well fertilized at seeding time. We must be careful, however, to make the application in a way that will achieve efficient use of the fertilizer without injuring the seed.

Band seeding is widely recognized as the best way to establish a legume seeding. The seeds are dropped on top of the soil directly over a band of fertilizer that is covered by at least 1 in. of fertilizer-free soil. The seed should be pressed down with a press wheel. No other means should be taken to cover the seed. This method has been used successfully on many kinds of soil. The method was recommended first by Ohio agronomists.

Topdressing is a satisfactory method of applying fertilizers for established stands of alfalfa. Lawton and Vomocil (1954) used radioactive phosphorus techniques to study the effects of placing phosphate fertilizers on the soil surface in established stands of hay and at various depths to 36 in. A glass tube was used to place the fertilizer below the surface. Where fertilizer phosphate was broadcast on April 24 on the surface of Berrien sandy loam soil, a sampling on May 17 showed that 44 percent of the plant phosphorus came from the fertilizer. On that same date, only 15 percent of the plant phosphorus came from fertilizer that had been placed 3 in. deep, also on

April 24. Where fertilizer had been placed 12 in. deep, the figure dropped to 7 percent and to 2 percent for the 36-in. depth.

These data show that the quickest results from the applied fertilizer occurred with surface applications.

Apparently, the effective feeder roots of alfalfa were largely in the surface 6 in. of soil. Perhaps in looser subsoils they might develop at greater depths. The supply of oxygen is probably the deciding factor.

Side Placement for Small Grains

Conventional grain drills place fertilizer in direct contact with seed. Because there are so many rows, and application rates are generally low, this method of application has been quite satisfactory for oats and usually for barley. Michigan experiments have shown that emergence of oat plants may be delayed by contact fertilizers applied during periods of low soil moisture, but good stands eventually result. Serious and lasting injury seldom occurs because drought periods at oats planting time are short and because rates of fertilizer application are relatively low.

Wheat is often planted during extended drought periods. This is especially true in Michigan, where winter wheat is an important crop, occupying about 10 percent of the cropland of the state. The crop is very responsive to fertilizer, and its acre value is high. Furthermore, the crop is largely grown on light-colored upland soil, relatively low in available plant food. Thus, rather large quantities of fertilizers are needed for satisfactory production. During seasons of adequate moisture, 500 to 600 lbs of high-analysis fertilizers are profitable. However, dry planting seasons, when such fertilizer treatments delay germination and injure stand, are all too common. Michigan data show that during such seasons, 120 lbs of total plant nutrients $(N+P_2O_5+K_2O)$ is all that should be applied in direct contact with wheat seed. More will be injurious, but twice that quantity is needed when moisture is adequate. Grain drills that place fertilizer in bands separate from the seed (an inch to the side and an inch below) make it safe to go to the heavier applications.

It is of interest to note that grain rows, either wheat or oats, need not be as close together as 7 inches. Preliminary results of Michigan experiments indicate that they may be as far apart as 11 in. without appreciable loss of yield. Such spacing might make it possible to construct side-placement drills as cheaply as the contact drills of the present and past. Such drills may then be safely used for planting soybeans, peas, beans, sorghum, sugar beets, and corn. By closing some of the openings, the drills can be used for planting crops to be cultivated. As the use of chemicals eliminates cultivation for weed control, an all-crop drill will be even more desirable.

Water Solubility and Particle Size

When a fertilizer granule containing water-soluble phosphorus is placed in moist soil, 50 to 80 percent of the phosphorus moves out of the granule in 24 hours, but it spreads into the soil very slowly. Thus, as long as the soil is not stirred, a highly concentrated "shell" of soluble phosphorus remains around the granule. Fixation of such phosphorus into unavailable form is slow, varying inversely for a given amount of phosphorus with the size of the granules.

The experiments that made possible these statements about the migration of phosphorus from granules into soil were conducted with tagged phosphorus (^{32}P). Further studies with this research tool have shown that a strong interaction exists between degree of phosphorus solubility, particle size, and method of placement.

When powdered fertilizer was well mixed with soil, the percentage of the phosphorus in water-soluble form had very little effect on yields or on the amount of fertilizer phosphorus that entered the plant. Apparently, the phosphorus from the fine particles of fertilizer contacted so much soil that very high fixation occurred. Depending on soil pH, phosphorus fixation is as iron, aluminum, or calcium phosphates.

When granular fertilizer was well mixed with soil, the percentage of the phosphorus in water-soluble form did have a marked effect on the percentage of fertilizer phosphorus absorbed by the plant and did have a marked effect on yields. To be satisfactory, granular material should have 40 to 60 percent of its phosphorus in water-soluble form. The experiments have shown that granular material containing but a small percentage of phosphorus in soluble form contact too little soil to be effective as starter fertilizer for rapidly growing plants.

When granular or powdered fertilizer was banded in soil, the uptake by plants of the fertilizer phosphorus increased directly with an increase in the degree of solubility of the phosphorus in the fertilizer. Likewise, the yields increased with the same increase in degree of solubility of the phosphorus. For maximum efficiency, fertilizer to be applied in bands should contain not less than 40 percent of its phosphorus in water-soluble form.

The experiments showed definitely that fertilizers that contain no water-soluble phosphorus should be thoroughly mixed with soil and should be powdered rather than granular.

Liquid versus Solid Fertilizers versus Placement

There are two ways of avoiding extreme fixation of the phosphorus in soluble solid fertilizers. Granulation provides the first method and banding the second. Both methods are effective because they reduce the area of fertilizer-soil contact.

It is not possible, of course, to granulate solutions. Then the only method of reducing fertilizer-soil contact is to place solution fertilizer in bands. Experiments have shown this lessens fixation. Radioactivity counts made on 8-week-old corn plants were about equal where liquids and solids were banded at planting time. Likewise, the counts were about the same where the two materials were thoroughly mixed with the soil before planting. However, the counts were three times as high for both liquids and solids where the fertilizers were placed in bands.

In this comparison, the same formulations were involved, since the liquids were obtained by dissolving the solid fertilizers in water. In a comparison of solid and liquid fertilizer, it is essential that the solid material have a high percentage of its phosphorus in water-soluble form and that the two materials be placed in the same location relative to the seed.

Liquid fertilizers applied to soil as spray cannot be expected to increase yields as much as the same plant foods applied in bands. In one experiment on a sandy loam soil, 50 lb of P_2O_5 was sprayed just as corn was emerging. On comparative plots, the same amount was injected in a band close to the row. The yields were 56.0 and 72.9 bu/acre, with the higher yield caused by band treatment. In spray treatment, much of the phosphorus must have been fixed into forms not available to the plants.

Polyphosphates in Liquids

The phosphorus in superphosphate is in the ortho form. In solution, ionization would form $H_2PO_4^-$, HPO_4^{2-}, and/or PO_4^{3-}, depending on soil pH.

During recent years, liquid fertilizers have increased in popularity. In Michigan in 1982, 36.6 percent of the mixed fertilizer tonnage was sold as liquids, 261,889 tons of a total of 716,487 tons. In high-grade liquid mixtures, polyphosphates make up a large part of the phosphate carriers. When mixed with soil, they form complex, *soluble* ions with iron and aluminum, whereas orthophosphates form *insoluble* compounds with iron and aluminum. This solubility difference suggests that polyphosphates might remain longer in forms readily available to plants. Considerable research, however, shows this advantage to be rather short-lived and polyphosphates soon convert by hydrolysis to orthophosphates, in some cases to the extent of 50 percent in 2 weeks.

Hashimoto and Lehr (1973) studied rate of mobility of the phosphorus of two orthophosphates (mono- and diammonium) and several polyphosphates in Hartsells fine sandy loam (Typic Hapludults) from Alabama. They found that "the distribution patterns of all the P sources were established within one week of their application to the soil columns, and only minor changes occurred thereafter." This would indicate that hydrolysis of polyphosphates to orthophosphate had indeed been rapid.

Gillian (1970) found that pyro- and orthophosphate solutions (no soil involved) were equally effective in supplying phosphorus to seedlings of P-deficient wheat *(Triticum vulgare)*, corn *(Zea mays L.)*, and barley *(Hordeum vulgare L.)*. They found that the roots absorbed some of the pyrophosphate ions *per se* and hydrolyzed 40 to 55 percent of the pyrophosphate to orthophosphate within 24 hours. They concluded that hydrolysis was caused by the presence of phosphatase on the root surfaces of the seedlings.

Studies have found that pyrophosphate and orthophosphate were equally effective when applied for corn *(Zea mays L.)* grown in warm soil (24°C). When a cool soil (16°C), is used, however, the pyrophosphate is a less effective form than was the orthophosphate.

FERTILIZER RECOMMENDATIONS

In the final analysis, the main objective in field plot and greenhouse experiments with fertilizers and the testing of soil and plant tissue is to gain the information needed for making fertilizer recommendations. Fertilizer grade and rate studies date back to 1913 in Michigan and to even earlier dates in certain other North Central states. Present systems of soil testing (the quick tests for available nutrients) were not in general use for another quarter century. The Spurway Simplex methods date back to 1933. In the meantime, field plot experiments and much valuable greenhouse work furnished a mass of information regarding the relationship between kinds of soil and crop response to fertilizer.

Early Field Plots

The early field plots were studies to determine the best grades to use on different series of soil and the most desirable rates of application. Grades were studied by holding two nutrients constant and varying the third. Rate studies were arranged to include several rates of the grade that the investigator believed was likely to be the most effective for the particular soil and crop. Statistics were not generally used, and many unfertilized checks or uniformly fertilized checks were used to indicate soil variability or uniformity.

Soil Phase versus Fertilizer Response

Fine-textured soils were found to be lacking first in phosphorus and second in potassium. Phosphate alone or a fertilizer having a 2:1 ratio with respect to P_2O_5 and K_2O usually gave the best results. This was especially true where stable manure was applied.

The same experiments showed that crops on coarse-textured soils responded more to fertilizers having equal parts of P_2O_5 and K_2O or, with some crops, to fertilizers containing an even greater proportion of potassium.

The early work on organic soils always pointed to a need for potassium first and phosphorus second. Usually, the best results occurred where the fertilizers contained more potassium than phosphorus.

As soils accumulate phosphorus from fertilizer residues, the needed ratios change. Soil test to be sure that the correct ratio is used.

Soil-Test Correlation

The 1940s marked the period of soil-test correlation work. Quick tests had been perfected to the point where agronomists felt that they could be used in making fertilizer recommendations. They believed the results of such tests could supplement the historical information that was needed but that was often hard to obtain. They needed, however, some "critical levels" with which to work, some points below which fertilizer should be applied and above which they would not expect crop response to fertilizer. Over the years, the results of research have established these critical levels for many crops.

Expected Yields

Adequate nutrients represent just one of many factors that help to bring about top yields. In the first place, soils are so characterized that they fall in a certain range as far as yields are concerned.

Soil Management Principle

The goal of soil management is to place the productivity of each soil phase in the top of its productivity range.

Management is the art of putting a certain area of the soil (field or farm) at the top of the range, from which unknown or uncontrollable limitations prevent further rise. Costs sometimes discourage practices that result in attaining the absolute top in yields. The economic crest is determined by the cost-price relationship.

Many factors other than nutrients affect production on a given soil. These include correct tillage, the use of good seed, the control of insects and disease, adequate liming, and timely operations. When the farmer does all these things in the best manner known to agricultural science, he has the stage set for high yields and full use of recommended fertilizers. Then and only then should the maximum quantities of fertilizer be applied.

Basis for Fertilizer Recommendations

When a farmer wishes assistance in deciding what fertilizer to apply, he should first obtain a soil test that truly represents the area to be fertilized. The sample should be taken as directed in Chapter 10. One pint of soil is enough for the laboratory technician.

The Farm Advisor (County Agent in some states) cannot make a satisfactory fertilizer recommendation unless he or she knows the soil series or soil management group (U.S. Soil Conservation Service Land Capability Unit). This is because low tests on one group of soils call for different fertilizers than do low tests on another group.

An existing map made by soil classification survey parties shows the soil series being considered, or the management group may be determined by considering location, texture, and color. Such determinations made by soil scientists who know the general area are quite reliable. The U.S. Soil Conservation Service has such scientists in practically every agricultural county in the nation. Many persons attached to the local Extension Service office are also able to catalogue soils in this manner.

The soil test result should next be considered. To illustrate how management group, crop variability, yield expectation, and soil test results are dovetailed to arrive at a specific recommendation, we use a Michigan recommendation bulletin, E-550 (Warncke, Christenson and Vitosh, 1985).

A Specific Example

Assume a farmer who wishes to grow corn has taken a properly collected soil sample to his County Agricultural Agent who decides the soil is Miami loam (Typic Hapludafs). Table 14-1 (from Table 11 in E-550) shows that corn (maize) on mineral soils may be expected to yield from 100 to 200 bu/acre. The farmer thinks he should be getting as much as 180 bu/acre. Assume now the soil has been tested and that the phosphorus level is 50 lb/acre. The P_2O_5 recommendation should then be 80 pounds per acre (see fifth line and fifth column). If the test had been 125 pounds P per acre, phosphate fertilizer would not be needed.

Now let us consider the need for potassium in the fertilizer. The soil texture is *loam,* so the recommendation will come from the lower part of Table 14-1. The soil test for K proves to be 260 lb/acre. Following across on the tenth line to column 5 we find that the K_2O recommendation should be 70 lb/acre. A still higher test would have meant that potash fertilizer was not needed. Many farmers who neglect soil testing are applying fertilizers that are not needed. On the other hand, if this farmer's test result had been less than 99 lb, he would have needed 290 lb of K_2O per acre.

What about nitrogen? Previous management practices largely determine the need for nitrogen. Table 14-2 illustrates the way this is handled in Michigan. Again, with a yield goal of 180 bu/acre, the farmer would need to apply about 110 lb N per acre where the crop is to follow a legume or 210 lb N following a nonlegume and no manure. With 20 tons manure but following a nonlegume, the rate should be 130 lbs.

On organic soil, nitrogen would not be recommended and the P_2O_5 and K_2O recommendations would be as shown in Tables 14-3 and 14-4. On such soil, the predicted yield of sweet corn is 160 bu/acre, as indicated in column two in both tables. Assume the soil test levels of the two nutrients to be 50 lb of P and 250 lb of K. As shown in the lower part of column 2 of Table 14-4, the recommended amount of P_2O_5 would be 120 lb/acre, and of K_2O, 40 lb/acre (Table 14-3).

Soil-series influence is indicated by the titles to Tables 14-1 through 14-4. Similar recommendations are available for other crops on the various soils. Note especially the fact that quantities of K_2O needed on organic soils are just as great as on mineral soils. In fact, organic soils are quite infertile unless heavily fertilized. More about that later.

Because soils, climate, and systems of management are so variable in different states and/or countries, you are urged to contact local extension personnel for specific recommendations. The preceding examples are only for the purpose of illustrating the factors that should be considered.

TABLE 14-1 Annual Phosphorus (P_2O_5) and Potassium (K_2O) Recommendations for Corn Grown on Mineral Soils

	Yield goal, bu/acre					
Soil test	100	120	140	160	180	200
Phosphorus Recommendation, lb P_2O_5/Acre						
10 lb P/acre	60	80	90	110	130	140
20	50	60	80	100	110	130
30	40	50	70	90	100	120
40	20	40	60	70	90	110
50	0	30	40	60	80	90
60	0	0	30	50	60	80
70	0	0	20	40	50	70
80	0	0	0	20	40	60
90	0	0	0	0	30	40
100	0	0	0	0	0	30
110	0	0	0	0	0	20
Potassium Recommendation, lb K_2O/Acre on Sandy Loams and Loamy Sand						
25 lb K/acre	220	240	260	280	300	320
50	200	220	240	260	280	300
75	170	190	210	230	250	270
100	150	170	190	210	230	250
125	120	140	160	180	200	220
150	100	120	140	160	180	200
175	70	90	110	130	150	170
200	50	70	90	110	130	150
225	20	40	60	80	100	120
250	0	20	40	60	80	100
275	0	0	0	30	50	70
300	0	0	0	0	30	50
325	0	0	0	0	0	20
350	0	0	0	0	0	0
Potassium Recommendation, lb K_2O/Acre on Loams, Clay Loams, and Clays						
25 lb K/acre	240	270	300	320	350	370
50	210	240	260	290	320	340
75	180	210	230	260	290	310
100	150	180	200	230	250	280
125	120	140	170	200	220	250
150	90	110	140	170	190	220
175	60	80	110	130	160	190
200	30	50	80	100	130	150
225	0	20	50	70	100	120
250	0	0	0	40	70	90
275	0	0	0	0	40	60
300	0	0	0	0	0	30
325	0	0	0	0	0	0

Source: Table 11 in Warncke, Christenson, and Vitosh (1985).

TABLE 14-2 Nitrogen Fertilizer Recommendations for Corn Grain

Yield goal (bu/acre)	Previous crop or manure application				
	No alfalfa no manure	100% alfalfa	60% alfalfa (acre/acre)	10 tons manure	20 tons manure
100	110	10	30	70	30
125	140	40	60	100	60
150	180	80	100	140	100
175	210	110	130	170	130
200	250	150	170	210	170
225	280	180	200	240	200

Source: Table 3 in Warncke, Christenson, and Vitosh (1985).

RESIDUAL NUTRIENTS

Nutrients applied as fertilizer are not all used by the currently growing crop. Some may be lost by leaching, and some are changed through chemical and biological reactions into forms not readily used by plants. Some are "fixed" in the bodies of soil organisms and in the tissues of higher plants, some are adsorbed by clay minerals, and some are chemically changed to unavailable forms.

Cover crops pick up some of the plant foods applied for the main crop of the season and carry them over to the next year. It is believed that residual effects of soluble nitrogen fertilizers come about largely through this means (see Fig. 14-5).

In some cases the effect from previously applied fertilizers may actually be harmful in that soils may be made acid, an unfavorable nutrient balance may be established, or excessive quantities of secondary nutrients may be left in the soil. This latter situation is common in greenhouse soils. As rates of fertilizer application are increased, nutrient imbalance becomes more common in field soils. As a result, greater care is advisable in selecting fertilizer ratios.

Phosphorus and Potassium Accumulation

The results of field experiments have been used to decide economical ratios and rates of application for various crops, soil series, and nutrient levels. The field experiments are still being conducted, and data are still scanty for some situations. As a result, the most satisfactory grades are not always recommended and applied, and crop removal of nutrients is not always correctly estimated. As a result, some nutrients accumulate. To obtain information on such nutrient accumulations, many field plots have been sampled and tested for phosphorus and potassium.

Hanway of Iowa applied P_2O_5 and K_2O at rates of 0, 60, and 120 lb/acre, each nutrient alone and in all combinations. Recovery of applied phosphorus in the hay crop ranged from 33 to 62 percent, whereas that of potassium ranged from 74 to 140 percent. Soil tests showed that considerable "available" phosphorus had accumulated in the top 2 in. of soil but that soil potassium levels had not increased.

A Michigan crop sequence experiment on Sims clay loam (Mollic Haplaquepts) covered a period of 16 years. The soil on all plots was sampled when the experiment started. Samples were taken again 16 years later. By that time, all plots had received a total of 3200 lb of 4-16-8 fertilizer.

Soil tests showed that even at these relatively low rates of fertilizer application over the 16-year period, both phosphorus and potassium levels did build up. This occurred under seven different cropping systems. However, the phosphorus levels rose relatively more than did the potassium levels. The results indicated that it was time to change the fertilizer grade to one containing a higher percentage of potash, perhaps to one having equal amounts of phosphate and potash. An alfalfa experiment was started in 1954 on Kalkaska sand (Typic Haplorthods). One objective was to determine the effect of fertilizer on the longevity of the alfalfa. First 0-20-20, then 0-13-39 were applied at three rates. In 1956, the soils were tested to determine nutrient accumulation.

The data showed that where 358 lb of P_2O_5 and 544 lb of K_2O were applied between 1951 and 1955, soil phosphorus levels rose from 24 to 64 lb/acre, whereas potassium levels had risen from 59 to 104 lb. These relative changes confirm the wisdom of changing from the 0 : 1 : 1 to the 0 : 1 : 3 ratio fertilizer. The relatively slow accumulation of potassium as compared to phosphorus is due to leaching and luxury consumption of potassium.

Current soil-test information shows that extractable phosphorus has rapidly increased in Michigan soils. In 1971, 25 percent of the samples tested exceeded 100 pp2m extractable phosphorus, but by 1979 50 percent of the samples tested exceeded that level.

If fertilizer recommendations are made based on soil test, accumulations of nutrients such as phosphorus result in less fertilizer being applied. But when high soil tests are ignored and fertilizer recommendations are based on past experiences, the farmer losses money.

Trace Element Accumulation

Copper accumulates in soil and may remain in available form for several years. At the Michigan muck farm, 6 lb of copper applied in 1953 and 1954 increased 1956 Sudan grass yields from 2.9 to 14.8 tons green hay per acre.

Boron remains available to plants for several months after application to acid soils but is quickly fixed in an unavailable form when applied to alkaline soils. Since it is on the latter soils that the element is needed, application must be on a yearly basis. Residual benefits are almost nil.

Manganese is quickly oxidized in neutral or alkaline soils. Highly oxi-

TABLE 14-3 Potassium (K_2O) Recommendations for Vegetable Crops Grown on Sandy Loam and Loamy Sand

Asparagus old beds	(20)[a]	Asparagus new beds	(20)	Cabbage	(400)	Broccoli	(70)
Lima beans	(30)	Carrots	(300)	Cucumbers	(300)	Brussels sprouts	(50)
Peas	(30)	Endive	(160)	Eggplant	(150)	Cauliflower	(140)
Pumpkins	(300)	Lettuce	(400)	Flower beds		Celery	(600)
Radishes	(60)	Sweet corn	(160)	Horseradish	(80)	Home gardens	
Snap beans	(80)			Muskmelons	(150)	Tomatoes	(400)
Squash	(300)			Onions	(400)	Market gardens	
Turnips	(400)			Parsnips	(200)		
Strawberries	(100)			Peppers	(200)		
				Rhubarb	(300)		
				Rutabagas	(350)		
				Spinach	(120)		
				Sweet potatoes	(180)		
				Swiss chard	(150)		
				Table beets	(250)		
				Watermelons	(150)		

Soil test lb K / acre	Potassium Recommendtion, lb K₂O / acre				
25	200	240	290	330
50	180	220	270	310
75	150	200	240	290
100	130	180	220	270
125	110	150	200	240
150	90	130	180	220
175	60	110	150	200
200	40	90	130	180
225	20	60	110	150
250	0	40	90	130
275	0	20	60	110
300	0	0	40	90
325	0	0	20	60
350	0	0	0	40

[a] Numbers in parentheses after the crop is the yield potential in cwt (100 lb)/acre.

Source: Table 21 in Warncke, Christenson, and Vitosh (1985).

TABLE 14-4 Phosphorus (P_2O_5) Recommendations for Vegetable Crops Grown on Mineral Soils

Asparagus old beds	(20)[a]	Carrots	(300)	Asparagus new beds	(20)	Celery	(600)
Lima beans	(30)	Endive	(160)	Broccoli	(70)	Flower beds	
Peas	(30)	Lettuce	(400)	Brussels sprouts	(50)	Home gardens	
Snap beans	(80)	Parsnips	(200)	Cabbage	(400)	Onions	(400)
Turnips	(400)	Pumpkins	(300)	Cauliflower	(140)	Tomatoes	(400)
Strawberries	(100)	Radishes	(60)	Cucumbers	(300)	Market gardens	
		Rutabagas	(350)	Eggplant	(150)		
		Spinach	(120)	Horseradish	(80)		
		Sweet corn	(160)	Muskmelons	(150)		
		Sweet potatoes	(180)	Peppers	(200)		
		Squash	(300)	Rhubarb	(300)		
				Swiss chard	(150)		
				Table beets	(250)		
				Watermelons	(150)		

Soil test *lb P / acre*		*Phosphorus Recommendation, lb P_2O_5 / Acre*		
10	130	170	210	250
30	110	150	180	220
50	80	120	160	200
70	60	100	130	170
90	30	70	110	150
110	0	50	80	120
130	0	20	60	100
150	0	0	30	70
170	0	0	0	50
190	0	0	0	20
210	0	0	0	0

[a]Numbers in parentheses after the crop is the yield potential in cwt (100 lb)/acre.

Source: Table 20 in Warncke, Christenson, and Vitosh (5985).

FIGURE 14-5 Barley grown after sugar beets on Sims clay loam. The two bundles at the right (3) were larger than the left two bundles (6) because alfalfa was grown on the plot 3 years earlier. There is no legume included in the number 6 rotation. Fertilizer is necessary for successful alfalfa production on Sims soil.

Residual effect from nitrogen is shown by the greater size of the left bundle in each pair. The increase was the result of 40 lb of nitrogen applied as a sidedressing for sugar beets. Yields of the plots represented by the bundles were, left to right, 43.9, 34.2, 47.6, and 39.1 bu/acre.

dized manganese is quite unavailable to plants, so it is advisable to apply only sufficient quantities of the fertilizer salts to satisfy the needs of the current crop. The residual portion becomes quite ineffective for the crops of the following years.

Zinc may be applied each year as a spray or in bands mixed with fertilizer. For highly responsive crops in Michigan, an application of 25 lb of zinc per acre remained sufficiently available for at least 7 years. Also, after several years of banding zinc where the total application was equal to 25 lb, rates can be greatly reduced or eliminated.

Fertilizer Technology, Past, Present, and Future

Fertilizer technology has changed much during recent years. Demands for high grades of dry fertilizers and an urge to save labor by using solutions that may be pumped and sprayed have resulted in marked changes in carriers. Formulas are greatly changed. The new materials contain much less calcium and sulfur and smaller percentages of impurities that in the lower-grade materials carried trace elements of appreciable quantities.

The importance of secondary and trace elements as plant nutrients has been discussed earlier in this chapter. Farmers should certainly be concerned when their fertilizers no longer contain appreciable quantities of essential nutrients. The nature of the soil actually determines when the omitted nutrients may begin to limit production and/or quality of crops.

In some instances, it is desirable to use high-grade fertilizers that do not carry unneeded anions such as sulfates and chlorides. By so doing, salt effect may be reduced and unwanted residues avoided. This is particularly important in greenhouse and vegetable soils where rates of fertilizer application are very high. In some intensive agricultural areas, streams contain so high a concentration of soluble salts that the water is not desirable for municipal use. Modern high-grade fertilizers are helping to prevent such occurrences.

REFERENCES

Anderson, A. J. (1956). Molybdenum as a fertilizer. *Advances in Agronomy,* vol. 8. Academic Press, New York.

Cook R. L., and C. E. Millar (1944). *Fertilizers for Legumes.* Michigan State University Agricultural Experiment Station Bulletin 328.

Davis, J. F., W. C. Hulbert, C. M. Hansen, and L. N. Shepherd (1956). The effect of fertilizer placement on the yield of onions and spinach grown on organic soils. *Mich. Exp. Sta. Quart. Bull.* 39:25–35.

Donahue, R. L., R. W. Miller, and J. C. Schickluna (1983). *Soils, an Introduction to Soils and Plant Growth,* 5th ed. Prentice–Hall, Englewood Cliffs, N.J.

Engelstad, O. P., and S. E. Allen (1971). Ammonium pyrophosphate and ammonium orthophosphate as phorphorus sources: Effects of soil temperature, placement, and incubation. *Soil Sci. Soc. Am. Proc.* 35:1002–1004.

Foth, H. D. (1984). *Fundamentals of Soil Science,* 7th ed. Wiley, New York.

Friesen, D. K., A. S. R. Juo, and M. H. Miller (1980). Liming and lime-phosphorus-zinc interactions in two Nigerian Ultisols: I. Interactions in the soil. *Soil Sci. Soc. Am. J.* 44:1221–1226.

Friesen, D. K., M. H. Miller, and A. S. R. Juo (1980). Liming and lime-phosphorus-zinc interactions in two Nigerian Ultisols: II. Effects on maize root and shoot growth. *Soil Sci. Soc. Am. J.* 44:1227–1232.

Gillian, G. W. (1970). Hydrolysis and uptake of pyrophosphate by plant roots. *Soil Sci. Soc. Am. Proc.* 34:83–86.

Hanway, J., G. Stanford, and H. R. Meldrum (1953). Effectiveness and recovery of phosphorus and potassium fertilizers topdressed on meadows. *Soil Sci. Soc. Am. Proc.* 17:378.

Hashimoto, I., and J. R. Lehr (1973). Mobility of polyphosphates in soil. *Soil Sci. Soc. Am. Proc.* 37:36–41.

Lawton, K., and R. L. Cook (1955). Interaction between particle size and water solubility of phosphorus in mixed fertilizers as factors affecting plant availability. *Farm Chem.* April: 44–46.

Lawton, K., M. B. Tesar, and B. Kawin (1954). Effect of rate and placement of superphosphate on the yields and phosphorus absorption of legume hay. *Soil Sci. Soc. Am. Proc.* 18:428–532.

Lawton, K., and J. A. Vomocil (1954). The dissolution and migration of phosphorus from granular superphosphate in some Michigan soils. *Soil Sci. Soc. Am. Proc.* 18:26–32.

Millar, C. E. (1954). Sulfur in crop production. *Farm Chem.* M. September:

Reisenauer, H. M. (1956). Molybdenum content of alfalfa in relation to deficiency symptoms and response to molybdenum fertilization. *Soil Science* 81:237–258.

Robertson, L. S., and R. E. Lucas (1981). *Zinc: An Essential Plant Micronutrient.* Bulletin E-1012. Cooperative Extension Service, Michigan State University.

Vitosh, M. L. (1975). *Fertilizers, Types, Uses, and Characteristics.* Cooperative Extension Service, Michigan State University Agricultural Facts 53.

Warncke, D. D., D. R. Christenson, and M. L. Vitosh (1985). *Fertilizer Recommendations for Vegetable and Field Crops in Michigan.* Bulletin E-550. Cooperative Extension Service, Michigan State University.

CHAPTER 15

Cropping Systems and Management Practices: The North Central States

The soils and climate of Kentucky resemble those of southern Ohio, Indiana, and Illinois. As a result, the state has often been included, as far as soil management is concerned, with the 12 North Central states. By the opposite line of reasoning, perhaps, the eastern, northern, and western fringe of the area should be excluded. The agriculture of eastern Ohio and the northern parts of Michigan, Wisconsin, and Minnesota is quite different from that of the major portion of the North Central states. The differences will be mentioned later.

The western parts of Kansas, Nebraska, and North and South Dakota are included in the Great Plains. Their soil management problems are discussed in Chapter 16.

CHARACTERISTICS OF THE AREA

The North Central region of the United States includes what has long been known as the "Corn Belt." It also includes much of our wheat acreage; in fact, most of that grown for pastry flour. The climate has been termed humid continental, although rainfall varies from less than 50 to over 100 cm (20 to over 40 in.) annually. In fact, that much variation occurs within the eastern three-fourths of Kansas. In general, the southern part of the region receives more rainfall than does the northern part.

Long summers, 5 to 6 months frost-free with hot nights and high relative humidity, favor corn (maize) production in much of the area. Maximum rainfall comes during the summer months when corn has a high water requirement. When it is too warm and humid to sleep, the Corn Belt farmer can realize some comfort from the knowledge that the weather is ideal for corn. From the date of emergence, corn plants have been known to grow 5 cm (2 in.) a day until tassels appeared.

During the last half century, soybeans have moved into the Corn Belt. In 1948, more than 67 percent of the U.S. soybean acreage was in the five states—Illinois, Iowa, Indiana, Ohio, and Missouri. This is a good illustration of the expansion of a crop into the areas where it is the best adapted. The crop was first grown in this country in the Eastern Seaboard states and was hardly known in the Corn Belt until after 1920.

As a result of the work of plant breeders who have developed adapted varieties, the fringe areas of the North Central region are gradually increasing their acreages of these two principal crops. The result has been even more uniformity in the cropping pattern of the region as a whole. The late-summer traveler who attempts to view the landscape from a car realizes that corn plays an important role in the agriculture of these states.

Oats occupy a much smaller part of the region's crop acreage than was the case in the early part of this century, a result no doubt of the trend toward monocropping. Lesser production, however has resulted in higher prices, and the crop still plays an important role as a companion crop for legume and grass seedings in the northern fringe areas where corn and soybeans are less important. A general picture of cropping in the entire region is shown by Fig. 15-1.

GENERAL CHARACTERISTICS OF THE SOILS

Soils in the North Central region are greatly variable. The northern part of the region was largely glaciated, although the glaciers missed several counties in southwestern Wisconsin, southeastern Ohio, and a part of Missouri. The driftless area of southwestern Wisconsin extended also into Jo Daviess County, Illinois. Swamps and peat lands occupy 10 to 20 percent of the land surface in this area, and topography ranges from level to steeply sloping. Sandy outwash plains are scattered through southern Michigan, northern Indiana, central Wisconsin, and parts of Minnesota. Millions of acres of submarginal sandy soils lie in northern Michigan, Wisconsin, and Minnesota. Clay lenses at relatively shallow depths make some of these soils sufficiently productive to give them agricultural value.

Soil Management Principle

Clay lenses at relatively shallow depths can make sand soils more productive.

Central and eastern Michigan and southern Wisconsin are occupied by glacial till and lake-bed (lacustrine) soils interspersed with sandy outwash and a few swamps. The till soils occur largely on moraines with gently rolling to steep relief and are intermediate to fine in texture. Much of the area requires artificial drainage, and some areas are so slowly permeable that tile are not

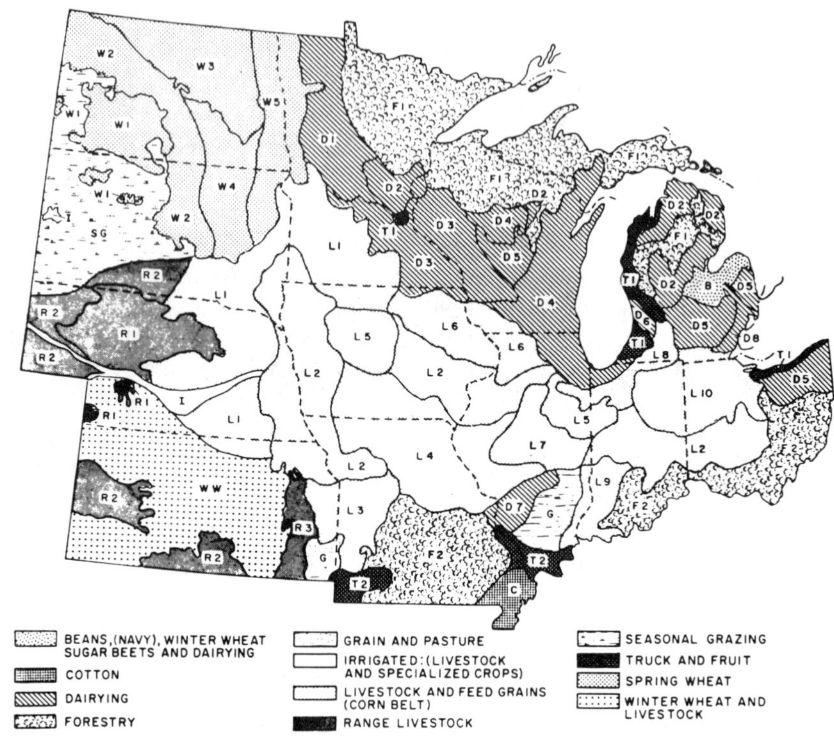

FIGURE 15-1 Major land use and types of farming areas in the North Central region. The meanings of the patterns and letters on the map are indicated in the figure. The letters and their numeric subdivisions are explained as follows: B. Beans (navy), winter wheat, sugar beets, and dairying; C. Cotton; D. Dairying: D1 Dairying, hay, and livestock, D2 Dairying and potatoes, D3 Dairying and livestock, D4 Specialized dairy (milk and cheese), D5 Dairying, small grains, hay, and corn, D6 Dairying, poultry, and truck, D7 Dairying, livestock, and poultry, D8 Dairying and truck; F. Forestry: F1 Forestry, dairying, and potatoes, F2 Forestry, livestock, and special crops (fruit, vegetables, tobacco); G. Grain and pasture; I. Irrigated: (Livestock and specialized crops); L. Livestock and feed grains (Corn Belt): L1 Corn and livestock, L2 Cattle feeding and hogs, L3 Livestock, grain, and dairying, L4 Livestock, pasture, and grain, L5 Cash grain (corn, oats, and soybeans), L6 Hogs and dairying, L7 Cash grain (corn, soybeans, and winter wheat), L8 Cash grain, dairying, and livestock, L9 Corn, winter wheat, and truck, L10 Livestock and cash grain; R. Range livestock: R1 Stock farms, R2 Range livestock and winter wheat, R3 Range livestock; SG. Seasonal grazing; T. Truck and fruit: T1 Truck, dairying, and fruit, T2 Fruit and mixed farming; W. Spring wheat: W1 Spring wheat and range livestock, W2 Spring wheat, W3 Spring wheat and flax, W4 Spring wheat and general farming (corn, flax), W5 Spring wheat, general farming (flax), and potatoes; WW Winter wheat and livestock. (From North Central Regional Publication 76, Wisconsin University Agricultural Experiment Bulletin 544, 1960.)

practical. Michigan, Wisconsin, and eastern Minnesota were largely forested. Light-colored, low organic matter soils are the result. A few small areas of prairie soil are interspersed.

The western part of the North Central region, including the Dakotas, Nebraska, Kansas, and the western portions of Minnesota, Iowa, and Missouri, are largely prairie with vast areas covered by silty loess. Prairie soil, dark in color, covers also a large part of Illinois and extends unevenly as far east as Lafayette, Indiana. Texture ranges from sandy loam to clay. Portions of South Dakota and Minnesota are covered with drift from the Wisconsin glacier, largely loamy in texture. Parts of southeastern Kansas and the larger part of Missouri south of the Missouri River were forested, and here the soils are light colored.

Sand hills cover much of northern and western Nebraska. Western South Dakota and North Dakota include the Badlands and Black Hills. Management of the agricultural soils of the western parts of these states and Kansas (roughly two-thirds) is discussed in the Great Plains section of Chapter 16.

The soils of northwestern Ohio, northern Indiana, and a part of southern Michigan, formed under forest vegetation, are light to dark in color and variable from clay to sand. Much of that portion of Ohio and sizable areas in northern Indiana and southeastern Michigan are lake plains. The soils are fine-textured, and drainage is very poor. Many of the soils of northern Indiana are at the opposite extreme, sandy loams and sands. The soils in the extreme northwestern part of Indiana, Lake and Porter Counties, and the southern part of Cook County, Illinois, are silty clay and clay. Loams and silt loams from glacial till occupy the remaining portions of northern Indiana, western Ohio, and southern Michigan. As in all glacial areas, local variations due to swamps and outwash are common. The soils of central Indiana and western Ohio are a mixture of light-colored silt loams and dark-colored silty clay loams.

The claypan soils of southern Illinois, southern Indiana, southwestern Ohio, and parts of Missouri are naturally acid, light-colored, and low in fertility. They vary in topography and drainage, and erosion presents serious problems.

The soils in central-eastern and northeastern Ohio are largely light-colored and are medium to strongly acid. Some have fragipan horizons, especially in northeastern Ohio. Southern and southeastern Ohio and parts of southcentral Indiana are unglaciated and have rolling to very steep relief. Practically all were developed under forest vegetation.

HISTORY OF SOIL MANAGEMENT IN THE NORTH CENTRAL STATES

Soil management in the North Central states has been influenced in no small degree by the results of the early field experiments in Illinois, Missouri, and Ohio. The Morrow plots in Illinois, started in 1876, were designed to compare rotations with continuous cropping. Three of the original systems have been maintained since 1876: Continuous corn; corn-oats (changed to corn-soybeans in 1967); and corn-oats-clover. Lime, phosphate, and manure treatments were introduced in 1904. In 1955, plots were again divided and treatments of lime, nitrogen, phosphorus, and potassium were added. Finally, in

1967, a high rate of lime, nitrogen, phosphorus, and potassium based on soil test was included.

Table 15-1 is a summary which shows that reduced yields under continuous corn are largely from nutrient depletion. Lime, nitrogen, phosphate, and potassium increased average yields from 32.3 to 117.8 bu/acre! Lime, manure, and phosphate, together with nitrogen, phosphate, and potassium, gave a small additional yield increase.

Although on the average the yield from fertilized corn grown in rotation was better than from fertilized continuous corn, examining yield from individual years is even more enlightening. For example, 1965 and 1966 were excellent years for corn. Fertilized continuous corn yields were equal to corn grown in a rotation during these years. But in 1976 and 1977 (poorer corn years), the fertilized continuous corn yielded only 64 and 75 percent as much as fertilized corn grown in a rotation. This illustrates the value of added soil organic matter in increasing yields through improved soil-water relations.

Soil Management Principle

Rotations improve the physical properties of soils as well as the chemical properties.

Missouri's Sanborn field, planned by J. W. Sanborn, was started in 1888. The primary purpose of the experiments was to determine the value of stable manure when used in continuous cropping systems and in rotations. One of the wheat treatments was N, P, and K fertilizers in amounts equal to the needs of a 40-bushel crop. Residues were not returned to the land. In general, the results showed that fertilizers were as effective as manure in keeping up yields but were much less effective in maintaining organic matter levels. Perhaps results would have been different if residues had been returned.

TABLE 15-1 Summary Data for Corn Yields from the Morrow Plots

		Additional treatments		
Rotation	None	LNPK	MLP and LNPK	MLP and high LNPK
			Bushels corn/acre	
Continuous corn	32.3	117.8	126.6	136
Corn-oats	37.9	133.4	134.4	156.5
Corn-oats-clover	59.1	139.5	139.8	153.6

Note: MLP is manure, lime, and phosphate since 1904; LNPK is lime, nitrogen, phosphorus, and potassium fertilizer since 1955; and High LNPK is lime, nitrogen, phosphorus, and potassium according to soil test since 1967.
Source: Odell, Walker, Boone, and Oldham (1982).

The Ohio experiments were established by Director C. E. Thorne in 1893. The primary objectives were to determine the effects of fertilizer and manure on the yields of crops grown continuously and in rotation. According to present-day standards, fertilizer rates were low compared to rates of manure application. Over a long period, manure treatments resulted in the highest corn yields, whereas complete fertilizers were the most effective for oats and wheat. Manure was especially valuable as a fertilizer for clover.

The value of lime in the production of legumes was evident at about the turn of the century. Within a few years after the Ohio experiments were started, clover seedings were failing, so a liming program was started in 1900. At Illinois, lime was applied on portions of each plot in 1904.

Mention might be made of the fact that lime investigations were included in the Jordan Soil Fertility plot (Pennsylvania State University) objectives when the work was outlined in 1881. That early work included studies with land plaster (gypsum), burned lime, and limestone. Over the years, applications of limestone to raise the soil pH from about 4.0 to over 6.0 resulted in marked increases in yields of all crops, including corn.

Considering the nature of these early experiments and the results, strongly positive for the use of rotations and lime, and the emphasis placed on the value of farm manure, it is not surprising that soil management recommendations were so strong for livestock, lime, and legumes. Where farm manure had been compared with fertilizers in the early experiments, residues were removed, rates of fertilizers were low, and, in many instances, fertilizer ratios were too low in nitrogen. As a result, the value of the manure was particularly outstanding for corn, a crop that recent work shows to be a particularly heavy user of nitrogen.

This general system was particularly successful in Illinois, Missouri, Indiana, Ohio, and parts of Iowa, Wisconsin, and Michigan because soils are generally acid, so lime was needed to grow clover, the basic crop in the system.

Further west, where rainfall is less, lime is needed to a lesser extent or not at all and soil-water management becomes all important.

Mechanization, expansion of the fertilizer industry, especially with respect to nitrogen, a broader research program in the experiment stations (see Fig. 15-2), and changes in the standard of living on the farms have worked together to greatly expand soil management possibilities on the farms of this region. No longer is it necessary to produce livestock to maintain soil fertility. In some people's opinions, there are easier ways, or at least more pleasant ways, to manage soils. A farmer may now have much more freedom of choice as to how to work the land.

Soil Management Principle

Soil management practices that bring about maximum sustained production are also those that conserve the soil.

FIGURE 15-2 An extension agronomist explains field plot results to a group of Michigan farmers. The soil is Sims clay loam, typical of the Saginaw Valley bean and sugar-beet areas.

MANAGEMENT FOR PRODUCTION AND CONSERVATION

Fortunately, the soil management practices that bring about maximum sustained production are also those that conserve the soil. The more soil fertility, soil management, and productivity are studied, the more truth there seems to be in this statement. The proper coordination of all known management practices is the keynote of soil conservation.

The soils of the North Central region are, in the final analysis, quite variable, as are also climate and crops. It seems well then to divide the area, for discussion purposes, into subregions. We have divided the North Central region into five subregions based on origin of soil, climate, and type of farming in a manner similar but not identical to other writers.

NORTHERN FOREST SUBREGION

Michigan, Wisconsin, and east-central and northeastern Minnesota comprise the Northern Forest subregion. Climate sets this subregion off from the remainder of the North Central region. Corn is highly profitable for grain in only the southern half of the area. Even there, corn on the organic soils, which make up 10 to 12 percent of the area, may be frosted during any month of the year. On the mineral soils of the southern half of the subregion, the frost-free season is only 110 to 130 days. As a result, frost prevention and the production of frost-resistant crops is important. Of importance, also, is early planting to ensure maturity.

An objective of plant breeders has been to develop short-season, high-yielding varieties; their success is evidenced by the gradual northward extension of corn and soybeans. More recently, cold-resistant varieties have made it possible to grow winter barley as far north as central Michigan.

Corn has been selected for quick germination in cold soil, and seeds are now commonly treated with chemicals that prevent mildew. As a result, it is safe to make earlier plantings and so make use of longer season varieties. Thus, the subdivision is gradually increasing in importance as a corn- and soybean-producing area.

Forages do well in this subregion. In fact, in the northern half of the area, they are by far the most important crops. Alfalfa, well adapted to the entire subregion, is a very valuable soil-building crop. The data in Table 15-2 show how valuable the crop may be in bushels of corn and tons of sugar beets when these crops follow the alfalfa in that order. The simple production of 1-year alfalfa-brome hay (two cuttings) brought about, as an average over a 15-year period, 29.3 bushels more corn and 1.4 tons more sugar beets. These results show that it may be profitable to grow the forage as a soil-building crop even if the hay is not used. However, in the same experiment, legumes grown as green manure catch crops, three times in the rotation, were almost as efficient in increasing corn yields as was the alfalfa-brome hay. The corresponding increases in yields were 21.9 bushels of corn and 0.9 ton of sugar beets.

Organic Matter Must be Maintained

As stated in Chapter 4, organic matter is essential for profitable crop production. Northern forest soils are notably low in this constituent. Percentages range below 1.0 in most of the sandy soils of central Michigan, central Wisconsin, and northeastern Minnesota. Amounts as high as 4 percent are common in the fine-textured, poorly drained soils. Surface soils throughout the area are naturally shallow. Water and wind erosion have taken and are taking

TABLE 15-2 The Effect of Legumes as Hay or Green Manure on the Yields of Corn and Sugar Beets that Follow in the Rotation. Sims Clay Loam Soil, 15-Year Averages. Michigan Experimental Station

	Yields per acre	
Rotation	Corn bu	Sugar beets tons
Alfalfa-brome, alfalfa-brome, corn, sugar beets, barley	63.4	12.5
Alfalfa-brome, corn, sugar beets, barley, oats	64.6	12.6
Wheat, corn, sugar beets, barley, beans	35.3	11.2
Wheat (G.M.), corn (Sw.C.), sugar beets, barley (G.M.), beans	57.2	12.1

G.M. = green manure; mixture sweet, alsike, red, mammoth clovers. Sw.C. = green manure; sweetclover.

heavy tolls on the sloping and sandy lands. Soil-conservation practices are badly needed.

Fortunately, the best soil-conservation practices are the best practices for raising yields and cutting unit costs of production. The production of sod-forming forages, with adequate use of lime and fertilizer, from 20 to 50 percent of the time, is a **must** on much of the land of the area. Some of the land should be covered with forage crops 75 to 100 percent of the time. Much of the abandoned sandy land of central Michigan and central Wisconsin can be brought under production of forage crops by the application of adequate lime and fertilizer, including the trace and secondary nutrients. The economy of such practices will depend on the value of the livestock that may be produced. Future population pressure in the area will have a bearing.

Irrigation

A word should be said about the role of supplemental irrigation in the agriculture of sandy soils. Although the region is considered humid, summer drought is always in the picture. On the whole, water supplies are good, either from lakes and streams or from groundwater, which in many places is near the surface. Flowing wells are not uncommon.

Proximity to centers of population makes the production of relatively high-value crops such as potatoes, vegetables, and small fruits a profitable enterprise. These crops do well on sandy soils under the right soil management with irrigation. Hoglund and Cook (1956) estimated that the farmer who keeps all other soil management practices in balance may expect, by intelligent use of irrigation (see Chapter 3), to increase average potato yields as much as 180 bu/acre.

Rotations

Some people have questions regarding rotations. Surely the blind following of a rotation is not wise, but the production of soil-building crops is **always** wise. We should try to fit the crop to the soil. On level, permeable land, capable of producing 100 bu of corn per acre, it may be possible with adequate use of fertilizers to do so continuously. However, it will be necessary to return or leave all residues, grow cover crops, use correct tillage practices, and follow all other good cultural practices. If by following all these good practices it is possible to maintain the 100-bu yield, all indications are that the soil is being conserved. A farmer who has such land to the extent of his usual corn acreage and all the remainder of his soil is sloping and subject to erosion is wise to grow his corn on those adapted acres and to confine his sloping land to the production of protective crops. In Tables 15-3 to 15-6 are listed the least-protective rotations that will keep soil losses within permissible limits. Soils are divided on the basis of texture and are described following. The assumed management practices are listed.

1. Permeable, productive soils, medium texture, good drainage, pH high

enough for legume cover crops, **erosion not a hazard.** Soils where corn with adequate nutrients and approved cultural practices would produce 100 bu/acre. Grow continuous corn on such land. Use minimum tillage and adequate nutrients according to tests,[1] including at least 100 lb of nitrogen per acre. Leave residues on the surface until planting time and grow cover crops.

2. Loam to clay loam soils, limed or naturally high pH, light to dark in color depending on natural drainage. Use tile drainage where needed, apply adequate nutrients according to tests, use minimum tillage, and return all residues. Use the rotations in Table 15-3. Note the equalizing effect of strip-cropping and terracing. On 1 to 6 percent slopes it is permissible, by strip-cropping, to use the same rotation on all except 400 ft slopes; with terracing the same rotation can be used on all the slopes up to 6 percent. Strip-cropping allows the production of row crops on slopes as great as 18 percent.

3. Sandy loam soils, limed or naturally high pH, light to dark in color depending on natural drainage. Use tile drainage where needed, adequate nutrients according to tests, and minimum tillage, and return all residues. Use the rotations in Table 15-4. Note again the effectiveness of the erosion-control practices in making it possible to use more depleting rotations.

4. Loamy sand soils or sand soils containing layers of finer materials, limed or naturally high pH, light to dark in color depending on natural drainage. Use adequate nutrients according to tests, minimum tillage, and drainage where needed. Return all residues. Use the rotations in Table 15-5.

5. Sand soils, limed or naturally high pH, light to dark in color depending on natural drainage. Use adequate nutrients according to tests, minimum tillage, and drainage where needed. Return all residues. Use the rotations in Table 15-6.

Legume seeding must be made alone on all loamy sand or sand soils except for those of very high productivity. On nonerosive soils, this may be done in late spring or early summer. On soils with more than 2-percent slopes, seedings should be made as soon after grains are removed as soil moisture is sufficient. The grain straw should be removed.

Grain Farming

The agriculture of this Northern Forest subregion has always been built around livestock production; accordingly, all the preceding rotation suggestions include provision for meadow crops. There are several areas, however, where cash-crop farming is largely practiced, particularly in southwestern

[1]It is suggested that you obtain fertilizer recommendation publications from local Cooperative Extension Service or USDA Soil Conservation Service personnel or local government agencies.

TABLE 15-3 Least-Protective Rotations for Loam to Clay Loam Soils in the Northern Forest Subregion

| Slope | | Erosion-control practices | | | |
Percent	Length (ft)	None	Contouring	Strip-cropping	Terracing
0	All	R*R	—	—	—
1–2	100	R*	R*R	—	—
	200	RW*	R*	—	—
	300	RROM	R*	—	—
	400	RROMM	RO*	—	—
3–4	100	RROMM	R*	RW*	R*R
	200	RROMMM	RW*	RW*	R*R
	300	R*RMMM	RROWM	RW*	R*R
	400	R*OMM	R*OMM	RW*	R*R
5–6	100	RRMMM	RROWM	RW*	R*R
	200	R*OMM	RROMM	RW*	R*R
	300	RWMMM	RROMMM	RW*	R*R
	400	RWMMMM	RWMMMM	RO*O*	R*R
7–8	100	R*OMM	ROMMW*	RW*	RO*
	200	RWMMMM	R*OMM	RROMM	RO*
	300	RWMMMM	RWMMMM	RROMM	RO*
	400	OMM	OMM	RROMMM	RO*
9–12	100	RWMMMM	RWMMMM	RROMMM	RROMM
	200	OMM	OMM	R*OMM	RROMM
	300	OMMM	OMMM	R*OMM	RROMM
	400	M	M	ROMMM	RROMM
13–18	60	OMMM	OMMM	—	—
	100	M	M	R*OMM	—
	200	M	M	RWMMMM	—
18+	All	Grass or trees — permanent vegetation			

R = row crop, O = oats, W = wheat, M = meadow, * = green manure cover crop. Dash indicates erosion control practice not needed or not recommended because it is not effective.

Wisconsin, southeastern Michigan, and the Saginaw Valley of Michigan. Furthermore, there is currently a trend away from livestock on scattered farms where the operator works part-time off the farm. In fact, the ratio of part-time to full-time farmers is increasing rapidly throughout the area. Livestock do not fit well into a part-time enterprise.

Experiments have shown that where erosion is not a factor or where it is controlled by contouring, strip-cropping, or terracing, meadow crops may not be essential for fertility maintenance. However, organic matter maintenence is essential and may be accomplished by careful use of residues, heavy fertilization, and the production of green manures as cover and catch crops. This is indicated by the experimental results shown in Table 15-2.

A farmer may gain a clue as to where it may be permissible to omit meadow from the rotation by considering the success that he has with green manure catch crops and cover crops. If he is able to establish good legume seedings *(Melilotus alba* or *Trifolium pratense)* in grains to be turned under for a second grain or row crop, and if he finds it possible to grow cover crops in corn (either legumes or grasses), he is relatively safe in changing to a cash cropping system. He will, of course, find it necessary to apply much more nitrogen fertilizer after such a change.

Melilotus alba is the best green manure and cover crop in much of the area. A leaf-eating insect makes it a questionable crop in some localities. Later seeding, such as in corn in late June, makes it possible to grow the crop in localities where early spring seeded crops are destroyed.

TABLE 15-4 Least-Protective Rotations for Sandy Loam Soils in the Northern Forest Subregion

Slope		Erosion-control practices			
Percent	Length (ft)	None	Contouring	Strip-cropping	Terracing
0	All	RO*	—	—	—
1–2	100	RO*	—	—	—
	200	RW*	RO*	—	—
	300	ROROM	RO*	—	—
	400	RROMM	RO*	—	—
3–4	100	RROMM	RO*	—	—
	200	RROMMM	RW*	—	RO*
	300	ROOMM	RORWM	RW*	RO*
	400	R*OMM	R*OMM	RW*	RO*
5–6	100	RRMMM	RORWM	RW*	RO*
	200	R*OMM	RROMM	RW*	RO*
	300	ROMMM	RROMMM	RW*	RO*
	400	RWMMMM	RWMMMM	RO*O*	RO*
7–8	100	R*OMM	RROMMM	RW*	RO*
	200	RWMMMM	R*OMM	RROMM	RO*
	300	RWMMMM	RWMMMM	RROMM	RO*
	400	OMM	OMM	RROMMM	RO*
9–12	100	RWMMMM	—	RROMMM	RROMM
	200	OMM	—	R*OMM	RROMM
	300	OMM	—	R*OMM	RROMM
	400	M	—	ROMMM	RROMM
13–18	60	OMM	OMM	R*OMM	—
	100	M	M	R*OMM	—
	200	M	M	RWMMMM	—
18+	All	Grass or trees — permanent vegetation			

R = row crop, O = oats, W = wheat, M = meadow, * = green manure cover crop. Dash indicates erosion-control practice not needed or not recommended because it is not effective.

TABLE 15-5 Least-Protective Rotations for Loamy Sands or Sands with Some Fine-Textured Layers in the Northern Forest Subregion

Slope		Erosion-control practices			
Percent	Length (ft)	None	Contouring	Strip-cropping	Terracing
0–6	100	RROMM	—	—	—
	200	RROMM	—	—	—
	300	RROMMM	RROMM	—	—
	400	RROMMM	RROMMM	—	—
7–8	100	RROMM	RROMM	—	—
	200	RROMMM	RROMM	—	—
	300	R*RMMM	R*RMMM	RROMM	RROMM
	400	ROMMM	ROMMM	RROMM	RROMM
9–12	100	R*OMM	R*OMM	RROMM	RROMM
	200	ROMMM	ROMMM	RROMM	RROMM
	300	ROMMMM	ROMMMM	RROMM	RROMM
	400	OMM	OMM	RROMM	RROMM
13–18	60	ROMMMM	ROMMM	—	—
	100	OMM	OMM	RROMM	—
	200	OMMM	OMMM	RROMMM	—
18+	All	Grass or trees—permanent vegetation			

R = row crop, O = oats, W = wheat, M = meadow, * = green manure crop. Dash indicates erosion-control practice not needed or not recommended because it is not effective. More intensive rotations than RROMM are not recommended for the more level of these soils (0–5 percent slope), even though they would hold soil loss to the permissible amount.

Winter grains and grasses are more satisfactory as cover crops on light sandy soils. With plenty of nitrogen fertilizer, they may be more satisfactory than legumes because they are more easily established. The grasses may serve better also on poorly drained heavy soils where winter heaving destroys small legume plants but does less damage to grasses. Ryegrass (*Lolium perenne*) is one of the best plants for this purpose. It is especially good as an erosion control cover because it makes a tremendous root growth.

As soil series are mentioned in the following pages of this chapter, please turn to Table 15-7 for their taxonomic classification.

EASTERN FOREST SUBREGION

Ohio, Indiana, and parts of Kentucky are included in the Eastern Forest subregion. The climate is favorable for corn, soybeans, and winter wheat, with 35 to 40 in. of rainfall and a 160- to 180-day growing season.

Most of the soils, formed under forest vegetation, are light-colored and relatively thin. They vary in texture from the very heavy Paulding clays (Typic Haplaquepts) of northwestern Ohio to the sands and sandy loams of northwestern Indiana. Considerable variation occurs in the rest of the subregion, but on the whole they are loams and silt loams.

Erosion is a problem in all parts of the area except in northeastern Ohio. Those lands not subject to erosion have another soil problem that is perhaps more serious. The soils are extremely fine, with very slow subsoil permeability. Poor drainage has inhibited legume production, and the soils produce relatively better yields of soybeans *(So ja max)* than of corn *(Zea mays)*. Thus, these areas in Ohio have become strong soybean-producing areas. Corn yields have improved, however, in recent years because of increased use of nitrogen fertilizer, a substitute to some extent for the building effect of legumes.

Rotations for the Soybean Soils of Northwestern Ohio

Grain or cash-crop farming is common in northwestern Ohio. Reed, Slipher, and Beard (1950) suggest the following rotations for the gray flat soils (Crosby silty clay loam in west central Ohio, Trumbull silt loam and silty clay loam in northeastern Ohio, and Clermont silt loam in southwestern Ohio). See Table 15-7 for a taxonomic classification of the soils mentioned in this chapter.

TABLE 15-6 Least-Protective Rotations for Sands in the Northern Forest Subregion

Slope		Erosion-control practices			
Percent	Length (ft)	None	Contouring	Strip-cropping	Terracing
0–6	100	R*ROMM	—	—	—
	200	RROMMM	R*ROMM	—	—
	300	R*RMMM	R*ROMM	—	—
	400	R*OMM	R*OMM	—	—
7–8	100	RROMMM	R*ROMM	—	—
	200	R*OMM	R*ROMM	—	—
	300	ROMMM	ROMMM	RROMM	RROMM
	400	ROMMM	ROMMM	RROMM	RROMM
9–12	100	ROMMM	ROMMM	RROMM	RROMM
	200	ROMMMM	ROMMMM	RROMM	RROMM
	300	OMM	OMM	RROMM	RROMM
	400	OMMM	OMMM	RROMM	RROMM
13–18	60	OMM	ROMMMM	—	—
	100	OMM	OMM	—	—
	200	OMMM	OMMM	—	—
18+	All	Grass or trees — permanent vegetation			

R = row crop, O = oats, W = wheat, M = meadow, * = green manure cover cop. Dash indicates erosion-control practice not needed or not recommended because it is not effective. More intensive rotations than R*ROMM are not recommended for the more level of these soils (0–5 percent slope), even though they would hold soil loss to the permissible amount.

TABLE 15-7 Classification of Soil Series Mentioned in Chapter 15:

	Family and subgroup
Aastad	Fine, loamy, Udic Haploborolls
Ava	Fine-silty, mixed, mesic Typic Fragiudalfs
Barnes	Fine-loamy, mixed Udic Haplogorolls
Beadle	Fine, montmorillonitic, mesic Typic Argiustolls
Bearden	Fine-silty, frigid Aeric Calciaquolls
Blair	Fine-loamy, mixed, mesic Aquic Hapludalfs
Bluford	Fine, montmorillonitic, mesic Aquic Hapludalfs
Bonilla	Fine, loamy, mixed, mesic Pachic Haplustolls
Bonnie	Fine-silty, mixed, acid, mesic Typic Fluvaquents
Brookston	Fine-loamy, mixed, mesic Typic Argiaquolls
Burchard	Fine-loamy, mixed, mesic Typic Argiudolls
Buse	Fine-loamy, mixed Udorthentic Haploborolls
Cavour	Fine, montmorillonitic Udic Natriborolls
Celina	Fine, mixed, mesic Aquic Hapludalfs
Cisne	Fine, montmorillonitic, mesic Mollic Albaqualfs
Clermont	Fine-silty, mixed, mesic Typic Ochraqualfs
Clyde	Fine-loamy, mixed, mesic Typic Haplaquolls
Crofton	Fine-silty, mixed (calcareous), mesic Typic Ustorthents
Crosby	Fine, mixed, mesic Aeric Ochraqualfs
Ellsworth	Fine, illitic, mesic Aquic Hapludalfs
Fargo	Fine, montmorillonitic, frigid Vertic Haplaquolls
Florence	Clayey-skeletal, montmorillonitic, mesic Udic Argiustolls
Grundy	Fine, montmorillonitic, mesic Aquic Argiudolls
Hamerley	Fine-loamy, frigid Aeric Calciaquolls
Hickory	Fine-loamy, mixed, mesic Typic Hapludalfs
Houdek	Fine-loamy, mixed Typic Argiustolls
Hoyleton	Fine, montmorillonitic, mesic Aquollic Hapludalfs
Kranzburg	Fine-silty, mixed Udic Haploborolls
Labette	Fine, mixed, mesic Udic Argiustolls
Mahoning	Fine, illitic, mesic Aeric Ochraqualfs
Marshall	Fine-silty, mesic Typic Hapludolls
Muscatine	Fine-silty,mixed, mesic Aquic Hapludolls
Newberry	Fine-silty, mixed, mesic Mollic Ochraqualfs
Paulding	Very fine, illitic, nonacid,mesic Typic Haplaquepts
Pawnee	Fine, montmorillonitic, mesic Aquic Argiudolls
Poinsett	Fine-silty, mixed Udic Haploborolls
Ravenna	Fine-loamy, mixed, mesic Aeric Fragiaqualfs
Richview	Fine-silty, mixed, mesic Mollic Hapludalfs
Sharon	Coarse-silty, mixed, acid, mesic Typic Udifluvents
Sims	Fine, mixed, mesic Mollic Haplaiquepts
Sharpsburg	Fine, montmorillonitic, mesic Typic Argiudolls
Sinai	Fine, montmorillonitic, mesic Typic Argiudolls
Sogn	Loamy, mixed, mesic Lithic Haplustolls
Summit	Fine, montmorillonitic, thermic Vertic Argiudolls
Trumbull	Fine, illitic, mesic Typic Ochraqualfs (Glossaqualfs)
Ulen	Sandy, frigid Aeric Calciaquolls
Vienna	Fine-loamy, mixed Udic Haploborolls
Webster	Fine-loamy, mixed, mesic typic Haplaquolls
Woodson	Fine, montmorillonitic, thermic Vertic Argiudolls
Wynose	Fine, montmorillonitic, mesic Typic Albaqualfs

1. Soybeans, oats *(Avena sativa)*, winter grain, clover seed.
2. Soybeans, oats, clover seed, winter grain.
3. Soybeans, oats, clover seed.

On the light brown soils (Celina in western Ohio and Ellsworth in northeastern Ohio); these authors suggest a split between corn and soybeans, as:

1. $\begin{bmatrix}\text{Soybeans}\\\text{Corn}\end{bmatrix}$ Small grain, legume sod, legume sod

2. $\begin{bmatrix}\text{Soybeans}\\\text{Corn}\end{bmatrix}$ Small grain, legume sod

On the somewhat more productive dark-colored soils (Brookston and Paulding), they recommend a sequence of corn and soybeans followed by small grain, legume sod, legume sod. A second grain crop may be substituted for one of the years of legume on the more permeable of these soils, provided satisfactory tillage, residue, and cover crop practices are followed. With adequate applications of lime and fertilizer, these soybean cropping systems should result in continued productivity or soil building. Lime and plant nutrients should be applied as need is indicated by soil tests and the results of experiments. Lime needs are heavy on the gray soils, medium on the light brown soils, and light on the dark soils.

On the gray sloping soils (Crosby silt loam in western Ohio and Ravenna and Mahoning silt loams in northeastern Ohio), the same rotations as recommended for the gray flat soils may be followed except soybeans are generally not recommended on soils more sloping than 4 percent for two reasons. They produce lower yields on those soils and are not adapted to livestock farming. Such soils can only be maintained by keeping them in meadow from 25 to 50 percent of the time, which means a livestock enterprise is needed to make use of the meadow. Grain is then needed for the animals, so the farmer should grow corn rather than soybeans.

Eastern Ohio and the Sloping Lands of Indiana

This area, including the sandy soils in northern Indiana, may be managed in much the same manner as that recommended for the soils of the Northern Forest subregion. The areas have much in common as far as soil characteristics and cropping practices are concerned. They differ to some extent with respect to soil nutrient status and fertilizer needs, but such differences are indicated by soil-test results; you are urged to consult local extension personnel or publications for specific recommendations.

Kentucky

The soils of Kentucky are quite sloping, so erosion control is a major soil problem. Most of the soils need lime unless adequate lime has been applied.

On most Kentucky soils, legume yields, including soybeans, cowpeas (*Vigna sinensis*), and lespedeza (*Lespedeza stipulaces*) are increased by liming. Tobacco (*Nicotiana tabacum*) yields are not increased by liming and may actually be decreased. In tobacco rotations, the lime should be applied immediately before the legume and in as small amounts as will result in satisfactory stands. Legumes help to aggregate soils and thus bring about better tobacco yields.

Kentucky soils, except for those of the Bluegrass region, are low in phosphorus. Grain crops generally need fertilizers high in this nutrient, especially if they are to serve as companion crops for legumes. Nitrogen and potassium are usually deficient and should be applied according to current recommendations of the Kentucky Cooperative Extension Service.

The production of crops in rotation is generally recommended, and the minimum rotations suggested for the Northern Forest subregion should be generally satisfactory. However, because of the extreme variations in slope within the boundaries of the same farm or field, strict adherence to a rotation system may be difficult.

Corn and tobacco are the main row crops. Most farm plans include one or both of these crops. Particularly in the eastern mountainous areas, a farm may include only a few acres of bottomland really adapted to the production of a row crop. An acre of such land may produce more tobacco or corn than several acres of sloping upland. In that case, it is better management to grow them continuously on the bottomland and produce meadow crops and small grains on the upland. If both row crops are grown on the bottomland, they should be alternated, and a good cover crop and fertilizer program should be followed.

A friable, well-aggregated soil is especially desirable for tobacco. Many farmers who grow tobacco produce livestock as well. Manure, which is then available for the tobacco land, should be applied in recommended quantities. Tobacco stalks may carry diseases, so good management from that respect may dictate their application to lands that will not grow the crop in the near future. On the hill farms this would be the sloping uplands. Unless disease makes it unwise, crop residues of all kinds should be carefully returned to the fields or should be spread and plowed under or mixed with the soil where the crops were grown.

The fertility of the upland soils may be maintained with legume meadow crops and adequate lime and fertilizer. Erosion should be controlled by contouring, strip-cropping, or terracing, as soil and slopes require.

Soil Management Principle

Minimum tillage practices allow water to penetrate rather than run off.

Excess tillage should always be avoided, especially on slopes.

SOUTHERN PRAIRIE-FOREST SUBREGION

Missouri, south-central Iowa, west-central Illinois, and southeastern Kansas comprise the Southern Prairie-Forest subregion. The steeper slopes near the major streams and some of the level areas in the northeastern part were forested, so the soils are light-colored and relatively thin. Many are badly eroded. The tall-grass prairie soils are generally level to gently sloping.

South-Central Iowa and West-Central Illinois

The soils of south-central Iowa were formed from silty loess; some of them are quite permeable, which makes them highly productive. Some, however, like those in west-central Illinois, are slowly permeable and only medium in productivity. They are not well-adapted to corn production because of slope and tight subsoil. Much of what is now in crops is subject to erosion. These conditions favor a livestock rather than a grain agriculture.

Meadow crops for erosion control are advisable on practically all the land in these areas. Contouring, strip-cropping, and terracing should be considered. The cropping systems and erosion-control practices recommended for the Northern Forest subregion apply equally well to these portions of this Southern Prairie-Forest subregion. Recommendations regarding the use of fertilizer and lime may be obtained from the Iowa and Illinois Extension Services and local U.S. Soil Conservation Service personnel.

Missouri

The soils and climate of Missouri are quite variable. The surface soils are rather uniform, ranging from loams to silt loams in 90 percent of the upland areas.[2] Bottomlands vary from clay to loose sand. Soils of the Ozark region are high in silt, but they contain so much chert that they are also gravelly and stony.

The majority of the upland soils have heavy clay subsoils, a condition that makes them slowly permeable and quite subject to erosion.

The soils of the northern half of the state were derived from glacial till or from loess. Much of the land is hilly, and erosion-control practices are necessary. Because of slope and tendency to erode, these lands should be kept in meadow crops from one-third to one-half the time. Under a livestock or general system of farming with intelligent use of manure, cover crops and crop residues and adequate use of lime and fertilizer, the soils of northern Missouri will remain highly productive. On the whole, they will not stand intensive cropping, except in small level areas.

---

[2]Much of this material regarding the soils of Missouri was taken from J. E. Collier, *Agricultural Atlas of Missouri*, Missouri Agricultural Experiment Station Bulletin 645, 1955.

Missouri contains 19 million acres of cropland, with 53 percent of the total land in farms. Carter County, in the southern part of the Ozarks has only 7 percent of its area in cropland. Throughout the Ozarks, much of the land is thin, stony, and strongly sloping. Most of the cropland is in the valleys. Soil management in the Ozarks should be intensive (perhaps continuous row crops) in the valleys, grain and meadow on the moderate slopes, and pasture on the stronger slopes, a system recommended in eastern Kentucky.

Cotton *(Gossypium hirsutum)* is grown extensively on the alluvial soils of the seven southeastern Missouri Delta counties. It is not grown elsewhere in the state. The alluvial sediments that formed the soils of this area (sometimes called the "'bootheel" country) came from the good soils of the Ohio and Missouri valleys and the best soils of the Ozark highlands. Poor drainage resulted in high organic matter. The land was originally covered with timber. Since it was drained and cleared, it has raised cotton, 396,000 acres in 1955 and 137,000 in 1979. The respective production figures were 390,000 and 157,000 bales. Yields have not appreciably increased, probably one reason why acreage has dropped. Cotton can no longer compete with the newer system of farming in the area.

Irrigation is needed for cotton even though annual rainfall is close to 127 cm (50 in.). Recent experiments showed that adequate nutrients, including starter fertilizer and nitrogen sidedressing, in combination with irrigation on soils well-supplied with organic matter, resulted in yields of 3 bales of lint cotton per acre. The same treatments on low organic-matter soil resulted in yields of 2 bales. Furthermore, heavy fertilization without adequate water did not bring about satisfactory results.

Southeastern Kansas and West-Central Missouri

The residual soils of southeastern Kansas and west-central Missouri were formed from shales, limestones, and sandstone. Tight subsoils (claypan) cause poor drainage and inhibit the use of tile. The soils are gray to light brown in color, low in lime and organic matter and generally in phosphorus and nitrogen. Surface drainage and more permeable subsoils make the sloping lands more productive, provided erosion is controlled. Precipitation averages close to 100 cm (40 in.) annually.

Winter wheat *(Triticum vulgare)*, corn, sorghum *(Sorghum vulgare)*, oats, soybeans, and Korean lespedeza are grown extensively in the area. Alfalfa and sweetclover can be grown on the better soils after application of lime and fertilizer. After liming, phosphorus is the next nutrient to limit yields and wheat needs nitrogen unless it is grown soon after alfalfa. Potash is usually needed.

Corn yields have averaged about half those of the best Corn Belt soils. Extra nitrogen fertilizer can help to even up this difference, but cash cropping to corn will never be very profitable because of the impermeable nature of the subsoil.

Wheat is well adapted to the area. On the whole, the best soil management system revolves around livestock production, either cattle or sheep. This is particularly true for soils such as those of Woodson, Wilson, and Montgomery counties, Kansas, where sandstone predominates in the parent materials and the relatively mature soils are hilly and low in productivity.

Cash-crop farming with adequate use of residues, green manures, and nutrients, including lime, may be advisable in limited areas on the best types of soil and where the personal preference is for systems other than livestock. Persons contemplating such a program should seek advice from local soils specialists.

ILLINOIS, INDIANA, AND MISSOURI CLAYPAN SUBREGION

All or parts of about 30 counties in southern Illinois are in this claypen area. Included also are extensive areas in southwestern Indiana and southeastern Missouri. The lands may generally be divided into nearly level uplands (0 to 2 percent slopes), rolling uplands (2 to 7 percent slopes) and bottomland soils. Nearly level soils include Cisne, Wynoose, and Newberry silt loams. See Table 15-7 for taxonomic description of these soils.

The rolling uplands include those gently rolling (2 to 4 percent slopes) such as Bluford and Hoyleton silt loams, those with medium slopes (4 to 7 percent) such as Ava and Richview silt loams, and those with more than 7 percent slope such as Blair and Hickory silt loams.

The two main bottomland types are Bonnie silt loam and Sharon loam. Bonnie is mostly in the large bottoms and Sharon in the smaller.

These claypan soils are naturally acid, low in organic matter, and deficient in nutrients. The first step in their build-up for profitable cropping should be the application of lime and complete fertilizer applied according to need, again as indicated by soil tests.

Wills (1956), at the University of Illinois, has suggested several cropping systems for claypan soils, as follows.

1. Flat upland soils not threatened by serious erosion.
   3-year rotation: corn-soybeans-wheat (sweetclover catch crop).
   4-year rotation: corn-soybeans-wheat-clover.

2. Rolling upland soils subject to erosion unless control practices are used.
   5-year rotation: corn-soybeans-wheat-meadow-meadow (recommended erosion-control practices should be followed on these soils.)

3. Steeply rolling upland soils threatened by serious erosion.
   Rotation: permanent meadow

4. Bottomland subject to serious flooding.
   2-year rotation: corn-soybeans (rye and vetch catch crop).

5. Bottomland not subject to serious flooding.
   5-year rotation: corn-corn-soybeans-wheat-clover

Many claypan soil farms include land in all these classifications. Where that is the case, the sloping soils may be used for continuous meadow or grain and meadow whereas row crops are confined to the bottomlands or flat uplands.

Soil Management Principle

Sloping soils should be used for continuous meadow or grain and meadow.

Special care should then be taken to make efficient use of residues and full use of green manures. Heavier applications of nutrients are usually needed for the row crops in such systems. This is especially true for nitrogen. Manure should be applied for the row crops on the flat uplands.

Tillage is a problem on claypan soils. They are plowed with difficulty when too dry and are seriously compacted when plowed while too wet. The "just-right period is usually short. Therefore, it is well to have plenty of power so that the job may be done in a short time. Excessive tillage after plowing should be avoided. Aeration and water intake are slow when the soils are packed. Minimum tillage (spring plowing and immediate planting) is effective on such soils.

CENTRAL PRAIRIE SUBREGION

The Central Prairie Subregion includes the level to gently rolling prairie from east-central Illinois to west-central Minnesota, including nearly half of Iowa. The soils, well-aggregated and permeable, were formed from medium-textured glacial parent material or loess and are dark-colored and high in organic matter. The Webster and Clyde soils of Iowa and Minnesota are examples of such soils. They have been extensively and satisfactorily tile drained.

Muscatine silt loam, found in eastern Iowa and central Illinois has been called the ideal corn soil. It has a deep dark-colored surface, good structure, occurs on very gentle slopes, and seldom requires artificial drainage. Its deep surface soil makes it drought-resistant and easy to work.

Over much of the Central Prairie subregion, early agronomists advocated a soil management system that was based on the production of legumes (largely red clover) with application of lime and phosphates, as needed. Generally the amounts of lime applied were sufficient to raise soil pH no higher than to 5.8 or 6.0 because the high percentages of organic matter caused the soil to be so highly buffered that pH remained relatively low. That is the reason why rock phosphate was fairly effective as a source of phosphorus for clover hay, corn, and oats.

More recently, cropping systems have changed to include less clover hay, a very small total acreage of oats, and much higher acreages of corn and soybeans. Let us take a look at the figures for Illinois.

Crop	Acres harvested	
	1950	1980
Hay	852,000	440,000
Oats	3,911,000	260,000
Corn	8,234,000	11,460,000
Soybeans	3,948,000	9,250,000

Soil studies show that many soils have lost one-fourth to more than one-third of their original content of organic matter. Nitrogen-deficient corn and oats are common throughout the area. We have seen more nitrogen-starved plants in the Central Prairie than in Michigan, a region of low natural soil organic matter. The explanation lies in the acreage of nitrogen-fixing legumes. What should be done? There are two alternatives. Switch to a greater acreage of legumes, or purchase more nitrogen and make better use of residues and green manures.

Corn and soybeans, especially corn, are so well-adapted to the soil and climate that the second alternative may be the better of the two. It may be necessary also to increase the use of mineral fertilizers, possibly some of the trace elements. More recent experimental results from the University of Illinois show that potassium as well as phosphorus must be applied as fertilizer if production is to be maintained, even on the most productive prairie soils.

Tillage and Erosion Control

More attention must be paid to tillage and erosion control. The large acreage of intertilled crops and the adherence to the old idea that crop yields may be increased by much tillage has resulted in losses of organic matter and soil compaction to the extent that crops are suffering. Yields of corn on the **ideal** soils of this region should be much higher than they are. We must believe the truth of this statement when we see individual farmers on land with **much lower inherent fertility** and in regions with less favorable climate, consistently obtaining yields twice as great as the average yields in such a corn state as Iowa.

Strict adherence to approved erosion-control practices, including contour strip-cropping and terracing, and the practice of minimum tillage (plow and plant the same day), conservation tillage, or similar tillage practice, would go a long way toward bringing corn and soybean yields up to the levels that should be expected on such good soils.

WESTERN CORN AND SMALL-GRAIN SUBREGION

Included in the Western Corn and Small-Grain area are the eastern portions of the Dakotas, Nebraska, and Kansas, the portions of those four states not included in the Great Plains, and western Minnesota. Roughly, the area lies east of the 98th meridian. Gradation into the Great Plains is, of course, gradual. In fact, Norum, Krantz, and Haas (1957) included all of North Dakota and all but a few counties in southeastern South Dakota in the Northern Great Plains. We, however, are inclined to agree with the North Central Regional-3 Technical Committee on Soil Survey. Members of that committee showed that although all of North Dakota is spring-wheat country, a strip along the eastern end of the state (roughly soils in the Fargo-Bearden association) is adapted also to the production of corn, sugar beets, and potatoes. This type of agriculture extends, of course, over into western Minnesota to join the typical dairy and hay region of that state (see Fig. 15-1). The soils range from silt loams to clays and are nearly level. They are the poorly drained soils of the Red River Valley of eastern North Dakota and Minnesota, developed from lacustrine deposits. In some places they are calcareous in all horizons.

The balance of eastern North Dakota, all of the eastern four columns of counties in South Dakota, all or parts of several counties in Minnesota and much of the eastern one-third of Nebraska and Kansas are covered with Mollisols. They are used for corn, wheat, and small grains, with sorghum important in southern Kansas. Ulen soils are in this group; nearly level to gently rolling and quite sandy in texture. They lie in North Dakota just east of the Fargo-Beardon area. They are adapted to small grains if wind erosion is controlled. (Turn to the section on the Great Plains (Chapter 16) for methods of erosion control.)

Aastad-Hamerly-Barnes soils are black to dark gray loams developed from calcareous till under grass vegetation. They are naturally productive of small grains but need drainage in low places. Erosion is a hazard, and fertility maintenance must be practiced. The soils lie west of the sandy Ulen area and extend into what surely may be considered as the Northern Great Plains.

Several soil associations occupy the portion of South Dakota that lies east of the Great Plains. The Barnes-Aastad association occupies parts of several counties in northeastern South Dakota and all or parts of at least 14 counties in west-central Minnesota. The soils are black or nearly black well-drained loams or clay loams developed from calcareous loam till. They are used principally for cash grain farming. Erosion control and fertility maintenance are important management problems.

A strip of Barnes-Buse association soils lies across the northeast corner of South Dakota. These soils are well- to excessively drained. They are black or grayish brown and were formed from calcareous loam till. They are used for general farming with livestock. Water erosion is a hazard, and organic matter and nitrogen maintenance are serious management problems.

A long strip of Kranzburg-Vienna association soils lies to the west and somewhat to the southeast of the Barnes-Buse association strip. Kranzburg soils are sloping, well-drained loams or silt loams, dark grayish brown in color with a slightly acid reaction. They developed from loess over calcareous Iowan or Tazewell till. Vienna soils developed from loam to coarse clay loam calcareous glacial till. Soils in this association are used for general and cash grain farming. Organic matter and fertility problems need continued attention. The association extends into Minnesota at Rock County.

An area of Crofton soils covers the three-corner boundary between South Dakota, Iowa, and Minnesota and a considerable area in northeastern

Nebraska. They are dark-colored silt loams developed from loess over a till plain. Water conservation and erosion control are problems. The soils are moderately productive for corn, oats, and hay. Phosphorus fertilizers are needed, and drought often limits yields.

A long strip of Poinsett-Sinai association soils stretches from south of the Nebraska line almost to the North Dakota line and just west of the Kranzburg-Vienna soils. They are level to undulating, dark grayish brown, slightly acid silt loams or silty clays. Poinsett soil developed from silty Cary drift and Sinai soils from lacustrine silts and clays. They are used for general farming with water erosion as a hazard. Organic matter and fertility maintenance are problems in management.

The 98th meridian passes through a narrow north to south strip of Beadle-Houdek soils in northern South Dakota. They range from silt loams to loams, are well-drained, and are dark grayish brown in color. Although slightly acid in reaction, they are well-adapted to the production of livestock, cash grain, and general crops. Nitrogen fertilizers are frequently needed.

All or parts of 10 counties in southern South Dakota and parts of two counties in northern Nebraska (lying just east of the Poinsett-Sinai association area) are covered with soils in the Bonilla-Cavour association. They are nearly level, moderately well-drained, dark grayish-brown, slightly acid loams. They developed from calcareous Mankato till. Nitrogen deficiency is an important management problem.

Several soil associations of Humic-Gley origin extend from Cuming County in eastern Nebraska down to Bourbon County in eastern Kansas. The soils are continuations from similar soils in Iowa and Missouri. The upper counties in this general area are largely covered with soils in the Sharpsburg-Marshall-Burchard association, which are dark grayish-brown, acid, silty clay loams. Sharpsburg and Marshall soils formed from loess, the Burchard from glacial till. They are fertile, responsive to good management, including erosion control, and are used largely for general crops and livestock production. They are very good soils for corn and small grain, particularly wheat.

Pawnee-Grundy-Burchard association soils lie to the south of those just described. These soils are used for general crops and livestock production, but crop yields may be limited by excessive rainfall or drought because the soils are slowly permeable. Grundy soils are especially well-adapted to corn, and all soils of the area produce good yields of wheat, oats, and alfalfa. Brown County ranks near the top among Kansas counties in oats production.

Summit-Woodson-Labette association soils occupy parts of about 12 southeastern Kansas counties. Roughly, this area lies above the lower three tiers of counties. It extends also into three or four adjacent Missouri counties. The soils are dark grayish-brown, gently sloping, well-drained, moderately acid, silty clay loams. They developed from limestone and shale residues and are adapted to grains and livestock. They need lime and fertilizers and erosion is a hazard.

The Flint Hills section of Kansas, an area ranging from one to two counties wide, extends north to south across the state, starting just west of

Manhattan. Soils included in the area are Sogn, Summit, and Florence. They formed from limestone high in chert, are thin, but are well adapted to blue-stem grasses. Wheat and sorghum do well in the bottomlands and areas of lesser slopes and deeper soils. Fertilizers and lime are needed, and erosion control is difficult.

Generally, this Western Corn and Small-Grain subregion is a rather highly productive area where water conservation is important, response to nitrogen and phosphorus fertilizers is common, and extremes in climate are expected. It is a transition area between humid and semiarid climates. Although southern Kansas enjoys a rather high rainfall, evaporation and transpiration are so high that drought is common. Wheat and corn production plays an important part in land use. Fertilizer use is growing in the entire area, but the major soil management problem is organic matter management.

Soil Management Principle

Residues should be kept on or near the surface for protection from both wind and water erosion.

Tillage and planting methods in this subregion are unlike those of both the Great Plains and the Corn Belt. Summer fallow is seldom practiced, but water conservation is very important. Corn is usually listed and should be planted on the contour for control of water erosion. Terracing and strip-cropping are recommended on many soils.

Four Agricultural Experiment Stations lie in this subregion — in North Dakota at Fargo, South Dakota at Brookings, Nebraska at Lincoln, and Kansas at Manhattan. Cooperative Extension Services are quartered at the same locations. You are urged to seek advice regarding soil conservation practices, including the use of fertilizers, from specialists at these Stations. U.S. Soil Conservation personnel, located in the same cities, are also anxious to suggest soil management practices.

Soil Associations

The soil association and main management problems for the entire North Central region are described in detail in North Central Regional Publication 76, Wisconsin University Agricultural Experiment Station Bulletin 544, 1960.

The taxonomic designations for the soil series mentioned in Chapter 15 are given in Table 15-7.

REFERENCES

Aandahl, A. R., J. K. Ableiter, H. F. Arneman, et al. (1960). *Soils of the North Central Region of the United States.* North Central Regional Publication 76. University of Wisconsin Agricultural Experiment Station Bulletin 544.

Hoglund, C. R., and R. L. Cook (1956). Estimated crop yield, costs and returns from using current and recommended levels of fertilizer and production practices, Michigan, 1956. Mimeograph publication of Michigan State University.

Longwell, J. H., and S. B. Shirky (1957). *Cotton Research.* Missouri University Agricultural Experiment Station Bulletin 684.

Noll, C. F., F. D. Gardner, and D. J. Irwin (1931). *Fiftieth Anniversary of the General Fertilizer Tests.* Pennsylvania State University Agricultural Experiment Station Bulletin 264.

Norum, E. B., B. A. Krantz, and H. J. Haas (1957). The Northern Great Plains. In *Soil — the 1957 Yearbook of Agriculture.* U.S. Department of Agriculture, Washington, D.C., p. 494.

Odell, R. T., W. M. Walker, L. V. Boone, and M. G. Oldham (1982). *The Morrow Plots, a Century of Learning.* Illinois Agricultural Experiment Station Bulletin 775.

Pierre, W. H., and F. F. Riecken (1957). The Midland Feed region. In *Soil — the 1957 Yearbook of Agriculture.* U.S. Department of Agriculture, Washington, D.C., pp. 535–546.

Reed, E. P., J. A. Slipher, and D. F. Beard (1950). *Putting Soybeans into Permanent Farming.* Ohio Agricultural Extension Service Bulletin 311.

Smith, G. E. (1942). *Sanborn Field, Fifty Years of Field Experiments with Crop Rotations, Manure, and Fertilizer.* Missouri University Agricultural Experiment Station Bulletin 458.

Thorne, C. E., et al. (1924). *The Maintenance of Soil Fertility, Thirty-years Work with Manure and Fertilizer.* Ohio Agricultural Experiment Station Bulletin 381.

Wills, J. E. (1956). *Increasing Farm Production and Earnings on Claypan Soils in Southern Illinois.* University of Illinois Extension Service Circular 762.

CHAPTER 16

Cropping Systems and Management Practices: Southern and Western States

THE SOUTHERN STATES

Perhaps agriculture in the Southern states has changed more during the past 40 years than in any other region of the United States. R. L. Cook remembers well a trip through Georgia, Alabama, Mississippi, and Louisiana in late November 1939. There was an appalling lack of green vegetation at that time, actually less green in the fields of those states than in those of Michigan just 3 days earlier.

Why? Cotton was the main crop being grown. By late November it was harvested and the fields were bare. There were few farm animals in the area at that time, so there was no need for winter hay and pasture crops, and the age of conservation farming had not arrived. As late as in 1950, cotton was still the most valuable farm commodity produced in eight states (Alabama, Arkansas, Georgia, Louisiana, Mississippi, South Carolina, Tennessee, and Texas). Tobacco was first in North Carolina. Under such intensive row-crop agriculture, in a relatively hilly region, erosion had been severe. Topsoils were entirely gone from many fields. Something had to be done if Southern agriculture was to survive.

Much *has* been done. Livestock have invaded the South. Perennial legumes, winter annual legumes, winter grains, and improved grasses are being grown for livestock feed. Manures and crop residues for organic matter maintenance and the application of needed fertilizers have greatly increased crop production. Furthermore, the production of winter crops keeps the soil covered and so helps to control erosion.

Cotton is no longer "king" in the South. The data in Table 16-1 show that after 1949 the acreage of cotton decreased consistently in all states except in Louisiana where the acreage seems to have leveled off at about

TABLE 16-1 Cotton Acreage (Harvested) in the Deep South 1938 to 1979

Years	Alabama	Arkansas	Georgia	Louisiana	Mississippi	South Carolina	Tennessee	Totals all seven states
Av. 1938–47	1,691,000	1,916,000	1,608,000	970,000	2,397,000	1,118,000	684,000	
1949	1,810,000	2,530,000	1,600,000	1,050,000	2,730,000	1,270,000	830,000	11,820,000
Av. 1950–59	1,129,000	1,570,000	963,000	665,000	1,847,000	794,000	636,000	
1961	905,000	1,360,000	693,000	535,000	1,580,000	585,000	538,000	6,196,000
1977	395,000	930,000	170,000	540,000	1,360,000	153,000	300,000	3,848,000
1979	310,000	530,000	150,000	465,000	1,030,000	109,000	230,000	2,824,000

one-half the figure for 1949. The totals for the seven states are interesting. The 1961 acreage was 52 percent of that of 1949, whereas acreage in 1977 was 62 percent of the corresponding figure for 1961. The acreage may still be going down since the 1979 total figure is only about three-quarters that for 1977.

Many crops have replaced cotton. Gradually since 1949, livestock have taken over the South. As a result, erosion is under much better control and crop yields have increased. Conservation farming is being practiced. Much of the credit for the changes should go to the U.S. Soil Conservation Service, the Tennessee Valley Authority, the State Cooperative Extension Services, and many private research and extension organizations. Seed, machinery, and fertilizer companies have played important roles.

CHARACTERISTICS OF THE SOILS

Soils of the Southern states (especially the Southeastern) are variable, generally sloping, and formed under humid conditions. They have certain characteristics in common that set them apart from other soils in the humid region, particularly from a management viewpoint.

The soils are generally fine-textured, with the clay minerals largely of the kaolonitic group. High temperatures coupled with erosion have resulted in soils low in organic matter. This, in conjunction with the kaolonite clay, gives soils that have a low cation-exchange capacity.

Bases have leached from these soils, so in their natural, unlimed state they are acid and many times high in exchangeable aluminum.

Management practices that must receive special attention on these soils are pH control (see Chapter 13), nutrient availability (see Chapters 10 and 11), water control to prevent erosion, and cropping systems to build up soil organic matter.

Water-Control Practices

After preparation of a soil map (soil survey), the next step in the rejuvenation of a run-down, eroded farm is the establishment of a whole-farm water-management system. As of 1950, this was needed on 75 percent of the farms in the Southern cotton states. Much has been done.

Field cover, contour farming, contour strip-cropping, and terracing are control practices for soils of progressively greater erodability. All practices are used on many farms. In fact, where terracing is needed, the other practices are usually used also (see Fig. 16-1). Waterways used in connection with terraces must be protected by permanent vegetation.

Terraces are needed on most farms of the Southeast. An illustration of how they are used in connection with permanently grassed waterways may be seen by referring to the *Manual on Conservation of Soil and Water*, USDA. Soil Conservation Service Agricultural Handbook 61, 1954, page 47.

A cotton farm of 124 acres in the Piedmont plateau in South Carolina was to be planned. Annual precipitation is 122 cm (48 in.). The farm includes

FIGURE 16-1 On this small farm in Greenville County, South Carolina, all applicable soil-conservation practices are employed. A strip rotation of row crops and small grains, terraces and a meadow outlet, a wildlife border, and a new tree plantation are advanced measures used on what was formerly a single field. (Photograph by USDA Soil Conservation Service.)

only a very small area of Class I land. It has a large acreage of Class II and Class III. The Class II land is gray, sandy soil with 3 to 5 percent slopes. Most of the Class III land is brownish-red sandy clay with slopes of 4 to 7 percent, already considerably damaged by erosion. By the establishment of a system of terraces, it is possible to safely farm this Class II and Class III land in a good rotation. In fact, one piece of Class IV land was included in the terraced cropland area. By use of the terraces and a long rotation, this was deemed safe.

All waterways on this farm were seeded to *Sericea lespedeza*, a perennial legume that gives full protection to the soil and furnishes several cuttings of hay each year. Commercial fertilizer is needed to maintain satisfactory growth on waterways. In this 80-acre rotated crop area. The *Sericea lespedeza* waterways occupy a total area of 5.0 acres in five long, narrow strips. Care must be exercized that the strips are not disturbed when adjoining soils are plowed.

Rotations, Manures, and Cover Crops

Soil Management Principle

Soil productivity is closely correlated with vegetative cover.

"Soil productivity is closely correlated with vegetative cover." This statement was made by J. C. Lowery and W. W. Cotney, Extension Agronomists in Alabama. It shows that soil improvement or maintenance practices in the Southern states are fundamentally similar to those found valuable in other parts of the country.

Soil is perhaps the only natural resource of the country that increases in value with increased use. Those who disagree with this statement are probably misinterpreting the meaning of the word "use." Soils are being **used** when they are covered with vegetation. The more dense the vegetation, the greater their degree of use. The more dense the vegetation, the more residues or manures will be left in or on the soil as a contribution to the organic content of the soil. Organic matter increases the productive capacity of the soil. Few people disagree with this statement. It follows, then, that greater yields lead to still greater yields the following year until finally the limit set by outside factors such as light and carbon dioxide is reached. This was nature's method of building soil before humans interferred. The only possibility of reclamation of devastated areas that once were soil, is for humans to make it possible for nature to do the job again. The encouraging part of the picture is that humans may assist in doing the job faster than nature did it the first time. This assistance is in furnishing more effective soil-building crops and in supplying plant foods at a more rapid rate than nature did by weathering processes. Nitrogen, particularly, may be supplied at a faster rate than ever occurred through natural processes.

To show how successful the soil conservation work has been in the cotton states, we may cite the results obtained on one test demonstration farm in Randolph County, Alabama. The following is a quotation from Lowery and Cotney.[1]

> UTD farm — Randolph County: This is a farm of 123 acres with 51 acres of crop land and 12 acres of open pasture. Changes in cover program were: 5 acres of perennial cover in 1937, 25 acres in 1943; 14 acres of winter cover (winter legumes and small grain) in 1937; 22 acres in 1943; 6 acres summer cover in 1937; 12 acres in 1943. Effect on crop yields was: cotton increased from 305 pounds of lint per acre in 1937 to 467 pounds in 1943; corn from 23 bushel in 1937 to 41 bushels in 1943; oats from 22 bushels in 1937 to 50 bushels in 1943. Effect on livestock was: 2.8 units sold in 1937; 14.7 units sold in 1943. The net income for labor and investment: $587 in 1937; $2,091 in 1943.

Soil management practices need not be appreciably different in the newly established livestock-producing Southern states than in the general farming North Central region. Approved practices are

1. Provide good drainage.
2. Control erosion, water and wind.

[1]J.C. Lowery and W. W. Cotney, *Soil Improvement in Alabama*, Alabama Polytech Institute Extension Circular 290. UTD farm is Unit Test Demonstration farm, sponsored jointly by the Alabama Extension Service and Tennessee Valley Authority.

3. Apply lime and fertilizers as needs are indicated by soil tests.
4. Maintain or increase soil organic matter through adequate use of crop residues, legumes, and cover crops.
5. Use recommended cultural practices.
6. Adapt the crop to the soil. This may or may not mean strict crop rotation.

THE SOUTHWEST

The American Southwest, including Arizona, New Mexico, and parts of California, Nevada, and Texas, is an arid, semiarid, and subhumid region. Differences in altitude, ranging from sea level to over 13,000 ft, cause marked variations in climate. Home gardens in Taos, New Mexico, elevation 7000 ft, are planted with approximately the same vegetables and during the same season as they are in Lansing, Michigan, where the elevation is one-tenth as high.

The Midwesterner likewise marvels at the vegetation changes seen as he travels from Tucson, Arizona, with its typical desert vegetation to the top of Mt. Lemmon, where the huge ponderosa pines flourish. The trip requires less than an hour, but temperature difference on a particular day may be 30°F or more. Such climatic variations mean marked differences in soil characteristics.

Ponderosa pine is the predominant forest tree at elevations above 7000 ft. Other important species are fir, spruce, and white pine. There is considerable lumbering in the area, particularly in Arizona around Flagstaff and McNary. Most of the timbered lands are publicly owned. Woodlands, areas covered with juniper, pinyon pine, and oak, cover the mountains at elevations ranging from 4000 to 7000 ft. These trees are used for fuel and fence posts.

Range lands, lying chiefly at 3000- to 5000-ft levels, overlap the woodlands in elevation. One problem in range management is to keep worthless shrubs and trees from encroaching on the grasslands. Machines and chemicals are making this a simpler task.

In general, irrigated crops are grown below 4000 ft, and mostly below 3000 ft. The upper watersheds, grass and timber lands, collect the water for the irrigation projects below. Water is collected in reservoirs, such as were formed by the Roosevelt Dam on the Salt River and the Coolidge Dam on the upper Gila River. Many other dams on these rivers and their tributaries have so completely collected floodwaters that for years at a time water does not run out of the mouth of the Gila, the principle drainage way for the state of Arizona. Water is also pumped directly from streams, but this is not a dependable source, as water can be taken only when levels are above certain points. As a result, there are seasons when water is not available and good land must stand idle. Even with reservoirs, however, water rights apply only when water is available.

Deep wells furnish much of the irrigation water used in the Texas Panhandle and parts of New Mexico and Arizona. Such a source is more dependable but in most instances is more costly. The initial cost of an irrigation well is very great, and pumping costs are high where wells are deep. Generally in these areas, the aquifers have limited recharge and water tables have dropped drastically in recent years. Costs go up as water levels fall, and the availability of water also becomes less as the aquifers are depleted.

CHARACTERISTICS OF THE SOILS

There is little leaching of arid-region soils. Consequently, they remain richer in the soluble substances that plants use in growth and that come from the weathering of rocks and minerals. Because of lack of moisture, chemical forces of weathering are less active than in humid regions. As a result, soils remain quite similar to the parent rock material. Changes are more the result of disintegrations brought about by erosion and temperature changes than they are the result of chemical activities.

Marbut called dry-region soils "pedocals" because they are characterized by an accumulation of calcium carbonate in the solum. Roughly, pedocals lie in regions of less than 50 cm (20 in.) of annual precipitation. Southwest soils are in the great soil groups known as Brown, Reddish Chestnut, Reddish Brown, and Desert Soils (aridisols and mollisols), except in the higher mountain areas where podzol-like soils (spodosols) may be found. As one travels from the northeast, southwestward into the desert area, grasses become shorter and sparser and the desert plants become larger. The yucca furnishes the best illustration of the latter. Because grasses are scarce or absent and total annual vegetation growth is small, soil organic matter levels are very low, ranging from 0.1 to 1.0 percent over the region as a whole.

As is true in all desert areas, geological erosion has been severe in the Southwest. Fine materials, silt and clay, are redistributed because there is little vegetation to offer protection from the wind. Rainstorms are infrequent but violent. Much washing occurs in quite unexpected places. Levees, dips, bridges, and culverts are common in dry and barren areas. Periodic erosion leaves soil surfaces strewn, sometimes completely covered, with boulders and pebbles that finally provide sufficient coverage to protect the soil underneath from further destruction. Kellogg (1941) called these surface covers "desert pavements."

Some Red and Gray Desert soils have well-developed horizons. A hard crust on the surface is underlain by a few inches of porous soil. Below that zone a layer of soil quite high in clay is often found. The clay is the result of leaching from the surface and of chemical weathering in a zone that has commonly been more moist than the soil above or below. Then at depths of 2 to 6 ft there are hardpan layers. The early Hopi Indians looked for sandy soils with hardpan layers at these depths because they had discovered that such soils were sufficiently water-retentive to produce a crop, especially if they could divert water from a stream to thoroughly soak the soils before planting time. They did not use clay soils in this manner because such soils did not produce a surface mulch that would protect the water from evaporation.

Saline Soils

Where soils are not leached, soluble salts accumulate. When the accumulation is sufficient to interfere with the growth of the crops, the soils are called **saline**. The U.S. Regional Salinity Laboratory defines saline soils as those for which the conductivity of the saturation extract is more than 4 millimhos/cm (see Chapter 13 for definition) and the exchangeable sodium percentage is less than 15. The pH is usually less than 8.5. These soils are approximately what Hilgard called white alkali soils and the Russians called "solonchaks." The salts are predominantly sulfates or chlorides of calcium and sodium, usually less of sodium. Nitrates are usually present.

Saline-Sodic (Alkali) Soils

Saline soils that contain appreciable quantities of exchangeable sodium are designated as **saline-sodic or saline-alkali soils**. The Salinity Laboratory defines them as soils that have a conductivity of the saturation extract greater than 4 millimhos/cm (0.4 siemens/m) and an exchangeable sodium percentage **greater** than 15. Soil pH is usually no higher than 8.5, and colloids remain flocculated.

Sodic (Nonsaline-Alkali) Soils

Soils that contain very small quantities of soluble salts and sufficient exchangeable sodium to equal more than 15 percent are called **sodic (or nonsaline-alkali) soils**. The conductivity of the saturation extract is less than 4 millimhos/cm (0.4 siemens/m). The pH values range between 8.5 and 10, and growth of most crop plants is depressed. These soils were called "black alkali" by Hilgard and "solonetz" by the Russians. As a result of the high alkalinity, the organic matter dissolves and coats the soil grains to give the soil a dark color. The high sodium carbonate and bicarbonate concentrations are very injurious to plants. Small local areas of these soils are commonly called "slick spots" or "hard spots," probably because they have a puddled surface and are free of vegetation.

Treatment of Saline and Saline-Sodic Soils

Saline soils can be safely leached to remove soluble salts. This, of course, may be impractical because of the shortage of water, but it is practiced in some instances. In some areas, only waters of poor quality, that is, a Na/Ca + Mg equivalent ratio of greater than one, exists. These water cannot be used for reclamation purposes with much hope of success.

Saline-sodic soils cannot be improved by leaching simply with water. They should first be treated with a calcium salt (gypsum is the cheapest) to remove sodium from the exchange complex. Otherwise, the leaching will remove salts from the soil solution and the sodium from the exchange will react with carbon dioxide to form sodium carbonate and bicarbonate. The resulting soil will be sodic, a less-productive condition than existed before the leaching. In actual practice, Chang and Dregne (1955) did not find that leaching changed a saline-sodic Gila loam to a sodic condition. They explained their results by the fact that the calcium carbonate supplied considerable calcium to the soil solution to keep salt concentration high. Refer to Table 16-3. Note that Gila is a coarse-loamy, calcareous soil. As such, the supply of calcium was probably great enough to maintain salinity during the period of leaching with water, so structure to maintain permeability was retained.

But such success is not always the case. One of the writers (Ellis) has seen in Swaziland large acres of land that had been dispersed by leaching a saline-sodic soil with excellent quality water (low in total salts). The soil was sealed, and water would not penetrate it even under a 15 cm (6-in.) head. Once this has occurred, reclaimation is expensive and difficult.

SOIL MANAGEMENT PROBLEMS

In many desert areas, brackish water, varying, or course, in degree of saltiness, is available from surface and underground supplies. The Israelis are successfully using such waters. Irrigation with water containing 2500 ppm dissolved salts resulted in 59-percent higher yields of cotton than did irrigation with fresh water. This was probably because the salts kept the soil flocculated so leaching could continue (it was still a saline-sodic soil).

Incidentally, seawater contains about 35,000 mg salts per liter. None of our agricultural plants can tolerate that degree of saltiness. Where drainage and leaching are adequate, several plants (cotton, barley, wheat, sugar beets, rice, ryegrass, date palms, olive, pomegranate, and pistachio) can tolerate water up to 1500 mg/l. Irrigation and *leaching* should be frequent. Refer to Chapter 21 for more information on irrigation of arid lands. Refer also to the National Academy of Sciences 1974 publication *More Water for Arid Lands*.

Soil management in the Southwest falls in two general channels, that connected with dry-land farming where water management is the main consideration, and that associated with irrigation farming where the problems are probably more complex than those connected with any type of humid-area agriculture.

Dry-Land Farming

For the most part, dry-land farming is conducted in New Mexico and the adjoining areas in Texas, except that range management is a problem throughout the region. The grazing land of Arizona, for instance, occupies more than 80 percent of the total land area. New Mexico has a long-time average precipitation record of 13.89 in./year. The average for 1940 to 1954 was almost an inch less than the long-term average. Drought periods of varying duration are normal for the state. A similar climatic situation exists in western Texas and in Arizona, which is reflected in the fact that rights to irrigation water are sometimes worthless because water is not available.

Range-Soil Management

The management of range soils revolves around organic matter maintenance and erosion control. When the range is grazed too heavily, there is very little vegetation left to die and become incorporated with the topsoil, especially in the areas of better forage species. Forage production is closely related to soil fertility and porosity. These soil characteristics are dependent on organic matter levels so low in desert soils that constant replenishment is imperative.

Organic Matter and Erosion

As organic matter becomes depleted, soil granules disintegrate and fine particles are dispersed to clog soil pores and slow up water penetration. Lack of surface litter on such land exposes the soil surface to the beating action of heavy rains, which serves to hasten the disintegration of the weakened aggregates. Water then runs off and is lost as far as the range grasses are concerned. Furthermore, running water carries away the thin topsoil and prevents the establishment of new forage plants. Some of the indications that a range soil is being depleted are as follows:

1. Active erosion is evident in the form of bare gullies.
2. Surface litter is absent or nearly so.
3. Forage plants lack vigor.
4. Seedlings of better forage species are absent.

   Range-soil depletion may be avoided by regulated grazing. Water always determines the amount of forage available on a given area. During drought periods, numbers of cattle must be decreased. Undesirable species usually take over during drought periods where there is little competition from desirable species.

   Range soils may be improved for grazing by clearing, reseeding, and undergrazing. You are urged to seek advice from local range specialists connected with the State and Federal Extension Services and the U.S. Soil Conservation Service.

Crop-Soil Management

In the semiarid regions of the Southwest, much of the land is limited to range land by climate and/or topography. Of the cropped land, dry-land crops are grown extensively in the eastern part of the Southwest. Much of this land is cropped to wheat and sorghum. In the production of wheat in humid areas, fertilizer usage plays an important role. It is not so in dry-land areas where the principal problems are water conservation and erosion control.

Soil Management Principle

The dry-land wheat farmer must farm with the weather.

Rainfall varies greatly from year to year. Figure 16-2 shows how great was the variation at Clovis, New Mexico, from 1941 through 1947. Particularly important is the fact that in several instances 2 or more dry years were in succession. It is also important that most of the rainfall came between May and October and that the winter, windy months were dry. It is during that season that wind erosion may be serious. The rainfall pattern at Amarillo, Texas, is almost identical to that for Clovis. The dry-land cropping area of Texas includes most of the land west of a line extending from Brownsville to a point 50 miles east of the eastern edge of the Panhandle. Some of the land in the extreme southwestern part of Texas is too dry for dry-land farming. Most of the wheat in Texas is grown under dry-land management in the Panhandle and nearby areas.

A flexible cropping system rather than a rotation system is recommended in dry-land farming. Plans must be varied with soil-moisture supplies and the amount of crop residue available for winter soil cover. When mois-

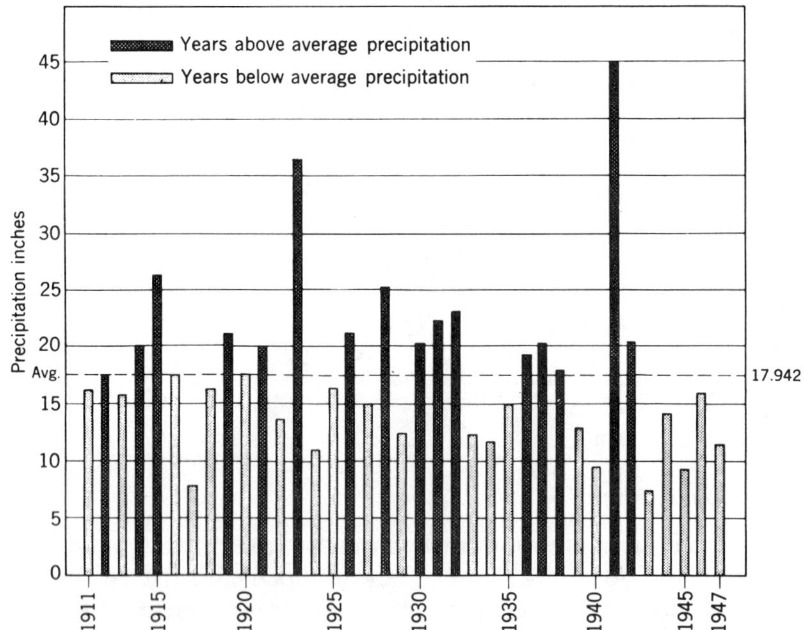

FIGURE 16-2 Annual precipitation at Clovis, New Mexico 1911–1947. (From L. R. Appleton. *Conservation Practices for the Wheat Lands of New Mexico.* New Mexico Extension Service Circular 221, 1949, p. 3.)

ture conditions are favorable, the wheat farmer should plant a crop, both because of the profit he may obtain and for the cover to his soil.

Wheat may follow wheat during normal or above normal rainfall years. When the fall season is very dry, and when the wheat crop just harvested left a good residue cover, it is better to hold off planting and use the next season for summer fallow. However, such a plan is not advisable if residues are not available because of the wind erosion that will occur during the dry winter months. Instead of summer fallow, it would then be better to plant a feed crop or plant a feed crop in alternate strips with fallow.

Water management may lessen the risks involved in dry-land cropping. Terracing and contouring may be used to hold water on the land so that it may be absorbed for later use by crops. These structures are needed more on wheat land than on row-crop land. They not only serve to conserve moisture but also to prevent erosion.

Sandy lands should be used for row crops rather than for wheat. Their storage capacity for water is low, and they do not resist blowing during the winter months. Sands are better for row crops because the deep furrows and ridges offer some protection from the wind. Furthermore, row crops need their water during summer when rains are more likely.

Although water is the most limiting factor for crop production in dry-land farming, fertilizers (according to K.G. Brengle, 1982) are becoming more commonly used. Occasionally, nitrogen may be profitable in the higher moisture years or in local regions of higher rainfall. The appearance of the plants will tell the farmer when nitrogen is needed (see Chapter 12). Some soils are low in available phosphorus, which may be detected by soil test. Iron and zinc have been found to be deficient on calcareous soils, particularly if they have been heavily fertilized with phosphorus.

Irrigation Farming

Irrigation is now practiced on several million acres in the dry-farming region of Texas, on almost a million acres in New Mexico, and on one and a quarter million acres in Arizona. In addition, large areas are under irrigation in southern California. The principal crops under irrigation, from the standpoint of acreage, are cotton, sorghum, alfalfa, barley, beans, and vegetables.

Water is scarce. The total irrigated acreage is limited by water supplies and not by land area. The expansion in irrigated acreage that has occurred during recent years has been largely the result of pumping from new wells. Underground water levels are falling in most areas. Future expansion will probably occur only through more efficient use of existing resources. Water waste may be avoided by the use of more scientific methods. Of particular importance is the use of a reliable method of determining soil moisture at the root level in order that irrigations may be controlled as to time and amount of water application. You are urged to consult local and national publications for advice on water management.

Drainage

Irrigated lands may need drainage, at least in portions of large areas. Excess water may come from excessive irrigation, impossible to avoid in certain situations, from seepage from ditches, or from subsurface flows. The last-mentioned type of water source may actually be at a considerable distance, outside the area controlled by the farmer.

Saline soils must be leached to remove soluble salts, which cannot be done without underdrainage. Deep, sandy soils may drain naturally, without a need for tile. On such soils it is possible to leach soluble substances below root zones by applying excess water during the preplanting irrigation.

Underdrainage is especially important on lands not properly graded for irrigation, where water from irrigation and rain may collect in low areas. Drainage helps to control erosion by keeping the soil more absorptive. Wet soils become puddled, so water enters more slowly.

Soil Tilth

Tilth refers to the structure or physical condition of soil, which determines aeration, rate of water penetration, drainage, and activity of soil organisms. Activity of soil organisms determines, to quite an extent, the availability of plant nutrients and the ability of plants to absorb the nutrients. Surely, the importance of soil tilth cannot be overemphasized.

Organic matter serves as food for soil organisms. In its various stages of decomposition, including those substances that are by-products of microorganisms, it makes up the binding material that holds soil particles together as aggregates. Desert soils are very low in organic matter content, ranging in Arizona, according to Fuller (1956), from 0.1 to 1.0 percent. Quoting from Fuller, "The continued productivity of the soils in Arizona will depend largely upon the replenishment and maintenance of the soil organic constituents." This statement can include all the soil of the Southwest, yes, all soils in the United States and perhaps all in the world. Variations occur only in the length of time required for depletion of this all-important constituent.

The desert soils of this region are not only low in total organic matter but also organic matter decomposition is very rapid, especially under irrigation. This means that constant replenishment is necessary. Much of the value of organic matter is exerted during its decomposition. Many years ago, Norman (1943) discussed the effect of organic matter on Iowa soils, stating that "The degree of aggregation is not necessarily directly related to the total organic matter present. It seems that there is a form of aggregation produced by the presence of **actively decomposing plant residues,** perhaps as a result of some product of microbial development." Norman attributed the remarkable productivity of Iowa soils largely to their high levels of organic matter. The words in the foregoing quotation, however, are an admission that, even in Iowa soils, continued replenishment of organic matter is necessary. Otherwise the system reaches a stage of decomposition where the remaining organic matter is quite ineffective. Decomposition must be active if the material is to function as it should.

The manager of desert soils may be encouraged by these truths in the characteristics of soil organic matter. Again, it is not the level of organic matter that is important but rather the supply of fresh material that may decide its effect on production. Fuller used a good illustration when he likened the organic matter in the soil to the water in a shallow horse-watering trough. The horse is not interested in the depth of the water but rather in the rate at which the water enters the tank. Fuller showed the influence of fresh organic matter on soil structure by citing the results of applying steer manure to an Arizona soil (see Fig. 16-3).

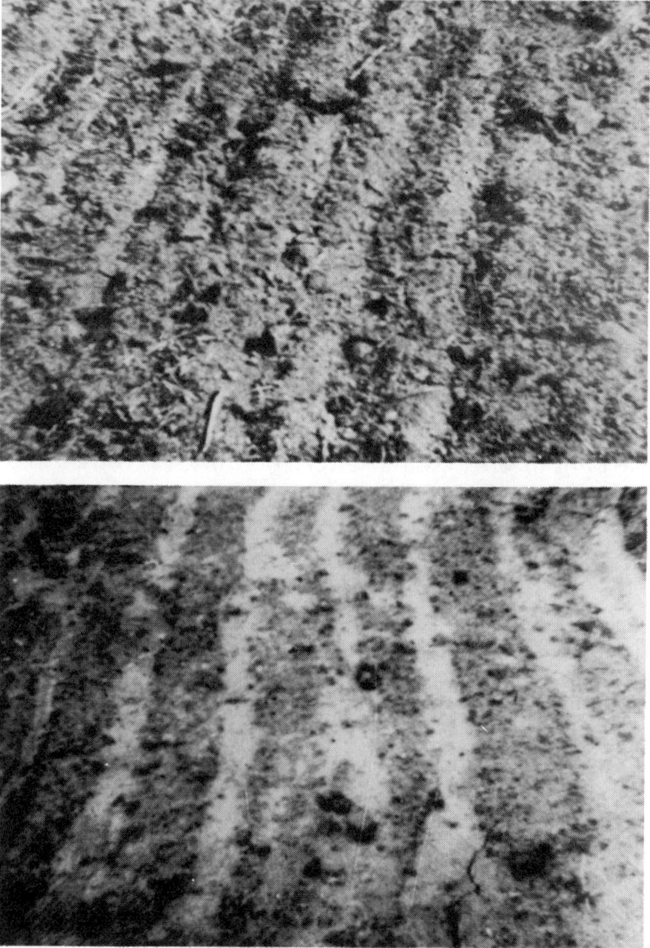

FIGURE 16-3 Influence of steer manure on the physical properties of the soil. *Upper:* Manure added to the soil. Note the structure and tilth is better than in the field shown below where no manure was added. Note how the unmanured land has washed until it has become smooth. (U.S. Soil Conservation Service photograph. Used by permission of the author from N. H. Fuller, *Soil Organic Matter,* Arizona University Agricultural Experiment Station Bulletin 240 (rev.) 1956.)

Management Practices

Management goals in this subhumid to arid region are not too different from what they are in the more humid North Central region. One simply must attain the goals by somewhat different means, depending largely on how much and what quality of water may be available. Organic matter levels are generally low, but they must be maintained.

Legumes and grasses should be grown as green manure for organic matter maintenance and erosion control. Animal manures serve the same purpose as, of course, they do in humid-soil management.

Lime is seldom needed except where is might be used as a source of calcium on alkali soils.

Fertilizers are needed less frequently in dry-land farming areas; but under irrigation farming, their need should be judged by use of soil tests, tissue test, and deficiency symptoms as explained in Chapters 10, 11, and 12.

You should contact the State Cooperative Extension Service, the U.S. Soil Conservation Service, or private agronomists in the area for detailed advice on specific soils.

THE GREAT PLAINS

The American Great Plains, as shown in Fig. 16-4, is a vast area bounded on the west by the Rocky Mountains and extending from the Canadian border almost to the Gulf of Mexico. Its eastern boundary is rather indefinite but is generally considered to be at the line east to west where dry-land farming begins. The southern and southwest boundaries overlap what has been considered in this book to be the **Southwest**.

Many articles and books have been written about the Great Plains. Explorers have called it a "region of lush grass," others called it the "Great American Desert." A quotation from Harold E. Myers (1943) (former Dean of Agriculture at the University of Arizona and formerly Associate Director, Kansas Agricultural Experiment Station) in his presidential address before the American Society of Agronomy shows vividly why the two opinions may have arisen. "The climate is at times most favorable for man, livestock, and crop production, followed with certainty by drought, hail, blizzard, wind and heat," Myers emphasized the fact that the Plains is an area of extremes. Average annual rainfall east to west varies from 20 to about 10 in., but the amount during a certain year has seldom been average. Floods and droughts are normal. During a 67-year period, annual rainfall at Hays, Kansas, varied from 35.4 in. to a low of 11.8 in. Furthermore, there have been periods of several continuous years when total annual rainfall has been much below the average figure. It is during such periods when "Dust Bowls" start. Soil management on the Great Plains can be so regulated, however, as to prevent the future occurrence of Dust Bowls. The purpose of this section is to review some of the important points in such management.

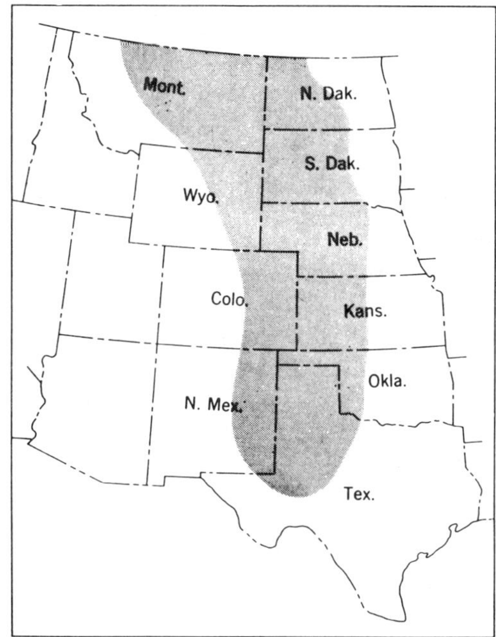

FIGURE 16-4 Area sketch of the Great Plains. (Taken from USDA Leaflet 394. Facts about wind erosion, and dust storms, on the Great Plains, by U.S. Soil Conservation Service.)

CHARACTERISTICS OF THE SOILS

The soils of the Great Plains are characterized by the fact that they are generally dark-colored and have a zone of calcium carbonate accumulation in the solum. That latter is the result of formation under too little rainfall to cause leaching, whereas the dark color is the result of formation under grass vegetation. Mollisols in the Eastern Plains give way to Aridisols as rainfall decreases from east to west. The Southern Plains (southwestern Kansas, southeastern Colorado, eastern New Mexico, and the Panhandles of Oklahoma and Texas) contain scattered areas of Aridisols and Entisols, those soils formed under prairie vegetation with low rainfall and warm climate. Although medium-textured soils are the most important agriculturally, the Southern Plains does contain large areas of very sandy soils. Similar soils are found also to occupy all or parts of 22 countries of central Nebraska, commonly termed the Nebraska sand hills.

Great Plains soils were originally well supplied with organic matter, especially those of the Northern Plains. As rainfall decreased from east to west, vegetation made gradually lesser growth, so organic matter accumulated at a slower rate from east to west. However, the soils of the Western Plains did develop under sufficiently dense grass cover to be fairly well supplied with this all-important material. The Southern Plains, although formed under grass, are in a region of much higher temperatures and have been exposed to greater winter erosion. As a result, organic-matter levels are lower and soil conservation problems are correspondingly more difficult. The Dust Bowl of the 1930s and more recently that of the 1950s were centered in this region.

Great Plains soils naturally have (or had) good structure. This is to be expected since they are grassland soils. They range in texture from very light sands to silt loams in the loess areas and varying textures in the High Plains. Soils in the High Plains were formed in place from the underlying parent materials. For that reason, they are quite variable. The loess soils range in slope from steep hills to level plains. Claypans have developed in some of the loess-plains soils to make them slowly permeable. Most loess soils of the Great Plains are deep, permeable, good-structured soils with very high productive capacity. They have high water-holding capacity, a characteristic of great value in a dry-land area.

THE DRY-LAND EXPERIMENT STATIONS

The first dry-land soils research was started in Colorado at Cheyenne Wells in 1894. However, the project was soon abandoned, and the next effort of the kind was started in Utah in 1901 by Dr. John R. Widstoe. Five stations were established in the state in 1903.

Professor E. C. Chilcott of South Dakota State College was also interested in dry-land investigations. He was responsible for selling Secretary of Agriculture James Wilson the idea that a research need existed in the Great Plains. Secretary Wilson presented the need to Congress and secured the appropriations for a new Office of Dry Land Agriculture. Professor Chilcott was placed in charge of the work in July 1905. The first stations were started in 1906, and a total of 25 were established, the last in 1916. They were well scattered over the area between Williston, North Dakota, and Big Spring, Texas, and as far west as Moccasin, Montana, and Tucumcari, New Mexico.

The well-known work of Briggs and Shantz on water requirement of plants is an example of the early fundamental research conducted by Chilcott's staff and cooperating Divisions in the USDA. Applied research included studies of tillage and seeding methods, crop rotation, manuring, summer fallow, and adaptation of crops. All phases of dry-land farming were studied at these stations. One problem that has an indirect influence on soil management is a study of the feeding value of silage after storage for several years. An experiment conducted at Colby, Kansas, involved silage that had been stored for 13 years. It is now known that the best investment a Great Plains cattleman can make is in silage during a wet year. In that way, he may be able to weather a drought period without sacrificing his cattle. The modern-day alternative is to produce silage on irrigated land so that a dependable supply of silage is available for feedlot operations. Such a balanced program means conservation of the soil, because loss of the cattle is invariably followed by an expanded wheat or sorghum acreage after the drought is broken.

One project of the dry-land stations that has led to a valuable method of predicting the best use of land during a coming year is on the relationship between soil water at seeding time and yield of wheat. Compton (1943) checked the results of these studies by measuring soil-moisture depth, precipitation during the growing season, and wheat yields on 2360 fields in western Kansas. He found a very close correlation between depth of soil moisture at wheat-seeding time and yield of that crop. Of course, rainfall during the growing season also had a marked effect on yields of wheat, but in only 9 percent of the fields did the yield exceed 10 bu/acre when the soil was wet no deeper than 6 in. On the other hand, 66 percent of the fields that were wet to 30 inches or more yielded more than 10 bu/acre. Compton arranged a chart (see Fig. 16-5) that may be used to estimate wheat yields from soil-moisture depth and rainfall during the period October 1 to May 31. Although the relationship between soil water at seeding and yield of sorghum yield is important, management may also exert an influence. The greatest water-use efficiency was obtained on conservation bench terraces.

The present agricultural possibilities on the Great Plains are much greater as a result of the work of the Dry-Land Experiment Stations and the State Experiment Stations that have been connected with them or have worked with them. As pointed out by Myers, for many years a leading agronimist in the area, most of the hard red winter wheat and much of the hard red spring wheat produced in the nation comes from the Great Plains. In addition, the area produced practically all the durum wheat, most of the grain sorghum, and a high percentage of the cotton and corn of the nation. That is not all. The Plains produce more beef than any other geographic area of the nation. The agricultural success of the Great Plains in future years surely depends on the establishment of a correct balance between grain and livestock production in order that soil fertility may be maintained. Better soil management will stabilize the agriculture in the area.

CROPS ADAPTED TO THE GREAT PLAINS

Buffalo thrived on the Great Plains. They made existence possible for the Plains Indians, who used them for food and clothing. Webb (1931) says that the Plains Indians could not have existed if it had not been for the buffalo.

The horse came to the Plains in the middle of the sixteenth century, supposedly abandoned by DeSoto or Coronado. The Kiowa and Missouri Indians are known to have had horses in 1682 and other tribes soon thereafter. The Plains country must have agreed with the horses, because by the time the American pioneers reached the Plains, horses were present in great numbers. At times, the Indians even used them for food.

The spread of the buffalo and the horse indicates that the Great Plains is good livestock country — good for any animal that can live on grass. Early agriculture in the area showed this to be true. The crops adapted to the Plains must then include those that may be used as cattle feed.

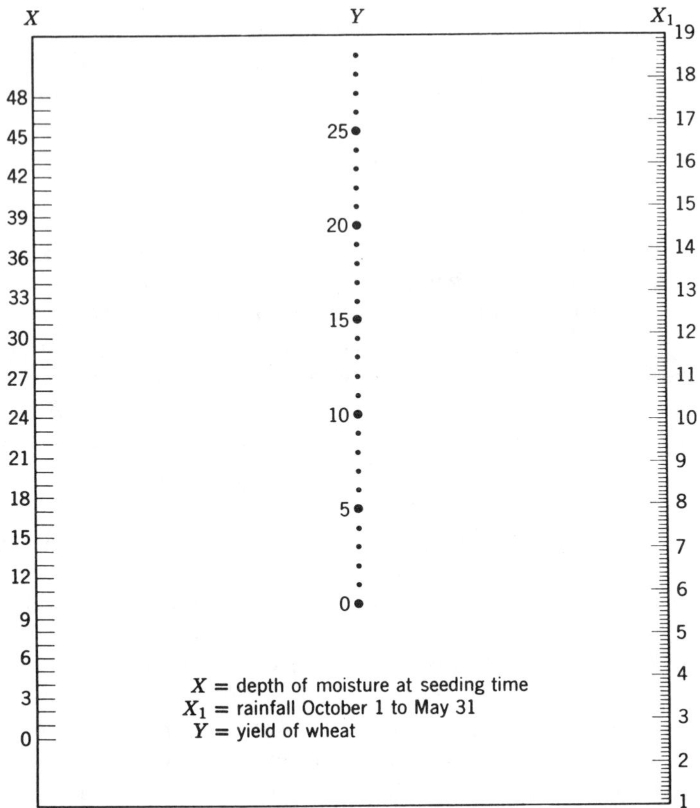

FIGURE 16-5 A chart for estimating wheat yield in western Kansas using depth of moisture at seeding time and rainfall from October 1 to May 31. Place one edge of a ruler at the point on the depth-of-moisture scale that corresponds to the depth the soil is wet at wheat-seeding time. Move the ruler so that the same edge is at the point on the rainfall scale that indicates the inches of rainfall between October 1 and May 31. The edge of the ruler intersects the yield scale at the expected yield of wheat.

At planting time, one does not, of course, know the inches of rainfall. However, a little manipulation of the ruler gives some idea of the slim chance for good yield that exists if soil moisture is at a shallow level. For instance, if the soil is moist to only 6 in., rainfall between October 1 to May 31 would need to exceed 16 in. to be assured of a 10-bu. yield. Rainfall records show there is little chance of that much rain. It would be better to make other use of the land during the next year. (This chart was taken from L. L. Compton. *Relationship of Moisture to Wheat Yields on Western Kansas Farms*). Kansas Extension Circular 168, 1943.

Water and Crops

Kearney and Shantz (1911) divided dry-land plants into four groups.

1. Drought-escaping plants — those that can grow to maturity while moisture is present.

2. Drought-evading plants — those that restrict growth during dry periods. They make efficient use of water.

3. Drought-enduring plants — those that make very little growth during dry periods. Mesquite is an example.

4. Drought resisting plants — those that store up water in fleshy tissue and resist water loss. Cacti are examples.

The dry-land crops must come from the drought evaders, those that use water efficiently. Sorghum is a better dry-land crop than corn is because it has a much greater secondary root system. History shows that sorghum starts where it is too dry for corn to be a dependable crop. Where moisture is plentiful, the two crops are about equal in water requirement. In the humid areas, farmers have preferred to grow corn, partly as a result of harvest problems brought about by excessive fall rains.

Approximately 25 percent of the Great Plains is adapted only to grasses. Much of this area is in the Sand Hill area of Nebraska, the Bad Lands of South Dakota, the extreme western part of the Northern Plains, and throughout the Southern Plains. According to the Soil Conservation Service land capability classification there were, in June 1955, 14 million acres of cropland in the Great Plains as a whole that should be returned to permanent grassland. At that time the same classification showed that a small acreage of potential cropland was still unplowed.

Legumes can be grown as forage on the entire Southern Plains and in the eastern part of the Northern Plains but only on irrigated land in the western third of Nebraska and corresponding portions of North and South Dakota and Kansas.

Wheat is the main cash crop in the Great Plains, hard red winter wheat in the portion south of central South Dakota, and hard red spring wheat to the north of that zone. Winter wheats are better adapted to dry-land farming than are the spring varieties because they mature earlier and fit a summer fallow system better. Alternated with fallow, winter wheat has been successfully produced in regions with less than 25 cm (10 in.) of annual rainfall. As little as 15 cm (6 in.) is sufficient in a part of South Africa. However, more rainfall is required for the crop in the southern Great Plains because of high evaporation losses. An amount of rainfall that results in desertlike conditions in New Mexico would make limited cropping possible in northern Colorado or Wyoming.

Other grains, principally oats and winter barley, may be grown in the eastern part of the Northern Plains. Corn may be successful during wet years, but the crop is not dependable, it is much safer to grow sorghum. Spring barley is more dependable than corn as a feed crop in northwestern Kansas.

WATER MANAGEMENT

Water is the limiting factor in the agriculture of the Plains. Crops are selected that can withstand long drought periods or that make their growth during the rainy season. Planting seasons and tillage practices are timed to take advantage of available soil moisture; the farmer must be ready to adjust cropping systems to the weather. Adherence to a strict cropping sequence would result in failure on the semiarid Plains.

Range-Land Practices

Water management on range land is largely a matter of controlled grazing. Enough forage must be allowed to remain on the land to maintain cover and soil organic matter. Soils will remain permeable to water, and the cover will prevent soil blowing.

Fences should be used for rotating cattle on pastures so that overgrazing in certain areas may be avoided. Gully control and brush clearing and reseeding are necessary on lands that have long been overgrazed.

Reserve supplies of feed are essential on a Great Plains cattle farm. Stored hay and silage make it possible to avoid overgrazing during drought years and so indirectly protect the soil. Furthermore, the farmers who are required to sell their cattle during drought periods are likely to increase their crop acreage when the rains come. Thus, protective soil cover is destroyed and the soil soon becomes depleted.

Cropland Practices

Water-management practices on cropland revolve around soil permeability, run-off control, soil water storage, and timeliness of operations.

Soil Permeability, Absorptive Capacity, and Runoff

Soil permeability is, of course, influenced to a marked degree by texture. The so-called "hard-lands" of the Southern Plains consisting of clay loams, silty clay loams, silt loams, and very fine sandy loams of the Richfield, Pullman, Colby, Abilene, Tillman, and Roscoe series are excellent wheat soils when moisture is sufficient. Wheat after wheat is common and is recommended when soil moisture is sufficient. (See Table 16-3 for the taxonomic classification of the series mentioned in this chapter). The soils have such high moisture-holding capacity near the surface that water is held in a too-shallow location so it may readily be lost by evaporation. This is probably why summer fallow has not proved to be as effective on these soils as on those in the Central Plains. Because the soil moisture is held so near the surface, the soils are more droughty than those in the sandier series — Pratt, Miles, Vernon, Amarillo, and Enterprise. However, since this is true, it is even more important that all precipitation enter the soil. Residues should be handled in such a way as to provide a stubble mulch. Such a soil surface is absorptive of rainfall and will not blow unless winds are very strong.

Soil Management Principle

Emergency tillage is sometimes necessary to stop soil from blowing.

On the sandy soils of the Southern Plains, water penetration is not a problem and it usually penetrates to sufficient depths that is not readily lost by evaporation. However, it is difficult to prevent wind erosion when such soils are cropped to wheat. The blowing is usually the most serious during the later winter months. Some control may be obtained by strip-cropping with sorghum. The sorghum stubble standing in the field during the winter is quite effective in protecting the strips of wheat, providing the wheat strips are not too wide.

Wheat does not do well after sorghum, so if the two crops are to be grown in rotation, the sequence should be sorghum-fallow-wheat. Wheat yields are likely to be increased as much as 30 to 60 percent over what they would be in a continuous wheat system, and sorghum does well after wheat. Furthermore, sorghum is much more likely than is wheat to produce a crop during drought years. Where wheat may be expected to fail 30 percent of the time, sorghum may fail only 10 percent of the time and perhaps not during the same years. Thus a more stable agriculture is assured, especially if the farm enterprise includes beef as well as wheat and sorghum. In fact, there are some years when sorghum makes a forage crop but fails to make a grain crop. Cattle may then make use of the feed.

Cover Crops

Throughout the Great Plains, organic matter maintenance, to ensure good soil structure, is a must if water intake and erosion control (water and wind) is to be efficient. *Residues should never be burned but should be retained on the soil surface as a mulch.* They should not be turned under as is recommended in the humid regions.

Cover crops are needed to give soil protection through the winter. Grain sorghums, sweet sorghums, Sudan grass, and millet are the best cover crops in the southern two-thirds of the Great Plains. Winter wheat, barley, or rye may be planted when fall moisture is sufficient. Cover crops are especially desirable after cotton.

Legumes should be used as cover crops and as sources of green manure on irrigated land. Mazurk, Zingg, and Chepil (1953) used a portable wind tunnel to determine the effect of alfalfa and manure on wind erodibility of an irrigated Mollisol (Tripp very fine sandy loam) at the Scotts Bluff Field Station at Mitchell, Nebraska. Two rotations, barley–potatoes–sugar beets and barley–alfalfa (3 years)–potatoes–sugar beets were conducted over a period of 39 years. The action of the wind caused a soil loss of 74,500 lb/acre where the 3-year rotation soil had not been manured. Where it had been manured with 12 tons/acre every 3 years, the soil loss was reduced to 2720 lb/acre. This surely shows that on these soils, farmers should make careful use of manure.

Alfalfa 3 years out of 6 did an even better job than did manure in lessening wind erodibility. From the land in that rotation, the soil loss was only 970 lb/acre. Wherever alfalfa or other legumes can be grown, farmers should take advantage of their beneficial effects in promoting soil stability and fertility.

Soil-Water Storage

There are several ways of increasing the stored water in a soil. This may be very important. Soil water research at Colby and Garden City, Kansas showed that it required 18.7 cm (7.37 inches) of soil water before any wheat grain was produced and that each additional 1.3 cm (0.51 inch) of water produced a bushel of wheat. Careful water management may increase the storage capacity of the soil by several cm and make the difference between a profitable crop or a failure.

Tillage Practices

Plowing loosens and aerates soil, which is one of the reasons for conducting tillage operations. The furrow slice, immediately after being turned, may hold as much as 20 percent more water than before. Sanborn showed in 1892 that crop yields increased with depth of plowing. Perhaps some of the reason was greater water-holding capacity. If soils are not packed again by immediate tillage after plowing, and if nutrients are not limiting factors, still deeper plowing or otherwise working of the soil may be beneficial. But it must also be realized that tilling is a drying operation. When tillage is used for weed control in a fallow system, seldom is there an advantage in tilling deeper than 10 cm according to Brengle (1982). Recent experiments indicate that subsoiling is beneficial on some soils. Deep plowing is recommended for sugar beets. Such deeply tilled soils take up more water before it runs off.

Tillage on the contour and the construction of terraces slows up the flow of run-off water. If fact, terraces are sometimes constructed entirely on the contour so all the water soaks in. Such construction, permissible where the terraces may be wide and the soil is quite permeable, are used where water conservation is important, as in the Great Plains.

Summer Fallow

Summer fallow is the practice of keeping land free of vegetation throughout a season for the purpose of storing a portion of the rainfall of that season for use by the next crop. Frequent tillage to prevent vegetative growth results in an additional advantage, that of noxious weed control. Also, there is some accumulation of available plant nutrients in fallowed soil. In the humid areas of the United States, summer fallowing for weed control is sometimes practiced. Otherwise, it is a practice peculiar to the semiarid sections, those areas that practice dry-land farming.

There is enough annual rainfall, on the average, to produce a wheat crop over all the Great Plains, except perhaps on the very sandy soils. The difficulty lies in the rainfall pattern, within seasons and between seasons. Precipitation on the Plains occurs mostly in summer, May through September, and is largely torrential. Much is lost by run-off and evaporation. Even during seasons of above average rainfall, there may be serious injury from drought because of unfavorable distribution through the summer. Yearly precipitation is seldom near normal. The variation at Hays, Kansas, has already been mentioned. At Dickinson, North Dakota (third county from Montana line), the 60-year average rainfall for April through July, the critical period for spring wheat, is 23.3 cm (9.17 in.). During that 60 years, 1892 through 1951, the variation was from 5.8 cm (2.3 in.) in 1936 to 54 cm (21.2 in.) in 1941.

Soil-water storage is very important at Dickinson, as is evidenced by the bumper crops of 1915 and 1942, both years that followed years of high rainfall. The critical 4-months total in 1914 was 42.6 cm (16.79 in.). It is interesting, however, that summer fallow is not recommended as a moisture-storage practice for spring wheat in the Dickinson area. The situation there is a good illustration of the fact that the advisability of fallowing depends on several factors and must be considered carefully for each locality. It is true that the highest yields of small grain are obtained after fallow, but they are not enough higher than after corn to pay for the loss of a year of cropping. This might not be true if winter wheat were grown, as there would be no water-storage period between corn harvest and wheat planting time. The further addition of hay and pasture to the rotation is valuable from the soil-improvement standpoint.

Continuous small grain is not a profitable practice in Dickinson County. Those who wish to grow only small grains should resort to fallow every third year, fallow-wheat-wheat or fallow-wheat-oats. In that case, early preparation for the second grain crop is important.

Fallowing is for storage, not for total water conservation. Evaporation losses are great during fallow but water removal by plant roots from lower depths is avoided, so it is possible to store a few inches of water for the next crop. The amount stored at three locations in western Kansas is shown in Table 16-2. There was definitely some advantage in early plowing but, according to the investigators, not enough, since in those areas more than 7.4 cm (2.9 in.) of available water at seeding time is necessary to have any assurance of a crop if rainfall should be of drought proportions. In other areas, where the value of fallow is less certain, early plowing as compared to late plowing might be the deciding factor as to the advisability of raising a crop every year or every other year or perhaps 2 years out of 3. As mentioned earlier, each inch of stored water may raise final yields by 2 bu/acre. That would be the case when October 1 to May 31 rainfall was inadequate for a good crop.

The need for fallow increases generally from east to west in the Plains area or, as stated by Brengle (1982) "as the precipitation and potential evapotranspiration decrease." For instance, in Cheyenne and Rawlins counties, Kansas, and all areas to the south in that state, alternate crop and fallow are recommended. In the next two-county strip, north and south across the state, fallow is recommended after 2 years of cropping; and in the third

TABLE 16-2 Effect of Time of Seedbed Preparation on the Quantity of Available Water in the Soil at Wheat Seeding Time

Location	Number of years	Average inches available water at seeding time		
		Late plowed	Early plowed	Fallowed
Hays, Kansas	23	1.54	2.90	7.96
Colby, Kansas	19	1.05	1.47	5.16
Garden City, Kansas	13	0.44	1.08	4.67

Source: A. L. Hallsted and O. R. Mathews, *Soil Moisture and Winter Wheat with Suggestions on Abandonment,* Kansas Agricultural Experience State Bulletin 273, 1936.

two-county strip, another year of crop may be grown between the years of fallow. The three strips make up the western half of Kansas, and fallow is not recommended for the eastern half.

Wind erosion limits the use of fallow in some areas and on some soils. This is especially true for spring-planted crops. Various control practices can be used to make fallow allowable where yields, without erosion injury, make it a very desirable practice. Fallowing may be done in strips, alternated with sorghum grown for summer control. Sorghum stubble left in strips may help during the winter. Deep listing in the fall is desirable after fallowing for spring crops. Summer-fallowed land should always be left rough-ridged where possible. Fallowing for wheat should be started in the spring rather than in the fall unless experiments in the area show that it is better to start in the fall. The most serious blowing usually occurs in late winter and early spring.

Soil series should be considered. To summarize, the soil must have ability to absorb more water than it is likely to receive during a continuous cropping system. This is true on all but the extremely sandy soils. The soil should be sufficiently permeable so that the water will move down to depths where it will be protected from loss by evaporation. Soils of intermediate texture fill this requirement.

Soil Management Principle

Summer fallow should be practiced on soils of intermediate texture.

Finally, the soil should be reasonably level or should be terraced so water may soak in rather than run off.

THE CONTROL OF EROSION

Dust is the scourge of the Great Plains. Dust bowls develop when soils of low rainfall areas are poorly managed. Overgrazing and cropping of grasslands were equally responsible for the dust storms of the past. Dust makes living conditions disagreeable, sometimes almost unbearable, but of even greater importance is the fact that soils are being destroyed, both by removal of surface soil and by deposition of infertile material on good topsoil. The future of Great Plains agriculture depends on the degree to which wind erosion is controlled.

Factors That Affect Soil Erodibility

Some soils blow more readily than do others. To arrive at the basis for a more intelligent wind-erosion control program, Professor W. S. Chepil, Kansas State University and the U. S. Agriculture Research Service, used a wind tunnel to study the characteristics of soils that affect erodibility. He found that the amount of soil erodible by wind is limited by the critical height of and distance between the nonerodible fractions (clods) that are exposed at the surface by the wind. The ratio of the height of the clods to the distance between them is the critical surface-roughness constant. This constant determines in part the volume of the nonerodible clods exposed by wind erosion and the volume of soil removable by wind. The amount of erosion will vary proportionately, other factors being equal, with the ratio of erodible to nonerodible fractions in the soil. This being true, the critical surface roughness constant determines how much soil will be removed by the wind, which means that soil should always be maintained in as cloddy a condition as is practical with the management system. Figure 16-6 shows an excellent soil surface for erosion control.

Several soil characteristics were found by Chepil to influence clod structure and erodibility.

1. *Soil texture.* With pure separates, the greatest degree of cloddiness and the greatest resistance to erosion occur with silt ranging in size from 0.005 to 0.01 mm in diameter. Highest erodibility occurs with fine sand, and clay is moderately erodible. The least-erodible soils are those that contain 27 percent clay and large quantities of silt. Erodibility decreases as clay increases up to 27 percent, then becomes greater as clay content increases beyond 27 percent. In clay soils, the shrinking and swelling of the colloids with drying and wetting causes the clods to disintegrate. Silt does not shrink and swell and is small enough to exert considerable adhesive influence in the clod, which probably accounts for its stabilizing effect. Sand or very fine sand weaken clod structure but decrease erosion because it acts as a pavement to protect the erodible fraction.

2. *Water stability.* Erodibility decreases with an increase in the percent of water-stable aggregates with diameter below 0.02 mm and above 0.84 mm. It increases as the percentage of water-stable particles between 0.05 and 0.42 mm increases. The effect of the water-stable aggregates on erodibility is arrived at by determining their effect on dry clod structure.

FIGURE 16-6 The lister did an excellent job of ridging and throwing up clods to stop wind erosion. Sandy loam soil in Ottawa County, Kansas. (Photograph from W. S. Chepil and N. P. Woodruff. *How to Reduce Dust Storms.* Kansas Agricultural Experiment Station Bulletin 318, 1955.)

Soils are more erodible in late winter and spring. This is apparently the result of frost on the size of the water-stable aggregates. Frost tends to break down those above 0.84 mm and to increase the size of those under 0.02 mm. In other words, the percentage of both fractions is decreased, which causes a decrease in soil cloddiness and an increase in erodibility.

Cultivated dry-land soils vary with respect to percentage of water-stable aggregates in the various size ranges. This would account for at least a portion of their great variability in wind erodibility.

3. *Organic matter.* Decomposing plant residues (wheat, straw, and alfalfa) cause a slight decrease in soil erodibility by wind. The effect is the result of an increase in water-stable aggregates greater than 0.84 mm in diameter, a decrease in the proportion of water-stable aggregates smaller than 0.02 mm, and a slight increase in the proportion of dry clods greater than 0.84 mm.

After 2 to 5 years of decomposition, the organic matter has the opposite effect. The tendency to increase water-stable aggregates greater than 0.84 mm has disappeared, the proportion of water-stable aggregates less than 0.02 has decreased, and the proportion of medium-sized aggregates has increased. The net effect is to decrease cloddiness and increase wind erodibility. This points to the desirability of making small, frequent additions of organic matter rather than infrequent large additions. It also emphasizes the importance of keeping residues on the surface where they will protect the soil particles from the wind rather than to mix them with the soil where they will decompose and eventually increase erodibility.

Practices That Reduce Wind Erosion

The practices particularly effective in reducing wind erosion (dust storms) on the Great Plains, especially if they may be conducted on an area-wide basis, are as follows.

1. Keep soil surfaces rough and cloddy. Be ready for emergency tillage when the wind starts to blow.
2. Keep soils covered with crops or residues as much as possible, especially during winter and early spring. As little as 500 lb of residues scattered evenly over an acre has been quite effective in controlling blowing. A **good** wheat crop is quite effective, but a **poor** wheat crop is ineffective.
3. Use stable manure, green manure, and rotations according to recommendations for the area, especially where irrigation is available.
4. Practice wind strip-cropping. Buffer strip-cropping may be used to advantage in certain instances.

Residues Management

Stubble-mulch farming may be defined as the practice of leaving crop residues at the soil surface where they serve to lessen the force of the wind against the soil particles and to protect the soil against the ravages of falling rain and moving water. In proclaiming the merits of this practice, Zingg and Whitfield (1957) quoted Lowdermilk as follows.

> Leaving crop litter, which is sometimes called stubble mulch or crop residues, at the ground surface in farming operations is one of the most significant contributions to American agriculture. Certain adaptations of the method need to be made to meet the problems of different farming regions, but the new principle is the contribution of importance.

The stubble-mulch system is particularly well adapted to alternate wheat and fallow. Tillage should start immediately after harvest and should be so regulated that crop residues are anchored but mostly exposed. Tillage should be just sufficient to control weeds and prepare the soil for the next crop. Traffic breaks the straw and induces blowing if excessive. Under careful management, wheat residues may be protective through the fallow year and into the wheat year until the young plants are large enough to furnish protection. Drilling is through the straw mulch (see Fig. 16-7). A minimum of 1 to 1 1/2 tons/acre at seeding time is necessary for adequate protection. This amount is usually available in a wheat-fallow system if careful tillage has been followed. Mulches save water and soil. A mulched soil surface readily absorbs rainfall. Once below the layer of mulch, the moisture is less subject to evaporation because of shading and protection from wind. Thus, more remains for

FIGURE 16-7 Drilling wheat through stubble mulch near Lincoln, Nebraska. Note the desirable combination of residue, roughness, and good soil structure. (Photograph by F. L. Duley. Used in A. W. Zingg and C. J. Whitfield. *Stubble-Mulch Farming in the Western States.* USDA Technical Bulletin, 1116, 1957.)

crop use. Furthermore, both wind and water erosion are lessened by mulches.

Wind Erosion

Research workers on the Great Plains have shown that anchored surface residues are very effective in controlling wind erosion but that amounts needed vary greatly depending on soil conditions. Chepil and Englehorn found that in 1950 in east-central Kansas that 1 ton of residue was required. In western Kansas in 1954, Chepil and Woodruff obtained stability with 500 lb of anchored wheat residue. Zingg (1950) working with a portable wind tunnel, found the erodibility of a land surface to be related to soil structure (dry-clod structure), roughness of ground surface, and the amount of vegetal cover in the form of residues or growing plants. His data provided the following formula:

$$X = \frac{A3.5}{2R0.8}$$

where X = soil material eroded in pounds per acre, A = percentage of dry surface soil particles less than 0.42 mm in diameter, and R = weight of wheat residue anchored in soil surface, in pounds per acre.

In Fig. 16-8, tons of wheat residue are plotted against dry-soil structure. The graph shows that dry-soil structure must be determined by sieving before the amount of residues required for soil protection may be prescribed. It is interesting that where the percentage of dry-soil particles below 0.42 mm in diameter is less than 40 percent, very little anchored residue is required to hold soil loss to a negligible amount. The importance of a cloddy structure is evident from the graph. Two tons of residue held soil loss to 1/4 ton/acre when the percentage of fine particles (below 0.42 mm) was 50 percent but allowed 1 ton loss when fines amounted to 70 percent. In fact, the data show that, under the wind tunnel conditions, it was practically impossible to add enough residue to hold the loss to the 1/4 ton figure. In other words, the effects of a cloddy structure and residues must be combined to effect satisfactory wind-erosion control.

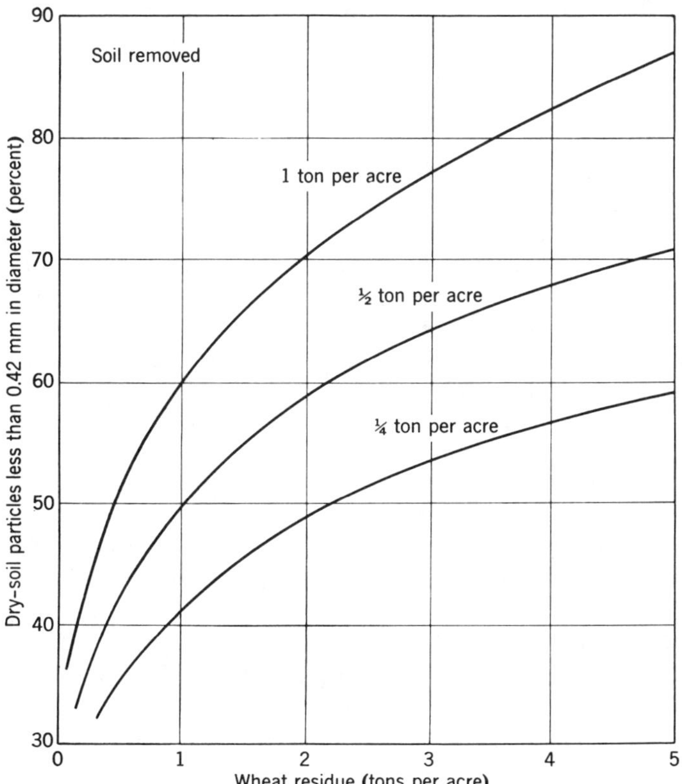

FIGURE 16-8 Tons of anchored wheat residue required to limit erosion to given amounts per acre on soils of varying structure, based on wind-tunnel studies. (From A. W. Zingg and C. J. Whitfield. *Stubble-Mulch Farming in the Western States.* USDA Technical Bulletin 1166, 1957.)

Water Erosion

Water causes much more erosion of the Great Plains, during most years, than does wind.[2] This is not commonly recognized because a small quantity of soil in the air is a nuisance to humans and so is always noticed. On the other hand, much greater losses by sheet erosion may go largely unnoticed, except by the conservation expert.

The tremendous advantage of leaving residues on the surface has all but eliminated the use of the moldboard plow in the Western Plains. Data obtained at Lincoln, Nebraska, are typical. During a 6-year period, land cropped in a corn-oats-wheat sequence was stubble-mulched as compared to being plowed with a moldboard plow. Annual soil losses from runoff were 1.26 tons and 6.01 tons, respectively, from the two areas, about five times as much loss from unmulched as from stubble-mulched soil.[3]

Residues on the surface protect the soil from the impact of the raindrops and lessen soil splash. This prevents sealing and compaction of the surface layer, which then continues to absorb water. Furthermore, the residues make the soil surface rough and mechanically obstruct water flow. Velocity is thus lessened and particles already suspended settle to increase the obstruction. Each bit of residue becomes a tiny dam and runoff is greatly lessened. The respective amounts of runoff that caused the erosion from the Lincoln experiment mentioned previously were annually 0.70 and 2.09 in. from stubble mulched and moldboard plowed areas. Water losses were 1 to 3 in. as compared to soil losses of 1 to 5 in. For the current crops, the water loss may be more important than the soil loss.

Surface residues may fail to reduce runoff when heavy rain falls on previously wet soil. This condition seldom occurs, however, in the Great Plains where rainstorms are usually violent and well scattered. In other words, stubble-mulched surfaces usually retain a high rate of water intake, as compared to bare surfaces, for all periods of rainfall.

Residues Save Water

It is logical that where less water runs off a piece of land, more will be left for crops. Soil-water studies, however, have pointed out several factors that tend to interfere with water accumulation within the root zone under mulches. In a very coarse-textured profile, percolation carries much of the water below root depths, perhaps to recharge groundwater reserves or to be lost through drainage channels.

In very fine-textured soils, storage is so close to the surface that evapora-

[2]Opinion of F. W. Smith, Kansas State University, given in a private conversation.
[3]Data obtained in 1954 by F. L. Duley and J. C. Russell. For source see Mazurk, Zingg, and Chepil (1953).

tion may occur before plants need the extra water, even though the mulch may slow the losses. The nature of the rainfall pattern-time lapse between rains has considerable bearing on the effectiveness of mulches in conserving water for plants.

The nature and amount of the residues must also be considered. Bright residues reflect more heat, and larger quantities tend to prevent radiation from the soil. Season is thus important, as residues affect soil temperature near the surface. During warm weather, when solar radiation is high, the temperature of the top few inches of mulched soil is lower than is that of bare soil, whereas during cooler weather, the opposite may be the case.

We may readily see that the immediate effect of stubble mulch on available soil moisture is dependent on so many factors that it may be difficult to evaluate. Many data have been published, however, to show the superiority of the system. Stallings (1957) showed that subtillage as practiced at Amarillo, Texas, during the period 1942 through 1948 resulted in higher yields of continuous wheat than was produced where the one-way plow was used. The difference was 1.9 bu/acre. There is little doubt that water made the difference, as yields in that area are always dependent on soil-water supplies.

Mulches Affect Soil Properties

Subtillage usually results in an accumulation of small water-stable particles, a condition that enhances water intake but that tends to produce a surface more subject to wind erosion. However, since wind erodibility is a function of completeness of vegetal cover and general surface roughness as well as surface-particle size, wind-erosion control may be obtained despite the higher percentage of wind erodible dry particles.

The tendency for an accumulation of small particles under stubble mulch may be minimized by inducing faster decomposition of the residues. Decomposing residues cause larger aggregates, which again disintegrate when decomposition ceases or slows. A desirable balance may be obtained by regulation of the rate of decomposition through tillage methods. Mixing the residues with the surface soil tends to increase decomposition rate. This may be accomplished by disking, whereas the exclusive use of sweeps avoids mixing and slows decomposition.

Excessive decomposition during a stubble-mulch system can be disastrous because excessive blowing may occur when the vegetal cover becomes inadequate. Emergency tillage to roughen the surface is then necessary.

Generally, stubble-mulch farming tends to conserve organic matter. Where relatively long-time studies have been conducted, at Amarillo, Texas, and Newell, South Dakota, mulched-plot soils contained more organic matter than did plowed soils. The differences are small but probably significant and are especially interesting in light of the general trend toward losses of organic matter with cropping.

LIME, FERTILIZERS, AND MANURES

Great Plains soils generally do not need lime. The soils were formed under too little precipitation to cause leaching and are characterized by a zone of calcium deposition above the parent material.

Saline and alkali soils are common in the Plains. They occur in small to large areas, usually in poorly drained bottomlands or in spots where the local relief is such that seepage occurs. Saline soils may result from irrigation with saline water unless drainage is provided and excess water is used. The management of saline and alkali soils was discussed earlier in this chapter.

Fertilizers

Fertilizers have not been extensively applied to the dry lands of the Plains. Field experimental results generally have failed to show a favorable yield response unless irrigation was practiced. Soils were originally well supplied with mineral nutrients and nitrogen. However, with losses in organic matter as a result of cropping and close grazing, total nitrogen levels are decreasing, which may account for the trend toward more favorable results from nitrogen fertilizers in recent years. Some results from western Nebraska illustrate what is happening.

Nitrogen Fertilizers

Rhoades (1956) reported some results from the North Platt Experiment Station. Soil nitrogen levels after cropping from 1908 to 1934 were as follows.

Cropping system	Loss of nitrogen (percent)
Continuous small grain	−15
1/3 intertilled, 2/3 small grain	−18
Continuous row crops	−27
1/2 fallow, 1/2 small grain	−30
1/2 fallow, 1/2 row crops	−36

These data show strikingly the effect tillage has on organic-matter depletion. As this continues, a need for nitrogen fertilizer will surely develop, perhaps only during years of favorable moisture. Fertilizer management should then be one of sidedressing and topdressing as need is indicated by tests and symptoms. The use of nitrogen fertilizers should result in an increase in crop residues, which will help to control erosion and compensate for organic matter losses resulting from cropping.

Still greater losses in organic-matter were reported by Rhoades to have occurred at the Box Butte Experimental Farm near Alliance, Nebraska. Continuous production of wheat from 1930 to 1955 resulted in a loss of 41 percent of the nitrogen, whereas continuous row crops depleted the nitrogen to the extent of 52 percent during the same period.

Experiments on irrigated land at Scotts Bluff showed that manure and legumes may prevent such losses. After 30 years of cropping without a legume, nitrogen losses amounted to 32 percent. With similar cropping, an annual application of 4 tons of manure per acre reduced the loss to 12 percent, and 6 tons/acre resulted in an increase in nitrogen amounting to 4 percent. In another series of plots, the Scotts Bluff experiments included the production of legumes 50 percent of the time. On those plots, the loss in nitrogen was 8 percent, and there was a gain of 2 percent where 2 tons of manure had also been applied each year. Thus it seems possible, on irrigated land, to prevent nitrogen losses by use of manure and legumes. Nitrogen fertilizers might have been even more effective.

Kansas State recommended from 0 to 50 lb of nitrogen per acre for small grains, corn, or sorghum on western Kansas unirrigated soils that had not recently grown alfalfa. On irrigated soils, use of 50 to 220 lb of nitrogen for the same crops was suggested, with the higher rates for corn and sorghum.

Other experiments and soils studies in western Kansas show that nitrogen topdressing will be profitable during years when late winter and early spring soil moisture levels are high. Nitrogen is not generally profitable for spring-seeded crops unless they are to be irrigated. In the Southern Plains, nitrogen is profitable only where irrigation is practiced.

Mineral Fertilizers

Great Plains soils are generally low in phosphorus and high in potassium. Sulfur is lacking in some areas, and there is considerable evidence that iron and zinc may be deficient. Iron and zinc deficiency will become more intense on irrigated soils when phosphate fertilizers are applied.

Phosphate fertilizers are especially needed for alfalfa on irrigated soils. The need should be determined by the use of soil testing, a service made available by the Extension Services in most of the states. Phosphate fertilizer applied for alfalfa will indirectly make more nitrogen available for wheat, sorghum, and other nonlegumes. This was shown by the Nebraska data from Scotts Bluff. On many irrigated soils, wheat, sorghum, and sugar beets also respond to phosphate.

Stable Manure

In dry-land management stable manures have not resulted in appreciable increases in crop yields when applied on the better soils, especially on those well supplied with nitrogen. It is suggested that such materials be applied on badly eroded spots and on alkali soils

On irrigated land, stable manure is very valuable, especially in the Southern Plains where organic-matter decomposition is rapid. Soil structure is a problem in the finer-textured soils, and the decomposing manure promotes aggregation. On the coarser soils it increases water-holding capacity, helps to prevent blowing, and furnishes much-needed nitrogen. Stable manure may be more valuable in these areas than in those where a greater abundance of water makes green manuring easier.

FLORIDA

Florida and the adjoining Flatwoods, including a large area in eastern South Carolina, parts or all of 20 counties in eastern Georgia, and a small area in southeastern Alabama, lie on a low coastal plain. Elevation increases from 25 ft along the Atlantic and Gulf of Mexico to about 300 ft in the central Florida ridge.

As the term "Flatwoods" indicates, drainage is generally poor and not well-defined. Several streams enter from the north and west, but streams in the central ridge area of the Florida peninsula are scarce. Surplus water seeps into cavities in the underlying limestone, into sloughs, and through pervious soils into numerous small lakes, and finally into the ocean or gulf.

Limestone underlies the entire region and is exposed in a few places. Unconsolidated materials, mostly sandy in nature, range in thickness from very thin to many feet over the limestone. Drainage has had a marked effect on the nature of the soils profiles. As in other parts of the country, soil characteristics have a marked effect on cropping patterns but climate is of even more importance in determining crop selection. The frost-free period varies from 250 days in the central northern part to 365 days at Key West.

Thornton and co-workers (1959) have shown that "economic crop production" is based on "the firm foundation of drainage, irrigation, and erosion control and that many soil management practices (building blocks) must be properly dove-tailed to make a satisfactory soil management structure" (see Fig. 16-9). In other words, the practices leading to satisfactory crop production in Florida are quite similar to those we have recommended elsewhere, even though at first thought Florida agriculture seems almost subtropical.

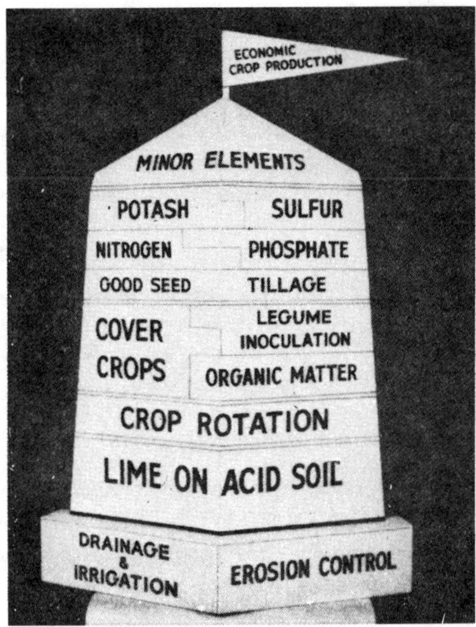

FIGURE 16-9 Economic crop production is based on good soil management practices. The foundation is drainage, irrigation, and erosion control. The building blocks vary with soil series, and they fit together differently in other climatic regions. (Courtesy Florida Agricultural Experiment Station.)

CHARACTERISTICS OF THE SOILS

The central ridge section of Florida is sandy, moderately high, well-drained, and medium acid. Because of the danger of frost, citrus, avocados, mangoes, and winter vegetables are mostly confined to the higher, well-drained soils of the central and southern parts of the state. Field crops, tung, and pecans are grown in northern Florida. The lower, poorly drained Flatwoods soils are mostly used for pasture and certain acid-loving crops such as potatoes, strawberries, and vegetables. Lime may be needed but should be applied according to test results to avoid overliming.

Florida soils are low in nutrients. Successful production requires careful application of fertilizers, the maintenance of organic matter, and irrigation of high-value crops. Irrigation may be valuable for frost prevention in low areas. Air drainage should be considered when lands are being planned for production of cold-sensitive crops. During the freeze of February 1981, citrus trees in pocket areas of central Florida were badly damaged while on the higher parts of the fields damage was slight or nil.

THE NORTHEASTERN STATES

The Northeastern states are broadly divided into mountains, plateaus, ridges and valleys, coastal plains, and lake beds. In terms of years of farming, it is the oldest area in the United States. Slopes are gentle to steep, soil textures vary from fine to coarse with loams predominating, and drainage varies from poor to excessive.

The climate of the Northeast is favorable for grassland farming and dairying, a type of farming favorable to an area where farms are small and hilly. The soil characteristics are favorable to dairying, and good markets for dairy products are close at hand.

GRASSLAND FARMING AND LAND USE

For many years, decades even, soil and crop research has shown that grasses and legumes, separately or in mixtures, are soil-building in effect on the soil. Some of the earlier and perhaps most outstanding work on soil-building versus soil-depleting crops was conducted in Ohio and was reported in 1928 by Salter and co-workers (see Chapter 6). Extensive crop-sequence (farming systems) experiments run in Michigan from 1940 to 1970 confirmed the Ohio results regarding the beneficial effects of grasses and legumes in erosion control, soil-structure maintenance, moisture-holding capacity, and nitrogen relationships.

Much of the soil in the Northeast is too steep to be cultivated more than 25 percent of the time. Some should remain in grass continuously. The humid climate is especially good for forage crops, and sheep make it possible to market them profitably (see Fig. 16-10). Intermediate slopes may successfully produce hay and pasture 75 percent of the time and grain during the other one year out of four. Thus the fields can be seeded back to the forage crop. Special erosion-control practices should be followed during the non-grass year.

Where a farm includes an area of level soil, the farmer may wish to grow a cultivated cash crop (potatoes are adapted to that climate) or corn for dairy feed. The corn can be grown year after year on the same land, as was suggested in Chapter 6 on fitting crops to the soils.

Dr. Ford S. Prince authored a very fine book entitled *Grassland Farming in the Humid Northeast* (1956). Most of the soil in the Northeast region is acid. Lime is generally needed for legumes and will be valuable for its effect in keeping phosphorus more available to growing plants.

Fertilizers are essential and should be applied according to need as indicated by soil tests, the crop to be grown, and the expected yield. Suggestions given for the Northcentral states, especially Wisconsin, where much grass and legume farming is practiced, may be studied. It is well to contact the Cooperative Extension Service and/or the U.S. Soil Conservation Service in your own state.

Taxonomic descriptions of the soil series mentioned in Chapter 16 are given in Table 16-3.

FIGURE 16-10 On very hilly farms, grass for pasture and hay may be the only crop grown. Sheep furnish an ideal means of selling grass from such farms. These young people say there are hidden pleasures in grassland farming. The soil conservationist agrees, but for slightly different reasons. (Photograph by 4-H Club Program.)

TABLE 16-3 Classification of Soil Series Mentioned in Chapter 16

	Family and Subgroup
Abilene	Fine, mixed, thermic Pachic Arguistolls
Amarillo	Fine-loamy, mixed, thermic Aridic Paleustalfs
Colby	Fine-silty, mixed (calcareous), mesic Ustic Torriorthents
Enterprise	Coarse-silty, mixed, thermic Typic Ustoshrepts
Gila	Coarse-loamy, mixed (calcareous) thermic Typic Torrifuvents
Miles	Fine-loamy, mixed, thermic Udic Paleustalfs
Pratt	Sandy, mixed, thermic Psammentic Haplustalfs
Pullman	Fine, mixed, thermic Torrertic Paleustolls
Richfield	Fine, montmorillonitic, mesic Aridic Argiustolls
Roscoe	Fine, montmorillonitic, thermic Typic Pellusterts
Tillman	Fine, mixed, thermic Typic Paleustolls
Tripp	Coarse-silty, mixed mesic Typic Haplaquolls
Vernon	Fine, mixed, thermic Typic Ustochrepts

REFERENCES

Appleton, L. R. (1949). *Conservation Practices for the Wheat Lands of New Mexico*. New Mexico Extension Service Circular 221.

Bonnen C. A., and B. H. Thibodeaux (1937). *A Description of the Agriculture and Type-of-Farming Areas in Texas*. Texas Agricultural Experiment Station Bulletin 544.

Bouyoucos, G. J., and A. H. Mick (1947). Improvements in the plaster of paris absortion block electrical resistance method for measuring soil moisture under field conditions. *Soil Sci.* 63:455–465.

Brengle, K. G. (1982). *Principles and Practices of Dryland Farming*. Colorado Associated University Press, Boulder.

Chang, C. W., and H. E. Dregne (1955). *Reclamation of Salt and Sodium Affected Soils in the Mesilla Valley*. New Mexico Agricultural Experiment Station Bulletin 401.

Chepil, W. S. (1950). Properties of soil which influence wind erosion I. The governing principle of surface roughness. *Soil Sci.* 69:149–162.

Chepil, W. S. (1953a). Factors that influence clod structure and erodibility of soil by wind I. Soil texture. *Soil Sci.* 75:473–483.

Chepil, W. S. (1953b). Water stable structure. *Soil Sci.* 76:389–399.

Chepil, W. S. (1954). Calcium carbonate and decomposed organic matter. *Soil Sci.* 77:473–480.

Chepil, W. S. (1955a). Sand, silt and clay. *Soil Sci.* 80:155–162.

Chepil, W. S. (1955b). Organic matter at various stages of decomposition. *Soil Sci.* 80:413–421.

Chepil, W. S., and C. L. Englehorn (1951). Report on Causes and Effects of Wind Erosion in East-Central Kansas in March, 1959. Agronomy Department, Kansas State College. Manhattan, Kansas.

Chepil, W. S., and N. P. Woodruff (1955). *How to Reduce Dust Storms*. Kansas Agricultural Experiment Station Bulletin, 318.

Cockerill, P. W. (1956). *Recent Trends in New Mexico Agriculture*. New Mexico Agricultural Experiment Station Bulletin 406.

Compton, L. L. (1943). *Relationship of Moisture to Wheat Yields on Western Kansas Farms*. Kansas Extension Circular 168.

Duley, F. L. (1948). *Stubble-Mulch Farming to Hold Soil and Water*. USDA Farmers' Bulletin 1997.

Fuller, W. H. (1956). *Soil Organic Matter*. Arizona University Agricultural Experiment Station Bulletin 240.

Humphrey, R. R. (1955). *Arizona Range Resources II, Yavapai County*, Arizona University Agricultural Experiment Station Bulletin 229.

Jones, O. R. (1975). Yields and water-use efficiencies of dryland winter wheat and grain sorghum production systems in the Southern High Plains. *Soil Sci. Soc. Am. Proc.* 39:98–103.

Kearney, T. H., and H. L. Shantz (1911). The water economy of dry-land crops. *USDA Yearbook* 10:351–362.

Kellogg, C. E. (1941). *The Soils that Support Us*. Macmillan, New York.

Mazurk, A. P., A. W. Zingg, and W. S. Chepil (1953). Effect of 39 years of cropping practices on wind erodibility and related properties of an irrigated Chestnut soil. *Soil Sci. Soc. Am. Proc.* 17:181–185.

Middleton, J. E. (1953). *Water Management*. Arizona University Agricultural Experiment Station Circular 205.

Myers, H. E. (1953). Romance of the Plains. *Agr. J.* 45:521–526.

National Academy of Sciences (1974). *More Water for Arid Lands*. Library of Congress Catalogue Number 74–10058.

Norman, A. G. (1943). *Organic Matter in Iowa Soils*. Iowa Agricultural Experiment Station—Agricultural Extension Service Bulletin.

Prince, F. S. (1956). *Grassland Farming in the Humid Northeast*. Van Nostrand, Princeton, N.J.

Rhoades, H. F. (1955–1956). *Commercial Fertilizers*. State of Nebraska Annual Report of the Department of Agriculture and Inspection, September 30, 1955, to October 1, 1956.

Richards, L. A., and W. Gardner (1936). Tensiometer for measuring the capillary tension of soil water. *J. Am. Soc. Agron.* 28:352–358.

Richards, L. A., L. E. Allison, L. Bernstein, et al. (1954). *Diagnosis and Improvement of Saline and Alkali Soils*. Arizona Handbook 60., U.S. Department of Agriculture, Washington, D.C.

Schwalen, H. E., K. R. Frost, and W. W. Hinz (1953). *Sprinkler Irrigation*. Arizona University Agricultural Experiment Station Bulletin 250.

Stallings, J. H. (1957). *Soil Conservation*. Prentice–Hall, Englewood Cliffs, N.J.

Swanson, N. P., and E. L. Thaxton, Jr. (1957). *Grain Sorghum Irrigation on the High Plains*. Texas Agricultural Experiment Station Bulletin 846.

Thornton, G. D., W. W. McCall, R. E. Caldwell, and F. B. Smith (1959). *Soils and Fertilizer*. State of Florida Department of Agriculture, Bulletin 137.

Webb, W. P. (1931). *The Great Plains*. Ginn, Boston.

Whitney, D. A. (1976). *Soil Test Interpretations and Fertilizer Recommendations*. Cooperative Extension Service, Kansas State University Bulletin C-509.

Zingg, A. W. (1950). Evaluation of the erodibility of field surfaces with a portable wind tunnel. *Soil Sci. Soc. Am. Proc.* 15:11–17.

Zingg, A. W., and C. J. Whitfield (1957). *Stubble-mulch Farming in the Western States*. USDA Technical Bulletin 1166.

CHAPTER 17

Soil Management for Gardens, Greenhouses, and Lawns

Loams high in organic matter, with pH ranging between 6.0 and 6.5, make ideal garden soils. The topsoil, that portion colored by decomposing organic matter, should be deep and underlaid by several feet of well-drained, permeable loam, silt loam, or clay loam. This "utopia" condition, however, is seldom found in the place where we would like to have a garden. In fact, the best garden soils are generally less desirable locations for homes. As a result, most prospective gardeners find themselves with anything but the soil condition suggested as ideal. Fortunately, most soil material can be improved or it may be picked up and traded for something better.

Greenhouse soil problems are not different in principle than are those encountered by gardeners or farmers. However, they must be dealt with in a somewhat different manner. The soil that produces the best corn in the field will also produce the best chrysanthemums on a greenhouse bench. Variations in intensity of crop production per unit volume of soil create different problems that must be solved. A small amount of soil in the greenhouse usually supports a tremendous amount of vegetation. Watering is usually with water high in calcium, perhaps also magnesium, so it is well to start with soil more strongly acid, say 5.5 to 6.0. The greenhouse manager usually has more choice of soil than does the gardner or lawn owner.

Greenhouse soils should be as high in percentage of clay as may be in keeping with the system of management to be followed. The method of watering to be used has some bearing on the most desirable amount of clay. It is essential that soils be highly permeable so that water intake may be rapid. This is especially desirable where water is applied to the soil surface through a hose. Rapid water intake is the result of a high percentage of coarse particles or a high degree of water-stable aggregation.

Total water-holding capacity increases with increasing percentages of clay and organic matter and with increasing degree of stable aggregation. In natural soils the latter characteristic is largely a function of organic matter, so the two usually increase together.

Sandy loams to clay loams should usually be selected. They should be free of weed seeds, harmful insects, and disease organisms. Clay loams high in silt and low in sand may be improved by addition of coarse sand. Clay loams and also loams may generally be improved by additions of fibrous, acid peat, especially if there is a need for lowering the pH.

If you find it difficult in your area to find a natural soil with suitable characteristics, you may wish to use a mixture of soil materials and acid peat as is recommended by some floriculture researchers. In most of the colleges of agriculture in our state universities and in the departments of agriculture in other countries, information on soil mixes is available. Soil survey maps can be used to locate suitable natural soils.

Most new homes are surrounded by subsoil materials quite unsuited for the establishment and maintenance of a good lawn. Steep slopes and terraces are difficult to maintain and mow, but in some cases, they cannot be avoided. Since erosion will always be a problem, retaining walls may be better.

Where topsoil is to be brought in, one must avoid abrupt texture changes in the profile. Otherwise, water percolation and capillary water movement may be inhibited. Turn back to Chapter 3 for a discussion of this problem. Perhaps it would be well at this time to review the article "How Water Moves in the Soil." See the Gardner (1979) reference of Chapter 3.

GARDENS

The gardener may be confronted with some of these problems—a buried rubbish heap or cement driveway, too much clay subsoil, too much sand, or just poor drainage of an otherwise good soil. After remedying such hazards he or she is ready to make a garden.

Organic Matter

The most important characteristic of a garden or greenhouse soil is that it be well supplied with organic matter. In fact, all *good* gardeners are "organic gardeners." Many, however, who call themselves by that name carry the philosophy too far. Unless they can bring in large quantities of animal manures and/or compost, they must also apply mineral nutrients from the fertilizer bag. See the discussion on organic farming in Chapter 22.

Soil Management Principle

Soils should never be worked or tramped when they are too moist.

Soil Water

When soil sticks to shoes, spade, or wheels, it is too wet. After being pressed firmly in the hand, it should crumble readily when the pressure is released. Then the gardener can go to work. The same test can be used by the lawn builder. Too much pressure on soil when wet can do long-lasting damage, even in winter when grass is not growing. The daily walk of the paper deliverer or mail carrier is injurious. It is not the damage to the grass, but the compaction of the soil that does the damage.

Careful attention should be paid to irrigation, either on the lawn or in the garden. If you like to be scientific about moisture control, try the Bouyoucos gypsum blocks and moisture meter, explained and pictured in Chapter 3.

Weed Control

Tillage and hand weeding is sufficient for weed control in the garden unless grasses that spread by stolons are present. Such grasses may be easily eradicated by spraying with "round-up" or another nonselective herbicide. Spray when the grass is 6 to 8 in. tall and is actively growing, then wait for a week or two before tilling the soil. Planting may soon follow.

The broadleaf weeds in lawns may be controlled by applying herbicides according to directions. Be careful about spray drift to sensitive plants. The herbicide 2-4-D is the one most commonly recommended.

Soil Testing

Soil tests make it possible to control plant nutrition in garden, greenhouse, or lawn. Much growth is expected in short periods of time, so plants must be well fed. Care must be exercised though to avoid excessive applications. This means frequent soil tests are advisable, perhaps three or four during the season in the garden and still more often in the greenhouse. Lawn soils should be tested at least once a year in the North and more often in the South where growth continues throughout the year.

Lime or Sulfur May be Needed

Soil-reaction control is important and not particularly difficult. A pH or lime requirement test is the first step (this subject is discussed in detail in Chapter 13). Garden crops vary in lime requirement, but a pH range of 5.5 to 7.0 is favorable and/or tolerable to a large number of the more common plants around a home. The desirable range for strawberries and potatoes is 5.0 to 6.5 and blueberries have a still lower preference range, between 4.0 and 5.0.

Azalea and hydrangea are examples of ornamental plants that grow well only in strongly acid soils, pH 4.5 to 5.5. The homeowner interested in these plants should start with a soil in this range, mulch with acid materials (peat or decomposed sawdust), and fertilize with ammonium sulfate as a source of nitrogen. The importance of this precaution varies directly with the base content of the irrigation water.

Liming Materials

Ground limestone is the safest liming material to use around the home grounds. The material is slowly soluble, so there is little danger of injury from excesses caused by uneven distribution. Dolomitic limestone furnishes magnesium as well as calcium. Refer to Table 17-1 for amounts (pounds or grams per 100 ft^2 or 9.3 m^2) to apply on different textured soils. These amounts are for fine material (all should pass a 10-mesh sieve with 50 percent fine enough to pass through a 100-mesh sieve) worked into the top 7 in. (17 cm) of soil.

Subsoils may have a pH quite different from the surface. If it is considerably lower, larger amounts of lime should be used and mixing should be to a greater depth. If soil organic matter is very high (except for muck, of course) increase the rates suggested in Table 17-1 by one-fourth, and if organic matter is very low, reduce the rates by a similar amount.

Sulfur

Most garden and ornamental plants do best when soil pH is below 7.0, even below 6.5. As already mentioned, medium to strong acidity is essential for several of our most common home garden plants. The desired reaction can be obtained in several ways: (1) complete soil replacement to the depth of most of the roots; (2) partial replacement by mixing acid peat with a portion of the original soil; or (3) treatment with sulfur. It is essential that sulfur-treated soils be well drained in order that sulfates formed from oxidation of the sulfur may leach away. In some instances, a combination of peat and sulfur is desirable. Mulching with acid peat will maintain a low pH after sulfur is used to obtain the initially desirable reaction.

Table 17-2 provides suggestions regarding the amounts (pints/100 ft^2 or 9.3 m^2) of sulfur to apply to lower soil pH from a specific original level to a level desired for a certain crop. The data are for sand and loam soils. If your soil is sandy loam, use an amount between the amounts recommended for sand or loam. If soil organic matter is very high, increase the suggested rates by 50 percent. Likewise, reduce the suggested rates if organic matter is very low.

Alkaline clayey soils, or any soils containing free calcium carbonate are not suitable for acid-loving plants. Remove them and substitute those having desirable texture and content of lime. Free calcium carbonate may be detected by using hydrochloric acid or vinegar. Bubbles of carbon dioxide indicate free lime. Ask for this test when you send your soil to the laboratory for a pH and/or plant-food test.

Mix sulfur thoroughly with soil sometime in advance of planting, several weeks if possible. If time permits, have the soil tested again to see that the desired pH has been reached. A second application may be necessary.

Fertilizer to Use

Adequate soil nutrients are essential for a successful garden or for successful plant production, flowers or vegetables, in the greenhouse. It is easy, however, to cause injury by excessive use of soluble fertilizers. Soil tests, though,

TABLE 17-1 Suggested Applications of Finely Ground Limestone to Raise the pH of a 7-in. Layer of Several Textural Classes of Acid Soils

	Finely ground limestone per 100 ft^2							
	pH 4.5 to 5.5				pH 5.5 to 6.5			
	Northern and central states		Southern coastal states		Northern and central states		Southern coastal states	
Textural class	(Pounds)	(Pints)	(Pounds)	(Pints)	(Pounds)	(Pints)	(Pounds)	(Pints)
Sands and loamy sands	2.5	1.9	1.5	1.2	3.0	2.3	2.0	1.5
Sandy loams	4.5	3.5	2.5	1.9	5.5	4.2	3.5	2.7
Loams	6.0	4.6	4.0	3.1	8.5	6.5	5.0	3.8
Silt loams	8.0	6.1	6.0	4.6	10.5	8.1	7.5	5.8
Clay loams	10.0	7.7	8.0	6.1	12.0	9.2	10.0	7.7
Muck[a]	20.0	15.4	17.5	13.5				

Source: C. E. Kellogg, Home gardens and lawns. In *Soil—the 1957 Yearbook of Agriculture,* U.S. Department of Agriculture, Washington, D.C., 1957, p. 678.

[a]Muck soils with pH above 5.5 do not require lime for garden crops provided they are well supplied with calcium and magnesium.

TABLE 17-2 Suggested Applications of Ordinary Powdered Sulfur to Reduce the pH of an 8-in. Layer of Sand or Loam Soil from an Original pH to a Desired Level

| | Powdered sulfur, pints/100 ft² | | | | | | | | | |
| | 4.5 | | 5.0 | | 5.5. | | 6.0 | | 6.5 | |
Original pH	Sand	Loam	Sand	Loam	Sand	Loam	Sand	Loam	Sand	Loam
5.0	$\frac{2}{3}$	2								
5.5	$1\frac{1}{3}$	4	$\frac{2}{3}$	2						
6.0	2	$5\frac{1}{2}$	$1\frac{1}{3}$	4	$\frac{2}{3}$	2				
6.5	$2\frac{1}{2}$	8	2	$5\frac{1}{2}$	$1\frac{1}{3}$	4	$\frac{2}{3}$	2		
7.0	3	10	$2\frac{1}{2}$	8	2	$5\frac{1}{2}$	$1\frac{1}{3}$	4	$\frac{2}{3}$	2

Source: C. E. Kellogg, Home gardens and lawns. In *Soil — the 1957 Yearbook of Agriculture,* U.S. Department of Agriculture, Washington, D.C., 1957, p. 678.

are a safe guide to successful nutrient control. Turn to Chapter 10 for a complete discussion of soil testing as a diagnostic tool. As mentioned in those pages, intensive cropping with the use of rather large amounts of fertilizers increases the need for frequent testing. We are so anxious to obtain the greatest yields possible on the small areas involved that we are inclined to use too much fertilizer. Thus soil testing becomes a means of avoiding excesses and of keeping nutrients in balance. All such soils should be tested each year before they are treated, then once or twice during the season.

Method of Application

Broadly speaking, two methods of fertilizer application are available to the grower of garden vegetables. These are broadcast and mixed in as compared to localized or band applications along the row. Usually, both methods are advised. In an intensive system, regulated by soil tests, one-half to two-thirds of the fertilizer should be distributed before plowing or spading unless soil tests show that nutrient levels are high and in balance. Then the broadcast application may be omitted.

Localized or band applications along rows or around plants are called starter applications when used at planting time, or supplementary applications when made later in the season as plants are growing. In fact, the supplementary (also called sidedressing) application is very important for crops such as tomatoes. They need so much fertilizer for top production that it may actually be dangerous to apply enough at planting time to furnish the needs of the crop through the season.

Frequent irrigations on sandy soil, with the natural rainfall coming intermittently, leaches soluble nutrients from the root zone. This may occur even in arid regions. If it does occur, you should be ready to apply additional nitrogen and sometimes potash. Extra soil tests during the growing season will tell you just when these extra applications should be made or you may tell by the appearance of the plants.

Soluble nutrients may be applied through irrigation water. Certain attachments for the hose make this possible. Much care is necessary to arrive at proper and accurate applications, and the nutrients must, of course, be soluble or be purchased in liquid form.

Localized Applications—Starter Solutions

Fertilizers applied locally should be close to but not touching seeds or plants. This is to avoid burning and dessication. Seeds in contact with fertilizer germinate slowly, and living plants may be killed with soluble fertilizer. Place the fertilizer for best results about 1 to 2 in. (2 – 1/2 to 5 cm) to the side and 2 in. (5 cm) below the seeds. The same distance should separate fertilizer from the roots of transplants. Certain garden seeders place fertilizer in this manner, or it may be done by hand. If distribution is being done by hand, an inverted flower pot makes a good tool for making a trench for fertilizer. Then after the fertilizer is covered, a transplant may be placed in the center of the circle.

Fertilizers used as starters should be soluble or should actually be in solution. Starter solutions may be purchased or prepared from soluble solid fertilizers. Add 1 to 1 – 1/2 oz (28 to 52 g) of solid high-phosphorus fertilizer to 1 gal (3.5 l) of water. Apply 1/2 to 1 pint (250 to 500 cc) of this solution around the roots of a transplant or along 10 ft (3 m) of row. In small home gardens, the use of starter fertilizers in this way give very satisfactory results, especially in cool seasons. It is a good idea to apply starter solution to plants in flats 2 or 3 days before transplanting. This lessens the shock of transplanting and aids in quick establishment of new roots.

Liberal use of soluble fertilizers, but under the control made possible by soil testing, makes it possible to grow tremendous crops of vegetables from small areas. The nine tomato plants, set 18 in. (1/2 m) apart, shown in Fig. 17-1 produced 4 – 1/2 bu (12.8 hectoliters) of very high quality fruit. The ornamental effect was pleasing throughout the season.

Soil Test Results and Fertilizer Rates

Your soil test results may be in absolute terms such as pounds (kilograms) of available nutrients per acre (hectare) or in parts per million parts of soil. On the other hand, it may be in relative terms such as low, medium, or high. This depends on the test method being used and the policies followed in the laboratory. Expression in absolute terms is usually accompanied by a statement as to the **range** for "low," "medium," and "high."

Use of these relative terms makes it possible to adapt results from various testing methods to general fertilizer recommendations. Table 17-3 suggests amounts and grades of fertilizers per 100 ft² (9.3 m²), per 10 ft (3 m) of 24-in.

FIGURE 17-1 These nine tomato plants, set 18 in. apart produced 4 1/2 bu of high-quality fruit. The fruiting period was from August 1 to October 15. The vigorous dark-colored vines and red fruit were highly ornamental.

TABLE 17-3 Amounts of Preplant and/or Planting Time Fertilizer for Home Gardens. The Amount Suggested for Each Method of Application is the Total Requirement for the Season

Kind of soil and ratio and grade of fertilizer	Amounts per 100 ft², 10 ft of 24-in. row, or per plant 2′ × 2′											
	Low soil test				Medium soil test				High soil test			
	100 ft² (lb)	(pints)	10-ft row (cups)	Per plant (tbl)	100 ft² (lb)	(pints)	10-ft row (cups)	Per plant (tbl)	100 ft² (lb)	(pints)	10-ft row (cups)	Per plant (tbl)
Sandy soils 1:4:4												
5–20–20	3	3.3	$1\frac{1}{2}$	$4\frac{1}{2}$	2.0	2.2	1	3	1.0	1.1	$\frac{1}{2}$	$1\frac{1}{2}$
3–12–12	5	5.6	$2\frac{1}{2}$	7	3.3	3.7	$1\frac{2}{3}$	5	1.7	1.9	$\frac{5}{6}$	$2\frac{1}{3}$
8–32–0[a]	2	2.2	1	3	1.3	1.4	$\frac{2}{3}$	2	0.7	0.8	$\frac{1}{3}$	1
Clayey soils 1:4:2												
5–20–10	3	3.3	$1\frac{1}{2}$	$4\frac{1}{2}$	2.0	2.2	1	3	1.0	1.1	$\frac{1}{2}$	$1\frac{1}{2}$
8–32–0[a]	2	2.2	1	3	1.3	1.4	$\frac{2}{3}$	2	0.7	0.8	$\frac{1}{3}$	1
Organic soils 1:2:4												

5–10–20	3	3.3	$1\frac{1}{2}$	$4\frac{1}{2}$	2.0	2.2	1	3	1.0	1.1	$\frac{1}{2}$	$1\frac{1}{2}$
Arid soils												
1:4:0												
8–32–0	2	2.2	1	3	1.3	1.4	$\frac{2}{3}$	2	0.7	0.8	$\frac{1}{3}$	1
All soils[b]		Cups	Tbl	Tsp		Cups	Tbl	Tsp		Cups	Tbl	Tsp
Ammonium sulfate	0.5	1.4	$4\frac{1}{2}$	$2\frac{1}{2}$	0.33	1.0	3	$1\frac{3}{4}$	0.33	1.0	3	$1\frac{3}{4}$
Sodium nitrate	0.6	1.0	$3\frac{1}{4}$	2	0.35	0.7	2	$1\frac{1}{3}$	0.35	0.7	2	$1\frac{1}{3}$
Ammonium nitrate	0.3	0.8	$2\frac{1}{2}$	$1\frac{2}{3}$	0.20	0.6	$1\frac{3}{4}$	1	0.20	0.6	$1\frac{3}{4}$	1
Urea	0.2	0.6	2	$1\frac{1}{4}$	0.15	0.4	$1\frac{1}{4}$	$\frac{3}{4}$	0.15	0.4	$1\frac{1}{4}$	$\frac{3}{4}$

[a]Use where potassium test is very high and where phosphorus tests are within the limits indicated, low, medium, or high.

[b]Apply these nitrogen fertilizers after season is one-third over. Avoid wet leaves and the axils and crowns of plants. Omit these summer applications of nitrogen and reduce preplant and planting time applications where stable manure has been applied at 1 bu/100 ft². Also, repeat or omit the recommended amounts as the need may be indicated by symptoms (see Chapter 12). Nitrogen may not be needed on well-drained organic soil. Do not apply nitrogen fertilizers to bearing strawberry beds until after harvest.

If the suggested grade is not available, use another grade with the same nutrient ratio, as for instance 6–24–12 for 5–20–10, and change the quantity to obtain the amounts of N, P_2O_5, and K_2O.

row, or per plant covering a 4 ft² (3600 cm²) area. Rates are given in convenient terms of pounds, grams, cups, or spoons, for sandy, clayey, and organic soils in humid areas and for all soils in arid regions. Your soil must be tested for available plant food if you wish to make full use of the data.

GREENHOUSE

In selecting and/or preparing soil for the greenhouse, one must be sure the pH is well on the acid side because watering with hard water will soon raise it. Be sure the organic matter content is high and that soil texture is medium, preferably loam. In many locations, suitable natural soil may be available. If the price delivered is not too high, such soil is better than a manufactured mixture. Be sure the soil does not contain soluble salts, not likely in a humid area but such might be the case. A soil test will tell.

Fertilizers and Soil Amendments

In the selection of fertilizer, particular attention should be paid to the acidifying or alkalizing effect of the material. **Generally**, acidifying fertilizers are the most desirable. Nitrogen fertilizers used for acidifying purposes should be applied in accordance with the need of the soil for nitrogen, **as determined by tests**. Suggested rates of application are to be considered as standard applications for soils very low in nutrients. It is assumed that such applications are to be made after soil tests show the need.

Nitrogen Fertilizers

> ***Ammonium sulfate.*** $(NH_4)_2SO_4$. About 21 percent nitrogen. Also supplies sulfur. Strong acidifying action.

> ***Ammonium phosphate.*** 12 to 21 percent nitrogen. Also furnishes phosphorus; 53 to 61 percent P_2O_5. Acidifying action. Apply on basis of nitrogen needed.

> ***Dried blood.*** 9 to 14 percent nitrogen. Acidifying action.

> ***Urea*** $CO(NH_2)_2$. 46 percent nitrogen. Acidifying action.

> ***Ammonium nitrate.*** NH_4NO_3. 33 percent nitrogen. Acidifying action.

> ***Sodium nitrate.*** $NaNO_3$. 16 percent nitrogen. Also adds sodium. Alkalizing action.

> ***Calcium nitrate*** $Ca(NO_3)_2$. 15 percent nitrogen. Also adds calcium. Alkalizing action.

Boron, Manganese, Iron, Zinc, Copper, and Molybdenum

Any of these elements except molybdenum is likely to be deficient in alkaline soils, especially if they were once acid. Molybdenum is made more available by liming. A study of the deficiency symptoms discussed in Chapter 12 will give you some idea as to a need for some of these trace or secondary nutrients.

Boron is likely to be needed for beets, turnips, cabbage, rutabagas, head lettuce, spinach, and cauliflower in alkaline (over pH 7.0) soils of the humid areas. Apply 1/6 to 1/3 oz (5 to 10 g) of borax per 100 ft^2 (9.3 m^2).

Manganese may be needed for beans, peas, beets, lettuce, cabbage, roses, and several flowering shrubs any place where soil pH is over 6.5. Apply 1/4 to 1/2 lb (110 to 220 g) manganese sulfate per 100 ft^2 (9.3 m^2).

Iron is widely needed as a fertilizer wherever soils are alkaline and/or where soil phosphorus is very high. Iron deficiency is especially common in trees and flowering shrubs in the arid and semiarid regions. Apply 1/2 lb (225 g) of ferrous sulfate per 100 ft^2 (9.3 m^2) or soak foliage with a solution containing 1/2 oz (14 g) ferrous sulfate per gallon (3.12 l) of water. The spray method is preferred.

Zinc, copper, and **molybdenum** may be needed, but the likelihood in any specific location is so slight that you are urged to seek local information before deciding on their use.

Phosphorus Fertilizers

Superphosphate. 18 to 20 percent P_2O_5. Supplies calcium but not in a form to raise pH. Contains calcium sulfate ($CaSO_4$), which supplies sulfur. May be considered as slightly acidifying because the calcium in the water may combine with the monocalcium phosphate to form dicalcium and tricalcium phosphates while releasing hydrogen ions.

Triple-superphosphate. 40 to 45 percent P_2O_5. Like superphosphate except does not contain calcium sulfate, so does not add sulfur. It is acidifying, particularly if banded.

Phosphoric acid (H_3PO_4). Used for conditioning water. The PO_4 ions react with the Ca_{2+} ions in the water to form relatively insoluble $Ca_3(PO_4)_2$. Thus the calcium ions become ineffective as far as pH is concerned. Plants are able to make use of the phosphorus. Soil tests will show when phosphorus levels are becoming dangerous.

Bonemeal. (Steamed) 20 to 25 percent P_2O_5. 1.6 percent to 2.5 percent nitrogen. Also supplies calcium. Alkalizing action.

Potassium Fertilizers

Potassium sulfate (K_2SO_4). 48 percent K_2O. Also adds sulfur. Neutral reaction.

Potassium chloride (KC1). 50 to 60 percent K_2O. Also adds chloride, which may not be desirable. Neutral reaction.

Potassium nitrate (KNO_3). About 46 percent K_2O and 14 percent nitrogen. Neutral reaction.

Manganese Fertilizer

Manganese sulfate ($MnSO_4$). Also adds sulfur. Acidifying action.

Iron Fertilizer

Ferrous sulfate ($FeSO_4$). Also adds sulfur. Acidifying action used to lower pH.

Magnesium Fertilizer

Magnesium sulfate ($MgSO_4$). Adds sulfur also. Neutral reaction.

Boron Fertilizer

Borax ($Na_2B_4O_7$).

Copper Fertilizer

Copper sulfate ($CuSO_4$). Adds copper and sulfur.

Acidifying Materials not Considered Fertilizers.

Aluminum sulfate ($Al_2(SO_4)_3$). Also adds sulfur.

Sulfur (S). Very strong acidifying action.

Sulfuric acid (H_2SO_4). May be used to neutralize alkaline water. Adds sulfur. Care must be taken to avoid excess of sulfur.

Alkalizing Materials not Considered Fertilizers

Limestone ($CaCO_3$). Should be finely ground. Use only if soil test shows the soil pH to be too low for the plant to be grown. Rate of application varies with buffer capacity of soil.

Dolomitic limestone ($CaCO_3 + MgCO_3$). May be substituted for limestone if it is known that magnesium is needed.

Hydrated lime ($Ca(OH)_2$). More quickly available than limestone. Apply only three-fourths as much.

Complete Fertilizers

It is usually impossible to obtain complete fertilizers containing the correct proportion of plant food for greenhouse soils. Furthermore, the possibility of using fertilizers to control acidity may be slight or nil if mixed fertilizers are used since most fertilizer mixtures have been treated to make them almost neutral. A fertilizer analysis indicates percentage of total nitrogen, (N), percentage of available phosphoric acid (P_2O_5), and percentage of water-soluble potash (K_2O).

Some Aids in the Calculation of Quantities of Fertilizer for Certain Rates of Application

1 oz contains 28.4 g.

1 lb contains 454 g.

1 acre contains 43,560 ft².

1 g/ft² equals approximately 100 lb/acre.

1/4 lb/100 ft² equals approximately 100 lb/acre.

0.2 g/gal of soil equals approximately 100 lb/acre.

1.6 g/bu of soil equals approximately 100 lb/acre.

1 oz/bu of soil equals approximately 1700 lb/acre.

Flower Pot Size Versus Quantity of Fertilizer

Flower Pot Size versus Quantity of Fertilizer

Diameter of pot (inches) 1 in. = 2 1/2 cm	Area of pot (ft²)[a] 1 ft² = 900 cm²	Grams fertilizer for several rates application per acre 1 acre = 0.4 ha			
		(100)	(300)	(600)	(1000)
3	0.049	0.049	0.15	0.29	0.49
4	0,087	0.087	0.26	0.52	0.87
5	0.136	0.136	0.41	0.82	1.36
6	0.196	0.196	0.59	1.18	1.96
7	0.267	0.267	0.80	1.60	2.67
8	0.349	0.349	1.05	2.09	3.49

[a]Calculations based on surface area of pot, not on weight of soil.

Leach Only when Necessary

Leaching is often necessary but should be avoided as a general practice. Excess water destroys soil structure and is wasteful of plant nutrients. Careful management may make it possible to leach quite infrequently or not at all. A few "don'ts" and "do's" may serve to illustrate ways of managing soil without excessive leaching.

1. **Don't** fill your benches with alkaline soil, particularly soil containing free carbonates, because it will then be necessary to use correctives that leave harmful residues. Sulfur, for instance, oxidizes to form sulfates.

2. **Don't** fill your benches with soil containing soluble salts. This may be hard to avoid in arid regions.

3. **Don't** make compost from organic materials that have been contaminated with salt. Leaves from streets or manures where salts were used as stable disinfectants are examples.

4. **Don't** use too much easily decomposed compost or nitrates may develop to toxic levels.

5. **Do** use a soil relatively high in clay and high in organic matter. Such soil will absorb a lot of residues before leaching is necessary. **Do** use acid peat for additional organic matter.

6. **Do** test soils frequently. Include tests for soluble salts. Avoid using fertilizers that supply more of the residue that is already high.

7. **Do** look for a salt-free irrigation water.

8. **Do** change soil frequently. Careful planning makes this practical. Sell it with potted plants.

9. **Do** something at once if an injurious salt level is detected. Delay may be costly.

LAWN PREPARATION AND SEEDING

Work lime (if needed) and fertilizer well into the top few inches of soil. Then alternately rake and roll until the surface is perfectly smooth and firm enough to bear up your weight without making depressions. This is necessary or during seeding and watering your feet will leave the surface rough and pitted.

Finally, rake very lightly, no more than to a depth of 1/4 in., and scatter the seed. Again rake very lightly (1/4 in. deep) and roll. If you plan to spread the seed by hand, you may get a more even spread if you mix the required amount of seed for a relatively small marked off area, perhaps 100 ft² (9.3 m²) with some bulky substance such as cornmeal or slightly moist sandy loam soil. The moist soil works very well but should be screened so that the seed and soil may mix thoroughly. Try to select a seeding time when there is little wind. Regulate rate of scatter so you can cover each small area several times. A uniform job of seeding is essential.

A light organic mulch as peat or straw ensures better germination and emergence. Avoid any material that may attract birds. Coarse burlap may be used on very steep slopes or in channels where there may be a flow of surface water. Scatter the seed and spread the burlap immediately. Pin the burlap to the soil with small wooden pegs. Wire may be used, but it must be removed, whereas the pegs will eventually rot. The young grass seedlings will penetrate the burlap readily. Do not remove the burlap. Cheese cloth may also be used, but coarsely woven burlap is preferred.

An alternative method of seeding is to cover the entire surface with grass mat, a mulch material containing the seeds imbedded in its fibers. Figure 17-2 shows a grass mat being laid on a home lawn. In this instance, the grass germinated and grew in an excellent manner (see Fig. 17-3). In fact, emergence was much faster where the mat was used than on an adjoining area seeded in the old-fashioned way. Late summer is the best time to seed a lawn in most parts of the United States. The grass is starting then during the cooler fall months, a favorable time for most grasses. Lawns started in the spring must be frequently watered throughout the summer. Zoysia grasses usually

FIGURE 17-2 Grass mat, a fibrous material containing embedded seeds, is being laid for a home lawn. Notice it is being wetted as fast as it is unrolled. This prevents trouble from blowing. Once the material is thoroughly wet, it sticks firmly to the soil. Germination is hastened by this cover, and erosion is lessened.

FIGURE 17-3 This excellent bluegrass-fescue lawn (foreground) developed between September 7, 1959, and June 5, 1960. Cover was almost complete. Seeding with grass mat is shown in Fig. 17-2. A Michigan location.

do better when started in spring or early summer; however, fall is satisfactory in the deep South. Sodding results in rapid lawn building. It is expensive but sometimes desirable, especially on slopes. Be sure to purchase sod grown from a desirable species of lawn grass, *not* from an old pasture. Sodding may be done at any time during the growing season, except late in the fall.

IRRIGATION

Watering of a new lawn is very important. The seeding should be done at a time when the entire soil profile is well supplied with moisture. The surface, of course, must not be wet enough to puddle during the seeding operations. If soil sticks to implements or shoes, wait for further drying.

Sprinkle lightly immediately after seed is raked in or at once after burlap, grass mat, or mulch is spread. Repeat frequently enough to prevent the surface from drying. The germinating seeds and young seedlings should be continuously moist until the new grass is well on its way. During hot, sunny days this will require at least two, perhaps three waterings each day. Mulch materials help to prevent evaporation so that watering may be less frequent. Set sprinklers to throw a fine spray at a rate that all water may penetrate the soil. Runoff will cause erosion and move the seed. Grass mat or mulch materials lessen danger of erosion.

Fertilizers

It is essential that lawn grasses be well fed. Otherwise, weeds will take over. It is well to add a complete fertilizer, according to soil-test results, in the spring. A herbicide may be applied at the same time. Later in the season, perhaps twice, nitrogen alone should be applied.

Lawn soils, depending on the degree of hardness of the irrigation water, are likely to become too alkaline. The annual soil tests will tell if this is the case. The tendency of the various fertilizers to prevent an increase in pH is indicated in their descriptions given in the preceding pages.

REFERENCE

Baker, K. F., et al. (1937). *The U.S. System for Producing Healthy Container-Grown Plants*, California Agricultural Experiment Station and Extension Service Manual 23.

Shifting Cultivation

We know that soils are formed largely by the many natural, geologic forces called **weathering**. Mineral expansion and contraction, chemical reactions, the activity of plant roots, and the forces brought about by soil organisms play major roles in nature's weathering process.

Soil Management Principle

Shifting cultivation allows a soil to rebuild itself.

Shifting cultivation is simply a way of allowing nature time to do what she is so well qualified to do—rebuild the soil that a period of cropping has partially depleted. Revegetation is fast in the tropics. That is one reason why nature has been depended on to rejuvenate the soil. Figure 18-1 shows regrowth of a woody stub 1 month after cutting and burning. The eroded condition of the soil is indicated by the almost complete coverage with pebbles. Depending on the characteristics of the soil, the climate, and the kind of vegetation involved, the length of the rebuilding period, called **fallow period**, is usually 5 to 10 times, sometimes 20 times, as long as the cropping period. Soil-building processes are slow, and the shifting-cultivation farmer does not help by furnishing inputs as one does in modern methods of soil management.

Shifting cultivation is in English called bush-fallow, grass-fallow, slash and burn, and natural fallow. By trial and error, the number of cropping years have been so balanced with the number of fallow years that yield stability has ensued. A stable agriculture exists. The farmer knows from

FIGURE 18-1 In the tropics, regrowth of woody vegetation is rapid. This picture was taken just 1 month after the area was burned. Note the many small pieces of plinthite (laterite), an indication of past erosion.

experience — his own or that of family members — that if the fallow period is sufficiently long a second cropping period will produce the same crop yields as did the first period. He has learned that a shortened fallow period will result in lower yields when the area is again cropped.

In Mexico and Central America, the term for shifting cultivation is *milpa* and the farmer is a *milpero*. In Spanish, the term milpa means cornfield. It is significant that the main crop is corn (maize), a very important food throughout Central America and Mexico. Many other crops are grown with the corn, usually beans, casava, yams, and sometimes bananas and plantain.

According to Wright and co-workers (1959), the average milpa family of British Honduras (now Belize) is made up of seven members. They rent about 5 acres (2 hectares) of new brushland each year, clear it by chopping down the bushes and smaller trees during February and March. As soon as the brush is dry, usually in April, they burn it, repiling as much as is necessary to get a good burn. A partial burn means poor crops (see Figure 18-2). They plant maize, coco, yams, beans, casava, and perhaps bananas, as shown in Figures 18-3 and 18-4. They can grow bananas only if they are to keep possession of the land more than 1 year. It requires 18 months from the setting of a cutting to the harvest of the first bunch of fruit.

The average farm family plans to do as much work off the farm as may be necessary to purchase the family needs they cannot grow on their rented acreage.

At the beginning of the year, the milpa farmer must locate and rent another area to be slashed and burned for another year's food production.

FIGURE 18-2 This shifting-cultivation farmer did not get a good burn. He should have waited until the wood was drier or perhaps done a better job of piling.

FIGURE 18-3 Maize and casava are commonly grown together in a shifting-cultivation system.

FIGURE 18-4 Sometimes the shifting-cultivation farmer plants many crops together—maize, yams, beans, casava, and bananas.

He knows that much depends upon his locating good land, covered with thick high brush, thus to furnish plenty of nutrient-carrying ashes to feed his crops. Contrary to the situation that exists in many countries where shifting cultivation is practiced, populated pressure is not great in Belize, so the need to discontinue the system is not pressing.

We may cite Sierra Leone as an example of a country where the practice of shifting-cultivation can no longer be afforded if the country is ever to feed itself. Land is so scarce and population increase so great, that farmers have shortened the fallow period to the extent that not enough ash is left after the burn and crop yields drop. Another reason for too little ash is that larger trees are removed and cut up for firewood to be sold for immediate cash (see Figure 18-5). Farmers of Sierra Leone *must* adopt modern methods of soil management if they expect ever to become self-sufficient in food production. Agricultural researchers are now experimenting to find out what inputs are necessary to allow the farmers to shorten the fallow period drastically or perhaps to change to continuous cropping. They cannot simply expand acreage because they are becoming short of land.

The majority of Sierra Leone respondents (81%) in a recent survey practice shifting cultivation because in no other way are they assured of satisfactory yields. By satisfactory, they mean sufficient yields to support their families on the commonly allocated 1 hectare of land. Land in Sierra Leone is owned and allocated by family groups or village officers. Currently, the West African Rice Research Institute is busy working out the practices necessary for a farmer to follow if he may successfully shorten his fallow period, in other words, resort to more years of cropping.

A great deal of land is necessary for shifting-cultivation to be a stable agricultural system. Ten years is commonly recognized as about the minimum safe length of the fallow (bush) period, although on very good soils a shorter bush period may suffice. Where 10 years is needed, only five to nine families may be accommodated per square kilometer. As population pressure reduces the area available to each family, fallow periods are shortened and some soil improvement inputs must be added. These may be dibble planting rather than broadcasting, better weed and pest control, application of fertilizers, and better seed. Research results will soon be worked out.

In any country, an increased percentage of the time in crops and better weed control means more erosion, so better control methods become essential. Contour planting, terracing, growing of tree crops, and use of cover crops are control measures. Figure 18-6 shows why erosion is such a problem in Sierra Leone.

A word about tropical grasslands (savannas). Contrary to their productivity in temperate regions, grasslands in the tropics are poor. They are not adapted to shifting cultivation. The vegetation becomes dry when it matures and has frequently been burned. Roots are shallow. Small quantities of nutrients are brought to the surface during the fallow period. After burning, heavy leaching takes place. Savanna lands are not good maize lands. See Fig. 18-7.

HISTORY OF SHIFTING CULTIVATION

Shifting cultivation is one of the oldest systems of agricultural management. It is said to have started in Northern Europe where mainly hunters inhabited

FIGURE 18-5 Sometimes the larger tree trunks and pieces of brush are removed for firewood. This is not a good practice because the ash from burning helps to bring about a stable agriculture.

FIGURE 18-6 Most of Sierra Leone's upland soil is hilly. Erosion is severe. Rows of tree crops on contour terraces could fairly well prevent soil loss. Then, with application of fertilizer, much shortened fallow periods, perhaps even continuous cropping, would be possible.

FIGURE 18-7 Savanna lands are not adapted to shifting cultivation. Grasses are shallow rooted and are dry when mature. Burning is frequent, so leaching is excessive. Thus there is little nutrient build up for the cropping year. This farmer applied fertilizer in back of where he stands, none in front where maize completely failed.

the land between 9000 and 3000 B.C. As wild food became scarce, the advent of farming created a fertility problem on cropped lands. This was initially solved by burning to clear a piece of land for 2 years of grain after which the land was allowed to recuperate by leaving it fallow (grass and trees) for 40 to 60 years. The long fallow period seemed necessary to allow plants to bring subsoil phosphorus to the surface where it became available to shallow-rooted plants. No supply of phosphorus or other nutrients were needed from off the farm. This was good because commercial fertilizers were not yet manufactured.

As settlements were developed and livestock were kept, manure became available. It was applied to small continuously cropped areas around the settlements. Phosphorus, in the form of hay and pasture, was brought in because the animals were returned to the settlements at night. "The meadow fattens the acre" was an old saying that originated at that time and has often been quoted in quite recent times. The system was stable as long as the meadow was large enough and the farmer did not export animals or crops from his land; that is, as long as subsistence farming was practiced.

By the end of the eighteenth century (1799), land reform started. Lands were parcelled out, so farmers lost access to the public pastures and, as the Napoleonic wars resulted in a possibility of export of grain and animals at high prices, lands became depleted of phosphorus. Farmyard manure was not sufficient to supply the phosphorus removed in the sale of grain and animals. This was eventually realized, and the fertilizer industry started in Denmark about 1850. The Danes started using bone but soon began importing rock phosphate from North Africa. Also, large quantities of feedstuffs were imported from overseas. By the turn of the next century, phosphorus levels in Northern European soils became positive. Since then, European countries have been strong users of commercial fertilizers and yields have been high.

In the United States, our early general farming operations, using such soil-building crops as grasses, clovers, and alfalfa rotated with small grains and row crops, were in effect a modified shifting cultivation. Think how that system varied from present-day monocropping with corn or wheat, a system made successful by all the inputs of modern soil management.

GREENLAND'S PHASES OF SHIFTING CULTIVATION

Greenland (1974), speaking before the FAO/SIDA/ARCN Seminar in Ibadan, Nigeria, in 1973 described "three phases" of shifting cultivation, with a fourth phase being continuous cultivation as practiced today in most developed countries. *Phase 1* is the simplest kind of operation, where the family dwelling is moved with the change in crop area. There is no thought of a second cropping period on the same land. Some of the criticism of the practice came as a result of the assumption that abandonment of an area occurred after the soil was destroyed for agriculture. This, of course, is not true, since failure to produce a satisfactory crop any longer is temporarily the result of depletion of readily available nutrients and/or prevalence of weeds and other pests that die off after cultivation ceases and soil rejuvenation is taking place.

Greenland's *phase 2* is where the farm dwelling stays put and only the cultivated area is shifted. In this arrangement, the family must have access to a rather large area of arable land, either through rental or ownership. In Sierra Leone, for instance, the land belongs to the village or total family unit. An immediate family — father, mother, and children — is each year assigned about 1 hectare for that year's operation. Other countries have different ownership arrangements. This phase varies from phase 1 in that after the soil is again ready for another cropping period, or when population pressure demands, the vegetation is again slashed and burned and cropping is repeated. The trouble with the system is that population increase continues to make shorter fallow periods necessary so there is not sufficient time for natural soil-building processes to bring the soil back to its original productive capacity. As a result, each cropping period sees smaller yields than were produced the time before.

Phase 3 may be a gradual development from phase 2 in which a portion of a family's land may be cropped continuously with productive capacity maintained by application of animal manures, domestic wastes, and organic composts. The manure may be from animals owned by the family, kept for meat and milk, or it may be purchased from cattle farmers who do not raise crops. Sometimes the animals owned by the family are grazed on public lands but are housed at home during the nights when, of course, manure would accumulate for use on the continuously cropped lands. The rest of a family's land area is then subjected to crop-fallow periods as in phase 2.

Greenland's *phase 4* is continuous cropping, but not necessarily continuous cultivation. Soil-building crops such as grasses and legumes may be grown, and animal manures and commercial fertilizers may be applied to maintain productivity. In other words, phase 4 is not essentially different from soil management in most developed countries.

A SHIFTING CULTIVATION AND SOIL CONSERVATION SEMINAR

There are in general but two methods of increasing world food production. First, by expanding the area of land under production and, second, by changing methods of management in such ways as to increase yields on lands already being cropped. To bring about such changes is the objective of the Food and Agriculture Organization of the United Nations. That organization (commonly referred to as FAO) has been long interested in studying ways of increasing yields of crops on shifting-cultivation lands or in determining what may be done to make continuous cropping successful on such lands. Accordingly, they were pleased to cooperate with the Swedish International Development Authority (SIDA) and the Agricultural Research Council of Nigeria (ARCN) in arranging a conference where observations and research results obtained by scientists in many countries might be presented and discussed. The work by Greenland already mentioned was presented at that 1973 conference.

Edouard Saouma, Director Land and Water Development Division, FAO, in his foreword of the shifting-cultivation report stated that 36 million km² of land representing 25 percent of the world's cultivated soils were under shifting cultivation. The figure today is probably just as great, but, it is hoped, some of the area is being improved by use of production inputs and by better selection of lands for cropping. For instance, 70% of Sierra Leone's rice land was, in 1980, still managed under bush-fallow methods — one crop year and five or more bush years. Rice is by far their most important crop. They are, therefore, attempting to improve their bush-fallow methods on the uplands and bring under production more swamplands for paddy rice.

Perhaps, worldwide, erosion is still our most serious soil-depleting force. In many countries, certainly in Sierra Leone, lands are hilly and easily eroded (see Figure 18-6). Any change to more years of cultivated crops must be accompanied by careful attention to erosion control.

SUMMARY — SHIFTING CULTIVATION

Shifting cultivation was long recognized as a stable system of agricultural land management in broad areas of Africa and Central and South America and was common in Southeast Asia and Oceania.

Hauck (1974) estimated that 36 million km² of land in the aforementioned areas is managed by this very inefficient system. Perhaps 250 million people are furnished a subsistence kind of living by such means. Unfortunately, increasing population is making it necessary for fallow periods to be shortened because available land is all allocated. As a result, productivity is gradually going down. The alternative is to apply the soil-management practices that will maintain or perhaps even improve the soil. Some of the needed information is available. Research being conducted at such institutions as the International Institute of Tropical Agriculture in Ibadan, Nigeria, and the West African Rice Research Institute at Rokupr, Sierra Leone, will, it is hoped, solve the unanswered problems.

REFERENCES

Greenland, D. J. (1974). Evolution and development of different types of shifting cultivation. In *Shifting Cultivation and Soil Conservation in Africa*, Seminar Report. FAO Soils Bulletin 24.

Hauck, F. W. (1974). Introduction to Shifting Cultivation and Soil Conservation Seminar, Ibadan, Nigeria, July 2–21.

Jones, R. A. D. (1980). *Rice Research for Self Sufficiency of Sierra Leone*. International Institute of Tropical Agriculture, Ibadan, Nigeria.

Turay H. (1980). Land Tenure Systems in Sierra Leone. A project report, Njala University College, Njala, Sierra Leone.

Wright, A. C. S., D. H. Romney, R. H. Arbuckle, and V. E. Vial (1959). *Land in British Honduras*. Report of the British Honduras Land Use Survey Team, Her Majesty's Stationery Office, London.

CHAPTER 19

Multiple-Cropping Without and With Interplanting

As the world population continues to increase, food production must likewise increase or some of us will go hungry. There are three ways, considering the world as a whole, in which world food production may be increased.

1. **Bring new land under cultivation.** This took place in the United States until about 1930. Since then, the conversion of farmland to other uses has more than offset the new lands cultivated. In much of the world (Western Europe and Asia, especially China and Taiwan), very little land is available for expanded production. China probably has less than one-quarter acre of arable land per person (as of July 1982). By the year 2000, the figure will be less than one acre of land to feed five persons. Another way of emphasizing this land scarcity situation—they must raise food for four times our population on one-third of the arable land *or* be in position (financially or politically) to import food. In fact, in fiscal 1981, China did import 7.7 million tons of wheat from the United States.[1]

2. **Increase yields of crops per unit area.** This is done through better cultural practices, timely cultural and water management, other production inputs, and sometimes through farming systems changes. In other words, by managing the soil in such a way as to cause it to produce at its maximum level, as illustrated by Fig. 1-1 (Chapter 1).

3. **Raise more than one crop in 365 days — simultaneously, in sequence, or by interplanting (*multiple cropping*).** Perhaps the most common examples of simultaneous crop production are seen in subsistence agriculture commonly called shifting cultivation or in Sierra Leone "bush fallow."

---

[1] Data from the *1982 Outlook on Foreign Agriculture*, USDA Foreign Agriculture Series, January 1982.

In Latin America, the most common cropping mixture is beans and maize. According to the CIAT, 80 percent of the beans grown in Latin America are produced on small farms and are mostly grown with maize.[2] These two crops account for much of the food in the Latin American diet. Fortunately, according to CIAT agronomists, maize yields are not depressed by the companion bean crop. Bean yields are about half as high as where they are grown alone. The agronomists do feel that much more research is needed on time of seeding each crop, irrigation management, fertilizer usage, cultural practices, soil management, and economics. The production of crops grown simultaneously involves much hand labor. That is where economics enters the picture.

With very short season varieties, double-cropping can be practiced in Michigan or other areas of similar climate. It is possible with a few short-season vegetables and by including a fall-planted crop that is harvested early in the next season. Actually, the seeding of a hay crop with small grains is a form of double-cropping. It requires 2 years to produce an alfalfa hay crop where seeding is alone. Thus a grain-hay sequence requires only 2 rather than 3 years. Navy beans have been grown after green peas and after an early harvested alfalfa crop. Irrigation may be necessary for this latter combination because alfalfa often dries out the soil too much for efficient bean-seed germination.

Double-cropping of wheat and soybeans is becoming common in southern Illinois, southern Ohio, and southern Indiana, a practice favored by the development of shorter-season varieties, the use of fertilizer to hasten growth, and irrigation to assure sufficient water for germination of the soybeans.

Developing Countries

A very high percentage of the World's peoples, including most of those in developing countries, live in the Tropics and/or Subtropics—roughly between latitudes 30° North and South. This makes multiple-cropping possible throughout most of the food-deficient portions of the world. Furthermore, labor is rather plentiful in most developing countries. This is a definite plus for multiple-cropping that involves simultaneous crop production and interplanting.

The production of leguminous trees and maize is one of the best examples of mixed cropping. In a January 14, 1980, lecture at Michigan State University, Albert Ravenholt,[3] formerly of the American Universities Field Staff, proclaimed the advantages of this cropping system in the Philippines. The practice is to set *Leucaena leucocephala* seedlings 1 m apart on the contour. After the Leucaena becomes well established, maybe 12 to 18 months, trim it back to 1 m in height, with successive cuttings every 3 months. Row

---

[2]CIAT—Centro Internacional de Agricultura Tropical, Series AE-1, 1976.
[3]P.O. Box 150, Hanover N.H., U.S.A.

spacing may vary according to land slope and the need for erosion control. Then plant maize between the Leucaena rows. Some growers have called this "alley cropping." The nitrogen fixed by the leguminous Leucaena may amount to as much as 500 kg N/ha, more than may be needed by the corn. Some of the leaves and fine twigs may even be removed for cattle fodder or nitrogen fertilizer application elsewere.

Research in Hawaii has also shown Leucaena to be a valuable fixer of nitrogen. With two plant types at different spacing and cutting frequencies, fodder yields varied from 11.9 to 20.8 metric tons/ha, with nitrogen percentages quite uniform at 4.3% in the fodder fraction. Total nitrogen in the fodder was nearly 600 kg N/ha/year in one plant type and about 500 kg N/ha/year in the other.

Taiwan

Perhaps the best way to illustrate multiple-cropping is to cite the activities of a very progressive country that passed through the developing stage (as of 1965, according to AID) to definitely be classed as "developed." In fact, the postwar history of Taiwan is one of the more progressive in the entire world. Agricultural progress is most outstanding. To quite an extent, the credit may go to their success with multiple-cropping.

The Tropic of Cancer passes through Taiwan at about its center, so the country is both tropical and subtropical. Because of its mountainous terrain, the island is very short of arable land, so they had no alternative as population increased but to raise more on the presently cultivated acres. They did it by good soil and crop management and by *multiple-cropping*. At present, their arable land-crop index is probably as high as 1.9. As labor becomes more expensive, this ratio will drop because interplanting requires hand work, which may become too costly.

Four Crops per Year

Rice is the main crop in Taiwan. The shorter-season varieties, maturing in 100 to 120 days, make it possible to grow three crops a year, but most growers prefer two crops of rice and two upland crops. In that way, the desirable soil-formation processes can go on normally during part of the year. Aerobic soil organisms cannot function normally in continuously saturated soil.

A Cropping System in South Taiwan

March 1	Transplant rice to be harvested June 20–30
June 1–10	Interplant melon transplants
June 20–30	Harvest rice
August 31	Harvest melons, transplant rice to be harvested December 20–30
November 25–December 1	Interplant sweet potatoes
December 20–30	Harvest rice
February 28	Harvest sweet potatoes and transplant rice for the start of another crop year

Diagramatically, the system appears as follows.

March 1   February 28

Rice   Melon   Rice   Sweet Potatoes

Double Cropping with Sugar

Monocropped sugarcane in Taiwan usually grows for 18 or more months. The plants remain quite small for several months. A crop of sweet potatoes, peanuts, or cabbage can be interplanted and harvested before the cane plants are large enough to be injured by the competition.

Another common practice is to interplant sugarcane with rice. Appropriate rows of rice are either left out or the plants are moved to adjoining rows, as shown in Fig. 19-1. Cane cuttings are then carefully planted in the open rows (see Fig. 19-2). The cane cutting cannot stand complete flooding, so they are placed on a ridge of such height that the water level will just reach but not quite cover them. A handful of mud over the node assures germination.

Some workers slope the cuttings into the soil so the end is above water (see Fig. 19-3). This kind of hand planting takes a great deal of time and can only be done where labor is plentiful or where the planted area is small, as is the case on most Taiwan farms.

FIGURE 19-1 A rice paddy where every fifth row of rice plants is omitted. Sugarcane cuttings will be planted at about the time the rice heads appear.

FIGURE 19-2 Sugarcane cuttings cannot stand continuous floodings, so they are carefully placed on a hand-constructed ridge where they are only partially covered. A handful of mud on the node aids germination by keeping that portion from drying if the water level drops for a short time.

FIGURE 19-3 Cane cuttings placed so one end is above water level. Germination is complete. This method of planting may be faster than the one shown in Fig. 19-2. It is probably just as effective.

Again, if the rice crop is planted March 1, the cane cuttings will be planted about May 15 and the rice harvested June 30. Sugarcane harvest then takes place at the end of February in time to again plant rice to start another year.

This rice-cane double-crop combination is a more profitable use of land than is the production of sugarcane over an 18-month period. It is necessary, of course, to have very good water-level control to make such a system successful.

Double-cropping to rice and wheat in Bihar, India, has tripled the wheat producing area without sacrificing rice production. The wheat is inter-planted with the rice. It can reach maturity after the rice is harvested and before the dry season is too severe. Much of Bihar's agriculture is only rain-fed with no irrigation

SUMMARY — MULTIPLE-CROPPING

Soil Management Principle

For successful multiple-cropping, water and fertility must both be care-fully planned and managed.

The practice of growing more than one crop a year is gradually increas-ing, even in temperate climates. The city gardener is becoming adept at making full use of the small amount of land available for a garden. Even Michigan farmers, especially those on small farms, are thinking of expanding their business by making more efficient use of their land.

Double- or multiple-cropping requires very careful planning and man-agement. Andrews and Kassam (1975) have presented an excellent review of the many ways in which multiple-cropping may be accomplished. Some of the needed precautions follow.

1. Year-round water control, both drainage and irrigation, is essential. Farmers in northern Taiwan cannot grow winter crops (December to February) on much of their rice paddy land because it is the rainy season and the land floods. In most tropical lands, rainfall is seasonal. Agriculture during the dry season is possible only where irrigation is available. Correct time of irrigation varies with the variety and other management practices.

2. One should be careful to raise varieties adapted to the area as far as hours of sunshine are concerned. Maize bred to thrive in the almost continuous sunny days of Colombia became too vegetative during Ghana's first corn season (July) because of continuous cloudy days.

3. Length of season and plant structure is important. The short, stiff straw, short-season rices bred by the International Rice Research Institute (IRRI) are excellent where interplanting is practiced.

4. Planting dates are important. Soybeans and sweet corn start slowly, then

grow rapidly. They can overlap each other to the extent of one-half their growing season without causing undue competition. In Latin America, where most of the edible beans are grown with maize, planting dates should be timed so harvest of the two crops is simultaneous, thus avoiding damage during harvest. This means that bean planting must be considerably later than maize planting. Research geared to variety, climate, and the kind of soil involved is needed.

5. Adequate nutrition is essential. We want fast growth but balanced growth, enough nitrogen but not so much as to result in excessive vegetation and lessened seed and/or fruit production.

6. Crop protection is highly important. Hot, humid weather hastens growth of crops, but also of weeds, insects, and diseases. Protection is a 12-month job if 365 days cropping is the goal.

7. Finally, multiple-cropping is good for the soil because it means continuous vegetative cover. Soil never needs a rest. It should always be growing something. If it was animate, we would say it is the happiest when it is working.

REFERENCES

Andrews, D. J., and A. H. Kassam (1975). *Multiple Cropping*, American Society of Agronomy Special Publication 27, Proceedings of a Symposium at Knoxville, Tennessee.

Guevarra, A. B., A. S. Whitney, and J. R. Thompson (1978). Influence of intra-row spacing and cutting regimes on the growth and yield of Leucaena. *Agron. J.* 70:1033–1037.

CHAPTER 20

Forests, Desertification, Erosion

"The forests of the world are being cut down at the rate of *50 acres* per minute." This statement is a quotation from former AID Administrator Douglas J. Bennet, Jr., who said further, "To stay even with present per capita fuel needs, we must plant 2.6 million more hectares of trees each year between now and the end of the century. This is a tall order perhaps, but we know how to plant trees, we have new varieties which grow rapidly, and tree-planting can be labor intensive, particularly in countries where labor is plentiful and cheap."

The foregoing statement by AID Administrator Bennet appeared in the July 28, 1980 issue (Vol. 3, No. 15) of *World Development Letter*, a biweekly publication of AID. In the March 1981 issue of *Agenda*, a publication put out 10 times a year by AID, the research zoologist Eugent S. Morton stated that 20,000 to 40,000 *square miles* of the world's forests are being destroyed each year. The larger figure is almost exactly in agreement with Bennet's "50 acres per minute."

FORESTS VERSUS FOOD

One-third of the people in the world depend on wood for fuel. In the rural areas of many developing countries, the figure runs as high as 90 percent. Much of the cooking and water heating is over open fires or on very poor stoves. A great deal of the heat is lost. Further, one government official is quoted as saying, "If at the end of the century our people are able to produce sufficient food, they will not have fuel to cook it."

As recorded in the November 1971 issue of *National Geographic*, Barry Bishop, an American geographer, asked a farmer in the Karnali zone of western Nepal how far he had to walk for a load of firewood. The answer was

"1 day." Asked about the time required when he was a child, the farmer answered "1 hour." One wonders how long the walk would take today.

In the September–October 1981 issue of the *Journal of Soil and Water Conservation*, Lester Brown, a noted world food authority, states that fuel wood cutting in Nepal is responsible for much of the increased erosion in that country. His estimate is that Nepal's rivers annually carry 240 million m³ of soil to India. The sad fact is that the soil deposited in India is doing more damage than good by adding sediment to rivers and reservoirs. In fact, the World Bank estimates (December 1979) that the beds of Nepal rivers flowing to India and Pakistan are rising at the rate of 15 to 30 cm/year. The Bank estimates further that at the present rate of tree cutting, the Himalayan ranges in Nepal will be denuded within 25 years. That means still more sediment into the rivers. Chronic flooding will spell disaster to millions of small farmers in eastern India and Bangladesh.

This Nepal pattern of firewood gathering is repeated throughout the developing world. Figure 20-1 shows a man in Guatemala just returning to the village with a load of firewood. Perhaps it was for home use, maybe to sell to villagers.

There is a close relationship between forest tree cutting, desert expansion (desertification), and erosion. The October 28, 1981, *World Development Letter* (published by AID) states that in West Africa's Sahel countries, trees are rapidly vanishing out to a radius of 50 miles around the eight capital cities, partly the result of firewood gathering and partly as a result of grazing by oversized herds of cattle, sheep, and goats. In fact, goats have been blamed for much of the loss of vegetation in many arid and semiarid countries. Pakistan has recently taken steps to eliminate goats as one way of resisting the spread of desertification.

Erik Eckholm and Lester Brown, researchers at Worldwatch Institute (Brown is also president of the Institute), stated that 630 million people live in arid or semiarid lands and that 50 million of them live where their pastures have turned to wastelands. Further, about 30 million live where the land is almost useless. They point out that overgrazing, the extension of grain farming onto lands that are not capable of its production, and firewood gathering in Morocco, Algeria, Tunisia, and Libya have combined to greatly overtax the environment. Range specialist H. N. LeHouerou says that an area equal to 100,000 hectares (250,000 acres) of range and cropland is lost to the desert each year.

This same thing has happened in many countries, including southwestern and southern Argentina, large areas of Mexico, much land in the southwestern United States, and parts of northeastern Brazil. In Chile's arid Coquimbo region, dividing Chile's Atacama Desert and the irrigated valleys of central Chile, cacti and less-productive annuals gradually replaced native pasture plants. As the pastures declined in quality and production, sheep replaced cattle and then goats replaced the sheep. One of the authors has seen goats climb trees to eat plums in Rumania, and Eckholm and Brown (1977) show in their article on "The Spreading Desert" a picture of a goat that died in a tree it had climbed to eat the leaves.

FIGURE 20-1 This man in Guatemala probably spent the greater part of a day walking to and from the area where he was able to gather his load of firewood. Note the machete in the crook of his arm. Perhaps the trees he cut should have been left for erosion control!

Desertification is taking place in Australia, where sheep overgraze the pasture around water ponds or tanks until for miles around the land becomes barren. Once that happens, the dry soil starts to blow and dunes are formed. After that, only irrigated trees adapted to the desert, such as acacia, can be made to survive. Likewise, large areas of our own Southwest (Texas, Arizona, and New Mexico) have become useless, badly eroded wasteland (definitely desert) by overgrazing with sheep. Wherever land cover, be it grass, trees, or shrubs, is destroyed, water and wind erosion increases until sand dunes and gullys destroy the usefulness of the land.

It is possible for forest destruction to affect food production in distant lands. Tropical forests are important in the life cycle of North America migratory birds. In the spring, the birds fly north to raise their young in the cooler climate. In so doing, they must have food. The swarms of insects that would destroy trees and crops are therefore kept in check by these hungry spring visitors. As an illustration, the spring webworms cannot produce a second generation because the migratory birds have arrived.

The tropical forests, especially in the Amazon regions, are being cut for forest products and for land clearing in shifting cultivation. Once the trees are gone, grass will take over and low productivity will result. Shifting-cultivation farmers knows that grasslands are not adapted to their method of soil rejuvenation. Refer to Chapter 2 for an explanation.

REFORESTATION AND EROSION CONTROL

In most of the developing countries, trees are being cut to produce cropland, timber, and/or firewood much faster than they are being planted. According to AID's October 28, 1981, *World Development Letter*, plantings in West Africa's Sahel region are "just 6 percent of the amount needed to stave off disaster."

According to Norman Myers, senior associate of the World Wildlife Fund, the yearly economic loss attributable to desert advance is greater than $6.5 billion. Is it any wonder that Lester Brown, a world food authority, stated in a recent article in *U.S. News and World Report* that the most serious constraint to hunger elimination in this century is soil erosion.

Reforestation will not only furnish fuel wood for the future, but may also furnish food for humans and animals and help greatly in slowing soil erosion — provided, of course, that the right trees are planted.

Soil Management Principle

Nitrogen-fixing leguminous trees can be valuable in soil management.

In lowland (below 1500 m) tropical regions the choice of a nitrogen-fixing leguminous tree might well be *Leucaena leucocephala*, a legume capable of fixing large quantities of nitrogen that has a very fast rate of growth — 10 cm (4 in.) in diameter in 18 months. The trees may then be cut back to 1 m (3 ft), after which the new shoots may be harvested every 3 months or may be pastured. Interplanted grasses, properly managed, may increase forage production for cattle and result in better food, as Leucaena alone should not make up the entire ration for long periods of time because the tissue contains mimosine, a toxic alkaloid. Cattle, sheep, or goats will thrive continuously on a ration containing 40 percent Leucaena or on 100 percent *Leucaena* for 2 or 3 months. Nonruminants cannot be fed a ration containing more than 5 to 10 percent Leucaena.

Leucaena is indigenous to Central America. It thrives best on neutral to alkaline soils. The CIAT is conducting research in Central America toward the adaptation of this very valuable plant to acid soils.

In the Philippines, where Leucaenea has long been grown for forage and soil improvement, especially erosion control, it is known as "Ipil-ipil." It is often planted on the contour on steep slopes in alternate rows or in alternate strips with maize (corn) where is furnishes all the nitrogen needed by the maize. Used in that way, it is kept trimmed back to about 1 m (3 ft). The clippings may be left entirely to benefit the maize, or some may be removed for livestock feed.

ACACIA ALBIDA

The leguminous tropical tree *Acacia albida* is sometimes called a "magic tree". It has been stated that "This is really a great tree. It's amazing how this tree survives and grows in areas of drought. It is a nitrogen fixing tree. Animals gather under this tree and fertilize the soil. During the rainy season, the leaves fall off and more fertilizer is produced."

Many thousands of these trees have been planted on the barren soil of Chad where so many cattle starved in the dry years of the early 1970s. About 85 percent of the trees survived. Millet planted under *Acacia albida* trees in Chad during the 20 years preceding the study had yielded 30 percent more than it did out in the open.

That deep-rooted trees may furnish enough shade to protect grasses against excessive transpiration and moisture loss by evaporation is shown by a picture by Georg Gerster on page 591 in the November 1979 (Vol. 156, No. 5) issue of *National Geographic*. Each tree is said to produce an "oasis" where the grass can survive and the soil is protected against wind erosion. If the tree is a legume such as *Acacia albida*, the grass is provided with the nitrogen it needs.

Other Acacias

Actually, there are about 800 species of the genus *Acacia*. They grow in the arid and semiarid tropics around the world, not excluding the United States. They grow where most other plants fail, furnishing browse for wild animals and cattle and protection to the soil. Usually, their canopy is not so dense but that grass can grow in the shade. Feed for animals is the result.

The species *Acacia senegal* is a main source of gum arabic, and its leaves and pods furnish fodder for animals. Besides production in Senegal, gum arabic is produced in Sudan, Mauritania, Mali, Nigeria, Niger, Chad, Tanzania, Ethiopia, and Somalia. It gets its name from the fact that 75 to 85 percent of its production is from the Arabic countries of Sudan and Arabia.

Acacia senegal is very deeply rooted (taproot) and has extremely wide-spreading side roots, probably explaining its ability to survive where rainfall is very low in a belt 300-km wide along the southern edge of the Sahara desert from Mauritania to Somalia. It cannot survive, however, against *overgrazing* by sheep and goats and *overcutting* by urban dwellers who need firewood. What is needed is extensive forestation and protection of the trees after they are planted. This would not only furnish fodder for animals and gum arabic for export but protection for the soil. On hilly land, trees are nature's best soil cover. With grass underneath, the protection is excellent.

EROSION

In countless locations throughout the world, erosion is greater than the allowable limit in tons per acre and it is increasing. As already noted, forest cutting and overgrazing are largely responsible. Improper soil and crop management is also responsible, particularly in our own country. What is happening in many states is illustrated by the situation in Michigan. In 1950, Michigan's total row-crop acreage was 2,419,000 — 30 percent of the total crop acreage. In 1978, corn alone occupied 2,660,000 acres, with the total of row-crop acreage amounting to 4,095,000 acres or 60.5 percent of that year's total planted acreage. Those of us who worked on soil conservation during the midcentury years remember well the fact that a soil conservation fieldworker measured the success of his endeavors on how many acres of depleting crops were that year converted to soil-building crops (legumes and/or grasses) in his district.

In addition, many of the proven soil-conservation practices are being disregarded. With four to eight row planters for corn and beans, contouring seems no longer practical and rye and ryegrass cover crops fail under the canopy produced by today's high-yielding crops. Attempts to prevent erosion losses by practicing minimum tillage (no-till on some farms) are only partly successful. Soil losses are commonly much greater than the recognized allowable limits. This means productive capacity will decline at a time when production increases are so badly needed as export food crops to third-world countries. We must not let this happen.

Dr. Sylvan Wittwer, former Director of the Michigan Agricultural Experiment Station, has called attention to the slowing of crop-yield increases during recent years. He says there is evidence that crop yields have reached a plateau, a break in the steady increases that were recorded in the years from the early 1940s to about 1970. There seems little doubt that the effects of erosion on crop production are coming into play.

Historical

One of our early soil-science teachers was Dr. Charles E. Millar. He pointed out the fact that crop yields in the United States did not increase appreciably during the later decades of the nineteenth century and through the first part of this century despite the introduction of fertilizers (on a small scale to be sure), pesticides, and other approved inputs. He explained it as being due to the fact that soil-maintenance efforts were not sufficient to balance the depreciating effects of cropping. We know now that the greatest depleting effect was loss of topsoil by wind and water erosion.

Then between 1940 and 1970, intensified research in all agricultural areas pointed out many improved production practices. Yields gradually and consistently increased until they reached the plateau pointed out by Dr. Wittwer.

Targeting Soil-Erosion Control Expenditures

In 1981, the U.S. Congress passed a farm bill that emphasized a need to target conservation expenditures to the nation's most erosive lands. Erosion exceeding 10 tons/acre occurs on only 17 percent of the nation's cropland. To confine assistance payments to lands where soil loss is greater than 10 tons/ acre would imply that losses less than 10 tons/acre can be disregarded. It further implys that the vast amount of erosion research, which shows that losses of 2 to 5 tons (on different soils) will eventually lower the productive capacity of the soil, can be disregarded. Such an assumption is an insult to the soil scientists who established the allowable erosion limits on a large number of soils.

Most of the good land in the world is being cropped. We must not allow it to lose its productive capacity. Instead, its capacity to produce food must be increased. Some of our most knowledgeable food experts say as much as a 100 percent increase must be made if we are to feed the world's population in the next century. Erosion *must* be controlled.

REFERENCES

Benge, M. D. (1981). *Leucaena leucocephala: An Excellent Feed for Livestock*. Technical Bulletin 25, Office of Agriculture Technical Assistance Bureau, Agency for International Development, Washington, D.C.

Brown, L. R. (1981). Next major problem for farmers: Soil erosion. *U.S. News and World Report* November 2.

Eckholm, E., and L. R. Brown (1977). The spreading desert, part I and The spreading desert, part II. *War on Hunger, A Report from the Agency for International Development*, August and September.

George, E. (1981). Magic trees in the desert. *War on Hunger*, July.

Morton, E. S. (1981). Silent spring revisited. *AID Agenda*, March.

Ogg, C. W., J. D. Johnson, and K. C. Clayton (1982). A policy option for targeting soil conservation expenditures. *J. Soil Water Conserv.* 37 (2):68–72.

CHAPTER 21

Management of Arid Lands and Water Management

In desert areas, soil management problems are largely confined to those connected with water supply. In areas where it never rains, as in the Atacama Desert in northern Chile, agriculture is ruled out. But where there may be minimal amounts of rain, say 100 to 200 mm (4 to 8 in.), limited agriculture is possible by making the most efficient use of what is available.

CROP SELECTION

Plants vary greatly in their ability to grow or even survive during extended periods of drought. Shallow-rooted grasses soon die when the top several inches of soil become air dry in an atmosphere almost devoid of moisture. We have already noted that grasses may survive under *Acacia albida* trees (and under many of the other 799 species of acacia) whereas they would die without the presence of the trees. This is because the trees are deep-rooted. They are legumes, and so supply nitrogen and bring up moisture from lower depths. The shade reduces transpiration from the grass and evaporation from the soil and attracts animals that eat fallen leaves and pods. The waste from the animals fertilizes the grass.

Cassia sturtii grows well in the Negev Desert, where annual rainfall is 150 mm, and without irrigation. It is a leguminous shrub, relished by livestock, and is quite resistant to grazing. Many similar plants are available in different parts of the world.

Efficient water users are as follows.

1. Plants that grow during the cooler seasons. Winter wheat is an example.

2. Rapidly growing, short-season plants.

3. Very slow growing perennial plants. Examples are cacti, oak, certain pines, and mesquite.

4. Grains with short straw.

5. Plants that close stoma during the day. Cacti, agave, and pineapple are examples. Also, those with stoma only on the lower side of the leaves.

6. Plants tolerant to salt so that extra water for leaching of accumulated salt is not necessary. Example are rice and sugar beet.

7. Plants that have deep, widespread roots. *Acacia senegal* is an example.

PRACTICE RUN-OFF AGRICULTURE

The principle involved in the Israeli Negev[1] is that of directing run-off from a relatively large area (20 to 30 ha on a desert slope) to a very small (1 to 3 ha) cultivated area in an adjoining valley. This represents a land area ratio of 1 to 20. Even then, it is usually necessary to select crops that classify as efficient water users, such as those mentioned previously.

Microcatchment farming in the Negev is described in the National Academy of Sciences publication *More Water for Arid Lands* (1976). Evenari, Shanan, and Tadmor (1968) also described the method of constructing small bunds for single fruit trees. The tree is placed in the low point of the bund so that all the rain that falls into the area will flow to the tree, thus assuring a sufficiency to supply that single plant.

EXCESSIVE GRAZING

The destruction of pastures by excessive grazing, particularly by sheep, goats, and camels, has already been mentioned as a practice that results in soil-destroying erosion. Avoidance is especially important in deserts because it is hard to get lands recovered with forage plants. Even in humid regions, vast areas of pasturelands have been destroyed by improper grazing methods. The uplands of Chile and large areas in our 17 Western states furnish good examples.

RAINWATER HARVESTING

In effect, rainwater harvesting is "runoff agriculture" except that it may involve more use of reservoirs and tanks for water storage, sometimes simply for livestock drinking water. In the arid regions of Australia, for instance, much of the pasture destruction results from excessive grazing in the vicinity of the water source. Sheep start eating as soon as they obtain a drink.

Various methods are used to increase the flow of water from the runoff to the reservoir or tank. Walls and ditches, strategically located, may do the job. Sometimes water penetration is lessened by applying powdered paraffin, which melts in the sun, or surfaces may be covered by butyl rubber or polyethylene film — anything to cause runoff instead of water penetration. Actually, desert soils do tend to seal over, so runoff is intensified.

[1]The Negev is a triangular area in the south of Israel, extending to the Gulf of Aquaba.

LESSEN EVAPORATION AND TRANSPIRATION

We have seen in a recent chapter how deep-rooted acacia (*Acacia senegal*) may make it possible for grasses to survive and furnish pasture for animals, and the protective effect of mulches is known even by agronomists in humid regions (see Chapter 7).

Much water is lost from reservoirs by evaporation that can be lessened by use of floating objects such as rubber sheets, wax, or lightweight concrete slabs. Losses are also reduced by filling reservoirs with sand. More reservoir volume is necessary, and, of course, the water must then be pumped as from a well.

Chemical antitranspirants may help by closing some of the stoma. The effect is rather short-lasting, perhaps 1 to 4 weeks.

Transpiration may be lessened by windbreaks or by enclosing plants within a plastic cover (partial enclosure) or by providing partial shade. Partial defoliation reduces transpiration. We take advantage of this when we transplant certain trees, sometimes by removing most or all of the branches. The tree becomes what the market terms a *whip*.

IRRIGATE SCIENTIFICALLY

<table>
<tr><td align="center">***Soil Management Principle***

Apply water only when it is needed.</td></tr>
</table>

Apply water only when soil moisture determinations indicate a need. Properly placed Bouyoucos blocks can make it possible to apply only the amount necessary for good results.

Lined canals prevent the loss of water before it reaches the plants. Closed-pipe transfer of water is best, and pinpoint application (trickle or drip) should be used where water is scarce and expensive.

Moisture Barriers

One of the newest moisture-saving ideas is that of reducing cropland percolation losses. Coarse and deep sand soils are generally unproductive because they do not retain sufficient moisture within the root zone to support a crop. This handicap may be eliminated by irrigation, but the necessary frequency of water application and the total amount needed are such that cropping may not be practical. Michigan alone has about 10 million acres of such soil. Many desert soils are of such nature, and, if water is available in the deserts, it is even more expensive.

The Michigan Experiment Station, under the leadership of Dr. A. E. Erickson, has successfully conducted research to solve this difficult problem. They had seen permeability slowed in certain deep Michigan sands by thin layers of soil considerably higher in clay than the rest of the profile. The result — better natural vegetative growth and better crops after the land was cleared, but where the thin, higher clay layer was absent — nonproductive soil.

The researchers tried several materials for artificial barriers at different depths from the surface. Clay (finely divided bentonite) and polyethelene film were quite satisfactory, but hot asphalt was easiest to apply and was the most efficient of the three materials. Erickson, Hansen, and Smucker (1969) insist that the method is recommended only on deep sand soils where irrigation water is available (see Fig. 21-1).

FIGURE 21-1 A deep sandy soil made productive by an asphalt moisture barrier. The barrier is in about the middle of the picture. Note the dark color above the barrier, in contrast to the light color of the dry sand beneath. This barrier was installed by a contractor for the Amoco Moisture Barrier Company.

REFERENCES

Erickson, A. E., C. M. Hansen, and A. J. M. Smucker (1968). The influence of subsurface asphalt barriers on the water properties and the productivity of sand soils. *Transactions of the Ninth International Congress of Soil Science*, Adelaide, Australia. 1:331–337.

Evenari, M., L. Shanan, and N. H. Tadmor (1968). Run-off farming in the desert, 1. Experimental layout. *Agronomy J.* 60:29–32.

National Academy of Sciences (1976). *More Water for Arid Lands, Promising Technologies and Research Opportunities*. National Academy of Sciences, Washington, D. C.

Rachie, K. O., J. P. M. Brenan, J. L. Brewbaker, et al. (1978). *Tropical Legumes: Resources for the Future*. National Academy of Sciences, Washington, D.C.

Organic Farming (Histisols)

Histosols (organic soils) constitute one of the 10 orders in the taxonomic classification of the soils of the United States, Puerto Rico, and the Virgin Islands. They form where organic-matter production exceeds decomposition over a long time. The soil must be saturated for at least 1 month of the year for this to take place. Suborders of the Histosols—Fibrist, Hemists, and Saprists—are divisions based on degree of decomposition of the organic matter, the Fibrists (fiber) being the least decomposed. Where organic soils grade into mineral soils, depending on the amount of clay present, Histosols range from 12 to 18 percent organic carbon.

Roughly 1 percent of the world's land area (0.50 percent of the U.S. land area) is in the order Histosols. In Michigan, the percentage is about 10.6 (1 acre in 10). Peter J. Lumbert, a 1982 M.S. graduate of Michigan State University, determined the Histosols acreage in Michigan, divided into soil management groups, to be as follows.

| | Area | |
Management groups	Acres[a]	Hectares
MC—deep high base status	2,031,971	822,660
MC-a—deep strongly acid	363,082	146,996
M/1C over clay	34,964	14,155
M/3C over loams	229,001	92,713
M/4C over sands	474,851	192,247
M/MC over marl	55,474	22,459
L-MC lowland[b]	107,083	43,353
Totals	3,296,426	1,334,585

[a]Acres/2.47 = hectares.
[b]Subject to flooding.

Perhaps the largest single area of Histosols in the United States is the Everglades of Florida. Deposition in that area started about 4400 years ago; by 1914 deposition of organic material had accumulated to a depth of 3.65 m. Time for accumulation varies for other areas. Perhaps an average of 15 cm/100 years is an acceptable figure.

FORMATION OF HISTOSOLS

Anaerobic conditions hasten the formation of Histosols. In tropical or subtropical areas, such formation occurs under water or where high ground-water levels keep the accumulating organic materials saturated. The Everglades of Florida illustrate this latter type of formation.

Blanket peats are formed in the cooler climates under moist conditions where continuous, complete saturation is not necessary. The muskegs of Alaska are formed largely from sphagnum mosses under high precipitation, high humidity, and low temperatures.

The climate of England, Wales, Scotland, and Ireland is favorable for the formation of Histisols. The cool climate with frequent and long-lasting rains results in luxuriant vegetative growth and very slow decomposition. As a result, an organic blanket builds up. On steep slopes in Northern England and Scotland, the writers have dug trenches to expose water seepage on the hillsides. The peat just a few inches from the surface was saturated. Decomposition would be very slow in such material, so the peat blanket continues to increase.

The Histosols of the northern humid sections of the United States were mostly formed in basins, lakes, and riverbeds that were formed as the glaciers receded (see Fig. 22-1). Such organic deposits vary greatly in nature, depending on the composition of the water in which the plants were growing. Water composition depends, of course, on the nature of the soil on the watershed. The pH of such deposits varies from strongly acid to alkaline. In some instances, they are underlaid by marl, clay, or bedrock — more often in Michigan by loamy or sandy materials.

Two textures, peat and muck, are arbitrarily recognized by farmers who manage organic soils. **Peat** (fibric) is that material in which the plant remains are readily identified. As examination readily tells whether moss, sedges, or woody plants gave rise to the deposit.

Muck (sapric) is organic material where the plant remains making up the deposit have decomposed so that they are no longer recognizable. Drainage and tillage bring about better aeration. This hastens decomposition, and peat soon changes to muck. The change may be confined largely to the plowed layer or it may extend deeper, depending on depth of drainage, the cropping system, and the amount and nature of the tillage. Rapid decomposition results in rapid subsidence and perhaps a short-lived organic soil. In places, of course, intermediate decomposition exists, as in humic materials.

FIGURE 22-1 Organic soil in the making.

PHYSICAL PROPERTIES OF ORGANIC SOILS

Organic soils are quite different from mineral soils in respect to physical properties. That extremes are the rule with organic soils is particularly evident when we consider water relationships. The effect of organic matter on the water-holding capacity of mineral soils was discussed in Chapter 4.

Water Relationships

Water-Holding Capacity

The water-holding capacity for organic soils is several times higher than it is for mineral soils. In fact, organic soils commonly hold water equal to several times their own weight. Water-holding capacities of 300 percent for wood-sedge peat, 1360 percent for sawgrass peat, and 3200 percent for sphagnum peat have been reported. An organic soil may contain a high percentage of water in a form not available to plants. Permanent wilting percentages as high as 80 are common.

Soil Management Principle

Available, not total, water is of importance in crop production.

Permeability

In the management of peat soils, permeability is perhaps one of the most important properties. Rate of water movement, vertical and horizontal, is greatly dependent on degree of decomposition and compaction. The latter varies with extent of tillage. Compacting the surface of a Houghton muck (Euic, mesic Typic Medisaprists) decreased the permeability rate from 30 cm/hour to 1.2 cm.

The rate of seepage through an Everglades peat (Euic, hyperthermic Typic Medihemists) was reported to be 0.5 cm (0.2 in.) an hour in the top 45 cm (18 in.) and 35 cm (13.6 in.) an hour in the 45 to 90 cm (18 to 36 in.) layer. The difference was the result of cultural practices. It is apparent that drainage and irrigation practices depend on rate of permeability.

Bulk Density

Dry Histosols are very light. They may weigh as little as 5 lb/ft^3. Bulk density depends on the nature of the organic material, the amount of mineral admixture, and the moisture content at the time of sampling. Dehydration results in shrinking and an increase in bulk density.

When Histosols become very dry, excessive shrinkage causes wide cracks that fail to close up completely when the material is again wetted, because of nonreversible structure changes. As a result, drainage conditions are changed and may become excessive, depending on the nature of the underlying strata. Organic material used on benches or in pots in the greenhouse sometimes become so dry that it is impossible to wet them and bring them back to their original physical condition. They should be managed in such a way as to prevent excessive drying. This may be accomplished through controlled drainage and irrigation.

CHEMICAL PROPERTIES OF HISTOSOLS

Histosols have a high exchange capacity and are highly buffered. Exchange capacities of 85 to 170 meq/100 g of air-dry soil (850–1700 mmol (p+)/kg) at pH 7.0 are not uncommon (determined by extracting with 1-normal ammonium acetate). Variations depend on the nature of the organic matter, degree of decomposition, and other factors. For example, two soils having the same pH (3.6) varied greatly in lime requirement. Four tons of limestone brought the pH of one to 5.0, whereas 8 tons was required to effect the same change in the other. In the first soil, 10 tons raised the pH to 7.0; but in the latter, 12 tons brought the pH to only 5.6. The second soil was much more highly buffered than was the first.

The high exchange capacity and the tendency for organic matter to otherwise absorb soluble substances results in a wide range of safety as far as fertilizer applications are concerned. In other words, excessive applications of soluble salts are less likely to cause injury to plants grown on peat or muck soils than would be the case for those grown on mineral soils.

WATER CONTROL

Histosols were formed in poorly drained areas. It is logical then that a radical change (excessive drainage) in the moisture content of the organic deposit will result in eventual complete disappearance of the soil formed from the deposit. The aim then must be to provide sufficient drainage to make cropping possible but to maintain the water table as high as the particular crop can tolerate. This will lengthen the life of the deposit. Dams like the one shown in Fig. 22-2 make it possible to hold the water table at any desired level.

FERTILIZERS ARE ESSENTIAL

> ### *Soil Management Principle*
>
> For nutrients other than nitrogen, fertilizers are more essential for organic soils than they are for mineral soils.

Fertilizers are more essential on organic soils than they are on mineral soils. As one might expect, this may not hold for nitrogen fertilizer, but it does for all the other nutrients. Potassium is likely to be the first nutrient to limit production, but phosphorus deficiency will usually be noticeable as soon as sufficient potash fertilizer is applied. Trace-element availability is less pH-dependent in Histosols than it is in mineral soils. It is true, though, that soil

FIGURE 22-2 Water-level control dams are necessary for control of groundwater levels.

reaction may be used as an indication of need. The likelihood of response is dependent on the actual amount in the soil, as determined by test, the crop to be grown, and the pH of the soil. Since organic soils usually contain lower levels of trace elements than do mineral soils, response to application of the element is more certain. Modern methods of soil testing and use of the emission spectrograph or plasma emission spectrograph for plant-tissue analysis make recommendations rather dependable. Refer to Michigan Cooperative Extension Bulletin E-486, 1981, "Secondary and Micronutrients for Vegetables and Field Crops," by M. L. Vitosh, D. D. Warncke, B. D. Knezek, and R. E. Lucas, Department of Crop and Soil Sciences, Michigan State University.

ORGANIC GARDENING/FARMING

Those who advocate organic gardening may be surprised that organic-soil farmers and gardeners need apply so much chemical fertilizer. It is because their soils (Histosols) were formed largely from the decomposition of vegetation very low in mineral nutrients. Potassium, particularly, leaches readily from dead plant materials (grasses, sedges, and woody plants). Likewise, the plant residues commonly used in composts may or may not be high in plant nutrients. In fact, they are likely to be low, especially if they were leached by rain before the compost pile was constructed. This means fertilizers should then be added to the compost pile as it is made up.

THE COMPOST PILE

Surely the organic gardener, now called by some a "biological agriculturist" (actually what kind of agriculture is *not* biological?), will wish to build a compost pile because garden and farm soils alike must be kept well supplied with organic matter if success is to result. As mentioned in Chapter 4, soil material without organic matter is not soil, and even the best garden soils can usually be made better by adding more of this critical soil amendment.

Typical organic gardeners, however, carry the *organic* philosophy too far. Just as surely as organic matter is essential, so also are the mineral nutrients and soluble nitrogen essential, and it matters not whether they are supplied from a compost pile, a cow stable, or a fertilizer bag. Plants do not to any extent take up organic compounds. The nutrients are liberated from the organic matter by decomposing microorganisms. They enter the plant roots in ionic form, such as NO_3^- or NH_4^+, $H_2PO_4^-$ or HPO_4^{2-}, and K^+. See Fig. 13-2 for an explanation of the effect of soil pH on phosphorus uptake by plants.

The materials from which gardeners commonly build their compost piles are mature crop residues such as straw and leaves, very low in the nutrients needed by the microorganisms that change the rough organic matter to compost. So we must add nutrients so that the materials will decompose rapidly and the final product will be good for the crops to be grown.

Table 22-1 gives some combinations of nutrients to be mixed with organic materials commonly used in the home gardener's compost pile. The fertilizers are measured by the cup and the organic matter by the bushel or other volume measure.

Building the Compost Pile

Arrange an open bin such as might be constructed with a snow fence formed from wire and wood pickets. The diameter can vary with the amount of compost to be made. The height should be at least 4 ft (120 cm), a common height for a snow fence.

Spread the organic matter in layers about 6-in. (15-cm) deep and scatter the needed amount of nutrients on top, according to the volume of refuse. An inch (2.5 cm) or so of garden soil spread over each organic layer provides microorganisms and nutrients that help to hasten decomposition. Enough water to moisten the pile should be added as it is built. It is good to add a layer of soil to the very top of the pile and leave the top slightly dished to catch rainwater. During a dry season and/or in dry climates, add water as needed. The material should always be moist but not saturated. Good compost cannot be constructed in a hole or tightly closed container because oxygen is necessary for the right kind of decomposition.

TABLE 22-1 Nutrients Needed for Composting of Assorted Organic Materials for the Home Garden

Nutrient materials	Rate in cups per tightly packed bushel
For acid-loving shrubs and for general use on alkaline soils	
Combination *A*	
Ammonium sulfate	1
Superphosphate (20 percent)	$\frac{1}{2}$
Epsom salt	$\frac{1}{16}$
Combination *B*	
Mixed fertilizer 12–6–6	$1\frac{1}{2}$
For plants not acid-loving and/or for use on acid soils	
Combination *C*	
Ammonium sulfate	1
Superphosphate (20 percent)	$\frac{1}{2}$
Ground dolomitic limestone	$\frac{2}{3}$
Combination *D*	
Mixed fertilizer 12–6–6	$1\frac{1}{2}$
Ground dolomitic limestone	$\frac{2}{3}$

Source: Taken with slight changes from C. E. Kellogg, Home gardens and lawns, *Yearbook Agr.* (USDA), 1957.

Some gardeners like to turn compost once during the period of decomposition. This helps to produce a uniform, well-mixed product, but it is not essential.

Compost may be mixed with garden soils or may be used entirely as a surface mulch, usually depending on the kind of crop being grown. Most home gardeners use the material in both ways, depending on how much they have. Some prefer to work compost into their soil and use sawdust, straw, or peat as mulch. This latter plan is recommended.

FARMING SYSTEMS

There is a move at present to break away from one- or two-crop cash farming (corn or corn and soybeans in the North Central region of the United States) and readopt the more general farm operations now being discussed under the term **farming systems**. Such a system would allow the production of cattle, the use of land for hay and pasture, and the production of small grains instead of growing such large acreages of row crops. Then rotations, once thought to be sacred (see Chapter 5), would help to control insects and plant diseases and legumes would fix atmospheric nitrogen so we would need to buy less commercial nitrogen fertilizers. The manure from the cattle barns can be applied for corn, a crop needing large amounts of nitrogen.

We are all for this system of farming, but we must go one step, perhaps two steps, further and also use commercial fertilizer to supplement the nutrients not supplied in the *organic system*. The amounts of fertilizer needed should be determined by soil tests, as explained in Chapter 10, just as recommended for the monoculture farmer. The *organic farmer* should not depend on charging excessive prices for organically grown produce to make up for low yields, because by so doing he is deceiving his customers. Such produce is no different from that treated with needed chemical fertilizers properly applied.

Utzinger and co-workers (1973), in a bulletin published by the Ohio Cooperative Extension Service, states "The use of organic material on soils does not change the nutrient composition of the crops. Plants, through the process of photosynthesis, turn simple chemical substances from the soil into complex carbohydrate molecules of food. Compost and manure have to be broken down by bacteria to form nitrate, potassium, and phosphate before they can be absorbed in the plant's root system. The same compounds may also be taken from chemical fertilizers. The nutrient composition of a plant is influenced primarily by the genetic composition of the seed and the maturity of the plant at harvest time."

We should not leave this subject without mentioning the use of pesticides. This is the second step organic farmers must take if they expect to raise enough food to more than feed their families; in other words, to do their part in helping to feed the world.

India has long been in need of improvements in the production of food. Recent research with wheat has shown average yields under irrigation in

Maharashtra state as 950 kg/ha. In the states of Punjab and Haryana, yields average 2000 kg/ha. Under an improvement program, the yields in all three states, again under irrigation reached 4000 kg/ha. Five main principles in effect were

1. Preparation of good seedbed and timely sowing.
2. Use of high-yielding variety.
3. Use of *fertilizers* in time, and supply them to soil according to *soil tests.*
4. Use of irrigation water when required by wheat plants.
5. *Plant protection* measures.

CHEMICALS—THEIR FUTURE?

Thirty percent or more of our present crop production in Michigan, and in many other states, can be credited to the use of chemical fertilizers. If we were not using them, we would have no food to export to other states or to developing countries. Perhaps we would not have enough for ourselves unless we changed our diets.

Wittwer (1981) says that annual losses of major food crops from pests, worldwide, approximate 35 percent, with the greatest losses occurring in developing countries where pesticide use is small, nil in some areas. Adequate and timely application of pesticides could greatly expand food production in the countries where it is most needed.

Integrated pest management with maximum attention to all known management and biological controls of diseases and insects, the production of legumes to supply some of the nitrogen needed in expanded world crop production, and careful use of all crop residues may *lessen* the need for increases in chemical use, but marked increases will still be needed in the developing countries. In the meantime, organic gardeners who plan to grow vegetables entirely without the use of pesticides should follow the directions given in Michigan State University Cooperative Extension Bulletin 529, entitled *Home Vegetable Garden.* The bulletin's number one suggestion is "avoid growing vegetables that are prone to attacks by insects (e.g. cabbage, cauliflower, broccoli and potatoes)."

REFERENCES

Clayton, B. S., J. R. Neller, and R. V. Allison (1942). *Water Control in the Peat and Muck Soils of the Florida Everglades.* Florida Agricultural Experiment Station Bulletin 378.

Dewan, M. L. (1982). *Agriculture and Rural Development in India, A Case Study on the Dignity of Labor.* Concept Publishing, New Delhi, India.

Feustel, I. C., and H. G. Byers (1930). *The Physical and Chemical Characteristics of Certain American Peat Profiles.* U.S. Department of Agriculture Technical Bulletin 214.

Feustel, I. C., and H. G. Byers (1936). *The Comparative Moisture-Absorption and Mois-*

ture-Retaining Capacities of Peat and Soil Mixtures. U.S. Department of Agriculture Technical Bulletin 532.

Foth, H. D., and J. W. Shafer (1980). *Soil Geography and Land Use.* Wiley, New York.

Lucas, R. E. (1982). *Organic Soils (Histosols), Formation, Distribution, Physical and Chemical Properties and Management for Crop Production.* Research Report 435, Michigan State University.

Shickluna, J. C., and J. F. Davis (1952). The chemical characteristics and the effect of calcium carbonate on the manganese status of five organic soils. *Mich. State Univ. Agr. Exp. Sta. Quart. Bull.* 34:303–319.

Utzinger, J. D., et al. (1973). *Organic Gardening.* The Ohio State University Cooperative Extension Service, Bulletin 555.

Wittwer, S. H. (1981). The further frontiers: Research and technologies for global food production in the 21st century. Unpublished Article.

Index